Advances in Intelligent Systems and Computing

Volume 514

About this Series

The series "Advances in Intelligent Systems and Computing" contains publications on theory, applications, and design methods of Intelligent Systems and Intelligent Computing. Virtually all disciplines such as engineering, natural sciences, computer and information science, ICT, economics, business, e-commerce, environment, healthcare, life science are covered. The list of topics spans all the areas of modern intelligent systems and computing.

The publications within "Advances in Intelligent Systems and Computing" are primarily textbooks and proceedings of important conferences, symposia and congresses. They cover significant recent developments in the field, both of a foundational and applicable character. An important characteristic feature of the series is the short publication time and world-wide distribution. This permits a rapid and broad dissemination of research results.

More information about this series at http://www.springer.com/series/11156

Javier Del Ser
Editor

Harmony Search Algorithm

Proceedings of the 3rd International Conference on Harmony Search Algorithm (ICHSA 2017)

Springer

Editor
Javier Del Ser
TECNALIA, University of the Basque Country (UPV/EHU), Basque Center for Applied Mathematics (BCAM)
Bilbao
Spain

ISSN 2194-5357 ISSN 2194-5365 (electronic)
Advances in Intelligent Systems and Computing
ISBN 978-981-10-3727-6 ISBN 978-981-10-3728-3 (eBook)
DOI 10.1007/978-981-10-3728-3

Library of Congress Control Number: 2017930144

Printed on acid-free paper

This Springer imprint is published by Springer Nature
The registered company is Springer Nature Singapore Pte Ltd.
The registered company address is: 152 Beach Road, #21-01/04 Gateway East, Singapore 189721, Singapore

Preface

This book presents the state-of-the-art technical contributions based on one of the most successful evolutionary optimization algorithms published to date: Harmony Search. The contributions span from novel technical derivations of this algorithm to applications in the broad fields of civil engineering, energy, transportation & mobility and health, among many others and focus not only on its cross-domain applicability but also on its core evolutionary operators, including elements inspired from other meta-heuristics.

The global scientific community is witnessing an upsurge in groundbreaking new advances in all areas of computational intelligence, with a particular flurry of research focusing on evolutionary computation and bio-inspired optimization. Observed processes in nature and sociology have provided the basis for innovative algorithmic developments aimed at leveraging the inherent capability to adapt, characterized by various animals, including ants, fireflies, wolves, and humans. However, it is the behavioral patterns observed in music composition that motivated the advent of the Harmony Search algorithm, a meta-heuristic optimization algorithm that over the last decade has been shown to dominate other solvers in a plethora of application scenarios.

The book consists of a selection of the best contributions presented at ICHSA, a major biannual event where leading global experts on meta-heuristic optimization present their latest findings and discuss the past, present, and future of the exciting field of Harmony Search optimization. It provides a valuable reference resource for researchers working in the field of optimization meta-heuristics, and a solid technical base for frontline investigations around this algorithm.

December 2016

Javier Del Ser
General Chair
ICHSA'17

Organization

ICHSA'17 is jointly organized by the OPTIMA (Optimization, Modeling and Analytics) research area of TECNALIA Research & Innovation and the Dept. of Communications Engineering of the University of the Basque Country (UPV/EHU), in collaboration with SERGOFI.

Executive Committee

Conference Chair

Javier Del Ser	TECNALIA, UPV/EHU and Basque Center for Applied Mathematics, Spain

Program Chair

Miren Nekane Bilbao	UPV/EHU, Spain

Publicity Chair

Andrés Iglesias	Universidad de Cantabria, Spain

Publications Chair

Sergio Gil-Lopez	TECNALIA, Spain

Steering Committee

Joong Hoon Kim	Korea University, South Korea
Zong Woo Geem	Gachon University, South Korea
Ali Kaveh	Iran Univ. of Science and Technology, Iran
Anupam Yadav	Korea University, South Korea
Ling Wang	Tsinghua University, China
M. Tamer Ayvaz	Pamukkale University, Turkey
Ioannis Kougias	European Commission, Italy
Nicolaos Theodossiou	Aristotle University, Greece
Osama Moh'd Alia	Univ. of Tabuk, South Arabia
Ponnuthurai Nagaratnam Suganthan	Nanyang Technological University, Singapore
Do Guen Yoo	Korea University, South Korea
Kusum Deep	Indian Inst. Technology Roorkee, India
M. Fatih Tasgetiren	Yasar University, Turkey
Reza Sirjani	Cyprus International Univ., Cyprus
Chung-Li Tseng	Univ. New South Wales, Australia
Ali Sadollah	Nanyang Tech. Univ., Singapore
Ardeshir Bahreininejad	Institut Tek. Brunei, Brunei
Rosni Abdullah	Universiti Sains Malaysia, Malaysia
Miren Nekane Bilbao	UPV/EHU, Spain
Sergio Gil-Lopez	TECNALIA, Spain
Iyad Abu Doush	Yarmouk University, Jordania
David Camacho	Univ. Autonoma de Madrid, Spain
Sancho Salcedo-Sanz	Universidad de Alcala, Spain

Organizing Committee

Conference Chair

Javier Del Ser	TECNALIA, UPV/EHU and Basque Center for Applied Mathematics, Spain

Local Chairs

Miren Nekane Bilbao	UPV/EHU, Spain
Cristina Perfecto	UPV/EHU, Spain

Organization

Jesus L. Lobo	TECNALIA and UPV/EHU, Spain
Ibai Laña	TECNALIA and UPV/EHU, Spain
Esther Villar-Rodriguez	TECNALIA, Spain
Ana I. Torre-Bastida	TECNALIA, Spain
Manuel M. Velez	UPV/EHU, Spain
Iker Sobrón	UPV/EHU, Spain
Izaskun Oregui	TECNALIA and UPV/EHU, Spain
Itziar Landa-Torres	TECNALIA, Spain
Diana Manjarres	TECNALIA, Spain

Contents

Harmony Search and Data Mining/Big Data

Harmony Search and Scheduling/Planning

Harmony Search and Telecommunications

Fundamental Analysis and Advances of Harmony Search

Sensitivity Analysis on Migration Parameters of Parallel Harmony Search

Ari Hong[1], Donghwi Jung[2], Jiho Choi[1], and Joong Hoon Kim[1(✉)]

[1] Department of Civil, Environmental and Architectural Engineering, Korea University, Seoul 136-713, South Korea
hoalee405@hanmail.net, y999k@daum.net, jaykim@korea.ac.kr

[2] Research Center for Disaster Prevention Science and Technology, Korea University, Seoul 136-713, South Korea
donghwiku@gmail.com

Abstract. Parallel Harmony Search (PHS) is a harmony search variant that employs parallel computing approach for improving the final solution quality of harmony search. A preliminary version of PHS was introduced in 2015 and applied to a highly nonlinear water infrastructure planning problem with extreme dimensionality. The application results showed that the PHS was promising and efficient in finding a feasible and near-optimal solution for such problem. However, sensitivity analysis on migration parameters of PHS should be performed to guarantee the best performance of PHS. Migration parameters of interest are migration frequency, migration topology, migration size, and so forth. In this study, PHSs with different migration frequencies are compared with respect to the final solution quality to identify the most efficient frequency in the water infrastructure planning problem.

Keywords: Parallel harmony search · Migration frequency · Sensitivity analysis

1 Introduction

Parallel computing is a kind of computation to make large computational tasks smaller, allocate them to multiple computing units, and perform the computations concurrently. Recently, parallel computing has been adopted to shorten computation time in meta-heuristic algorithms [1]. Some studies conducted fitness calculations of an existing meta-heuristic algorithm in a parallel manner while other studies suggested a coarse-grained parallel algorithm in which a group of population is evolved independently under a computing unit (e.g., a core) and search information is shared only at migration phase. Artina et al. [2] proposed a parallelization method of non-dominated sorting genetic algorithms-II (NSGA- II) [3] for multi-objective optimal design of water distribution systems. Abu-Lebdeh et al. [4] compared coarse-grained parallel genetic algorithms (PGAs) and cellular PGA in terms of their performances on a transportation

J. Del Ser (ed.), *Harmony Search Algorithm*, Advances in Intelligent Systems and Computing 514, DOI 10.1007/978-981-10-3728-3_1

system optimization problem. Jung et al. [5] was the first to propose a coarse-grained parallel harmony search (PHS), which employed a hierarchical migration topology. That is, the temporal global best solution identified at the migration phase is broadcasted/transferred to individual harmony searches.

There exists few more questions to answer on the best condition for PHS. For instance, how often migration phase should be exhibited, what kind of migration topology should be set up, and how many and what kind of solutions should be migrated.

In this study, we perform a sensitivity analysis on the migration frequency (MF) of PHS to investigate its impact on the final solution quality. The PHS with three different MFs are compared with that without migration (i.e., zero MF) in a water and wastewater system planning problem.

2 Parallel Harmony Search

Harmony search (HS) [6,7] which was first introduced in 2001 is a novel meta-heuristic algorithm inspired by improvisation of musical instrumental players. In HS new solution can be generated by three operators such as harmony memory consideration, pitch adjustment, and random selection. Harmony memory (HM), which is a space for storing good solutions, keeps updated by replacing the worst solution with the new solution if it is better than the worst one.

In PHS with migration, HSs are evolved independently and concurrently until migration phase. Migration phase includes both identifying the overall best solution and broadcasting it to other HSs except the HS which it is from, then serial evolutions are carried out until next migration phase. The PHS's efficiency can be affected by how often migration phase is exhibited (migration frequency, MF), what kind of migration topology should be set up (migration topology), and how many solutions should be migrated (migration size). In this study, PHS with zero MF and three MFs are examined and compared in terms of the solution quality to identify the most efficient migration through the optimization of a multi-scenario planning problem.

3 Application

3.1 Study Area

PHS with migration is examined through the planning problem of a developing area in southwest US which finds the most cost effective water and wastewater infrastructure design and expansion over staged planning periods. Total number of pipes is 333 per period and commercial pipe size are 0, 6, 8, 10, 12, 16, 20, 24, 30, 36, 48, 54, 60, 72 inches. The design flow, the design head, and the number of pump in operation under peak and average demand conditions should be determined for each 36 pump stations. In addition, plant capacity is determined to a potential location of each 6 satellite wastewater plants. The decisions are made at three phase: 2010–2020, 2020–2030, and 2030–2050. Total number of decision variables is 1449 ($= 3 * (333 + 4 * 36 + 6)$).

3.2 Results

Three performance metrics, the average, best, and worst solutions, of PHS with and without migration were compared. Three MFs are examined for the latter: migration every 25, 100, and 200 iterations until total 500 iterations, resulting in the migration phases of 20, 5, and 2 times, respectively. Note that the identical HS parameters were used for each HS in PHS. The performance metrics were obtained from 20 independent optimization runs of each PHS.

Figure 1 shows the average total cost changes of the four approaches over iterations. The solution by PHS with migration every 25 iterations is always worse than any other PHSs during total computation. PHS with migration every 100 iterations shows the best solutions until approximately 200 iterations, however, after that the solutions get worse than before. At 500 iterations, which is termination criteria, PHS with migration every 200 iterations has the best solution, 1.688 (109 USD), and PHS with migration every 25 iterations has the worst solution, 1.767 (109 USD). Final solution found by PHS with no migration was good rather than PHS with migration every 25, 100 iterations.

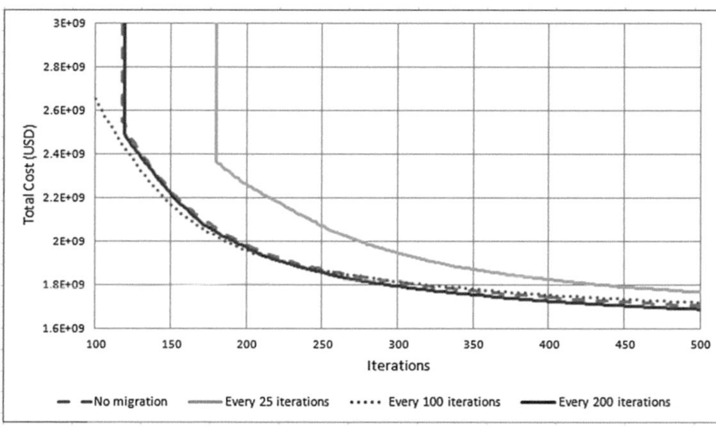

Fig. 1. Average variation of solution costs.

In Table 1, the PHSs without and with migration every 200 iterations are compared with respect to three performance metrics (Table 1). The worst solution cost of the latter was 0.23 percents less than that of the former. More significant gap between the two algorithms was found in the best solution. That is, PHS with migration every 200 iterations obtained 1.79 percents less cost. It was found that PHS with the migration produce better solution than PHS without migration overall.

Table 1. Performance metrics (Unit: 10^9 USD)

	Mean solution	Worst solution	Best solution
PHS without migration	1.703	1.764	1.623
PHS with migration	1.688	1.760	1.594
(every 200 Iterations)	(0.88% reduction)	(0.23% reduction)	(1.79% reduction)

4 Summary and Conclusions

This study conducted a sensitivity analysis on the migration frequency of parallel harmony search. In PHS with migration, individual HS is computed independently under processors in a workstation and every certain iterations migration phase occurs to identify the overall best solution and broadcast it to other HSs. The PHS is applied to multi-period and multi-scenarios planning of water and wastewater infrastructure in a developing area of southwest US.

First, PHSs without and with migration were compared in terms of three types of results: the mean, worst and best solution. It is found that overall PHS with migration has better solution than PHS without migration. Second, PHS with no migration and PHSs with migration every 25, 100, and 200 iterations are compared in terms of the average evolution of solution cost by iteration. The result shows that very frequent migration does not guarantee the best performance, in other words, it is not that the more frequently the migration phase occurs, the better the final solution is. In conclusion, migration frequency is a problem-specific parameter which should be calibrated for the problem of interest.

In the future studies, sensitivity analysis on the migration topology and migration size should be carried out to find the best efficiency of migration. Ensemble of various algorithms not only harmony search should be conducted for parallel computing.

Acknowledgements. This subject is supported by Korea Ministry of Environment as Global Top Project (2016002120004).

References

1. Elfeky, E.Z., Sarker, R., Essam, D.L.: Partial decomposition and parallel GA (PD-PGA) for constrained optimization. In: IEEE International Conference on Systems, Man and Cybernetics, pp. 220–227 (2008)
2. Artina, S., Bragalli, C., Erbacci, G., Marchi, A., Rivi, M.: Contribution of parallel NSGA-II in optimal design of water distribution networks. J. Hydroinformatics **14**(2), 310–323 (2012)
3. Deb, K., Pratap, A., Agarwal, S., Meyarivan, T.A.M.T.: A fast and elitist multi-objective genetic algorithm: NSGA-II. IEEE Trans. Evol. Comput. **6**(2), 182–197 (2002)

4. Abu-Lebdeh, G., Chen, H., Ghanim, M.: Improving performance of genetic algorithms for transportation systems: case of parallel genetic algorithms. J. Infrastruct. Syst. A4014002 (2014)
5. Jung, D., Choi, J., Choi, Y.H., Kim, J.H.: A new parallelization scheme for harmony search algorithm. In: Kim, J.H., Geem, Z.W. (eds.) Harmony Search Algorithm. AISC, vol. 382, pp. 147–152. Springer, Heidelberg (2016). doi:10.1007/978-3-662-47926-1_15
6. Geem, Z.W., Kim, J.H., Loganathan, G.V.: A new heuristic optimization algorithm: harmony search. Simulation **76**(2), 60–68 (2001)
7. Kim, J.H., Geem, Z.W., Kim, E.S.: Parameter estimation of the nonlinear muskingum model using harmony search. J. Am. Water Res. Assoc. **37**(5), 1131–1138 (2001)

Multi-layered Harmony Search Algorithm: Introduction of a Novel and Efficient Structure

Ho Min Lee[1], Do Guen Yoo[2], Eui Hoon Lee[1], Young Hwan Choi[1], and Joong Hoon Kim[1(✉)]

[1] School of Civil, Environmental and Architectural Engineering, Korea University, Seoul 02841, South Korea
{dlgh86,jaykim}@korea.ac.kr, hydrohydro@naver.com, okgohouse@hanmail.net
[2] Software Development and Engineering Department, Software Center, K-water Research Institute, Korea Water Resources Corporation (K-water), 200 Sintangin-Ro, Daedeok-Gu, Daejeon 34350, South Korea
dgyoo411@kwater.or.kr

Abstract. The harmony search algorithm (HSA) is one of the most widely used meta-heuristic optimization algorithms. During the last two decades, many improved versions and variants of HSA have been proposed to improve the algorithms efficiency and usability. In this study, a HSA variant with unique structural characteristics is proposed, named multi-layered harmony search algorithm (MLHSA). The multi-layered structure is specifically designed for the effective improvement of exploration and exploitation capability. The proposed MLHSA is applied to a set of benchmark problems to test and verify the efficiency. The application results show that MLHSA outperforms other meta-heuristic algorithms, indicating the competitiveness of the algorithm. The multi-layer concept can be easily employed to other algorithms, as a helpful tool for the improvement of existing algorithms convergence.

Keywords: Optimization · Meta-heuristic · Harmony search algorithm · Multi-layered harmony search algorithm

1 Introduction

Optimization is the process of selecting the best element from some sets of available alternatives under certain constraints. Meta-heuristic algorithms are well known approximate algorithms which can solve optimization problems with satisfying results [1]. The harmony search algorithm (HSA) [2] is one of the widely used optimization algorithm and at a same time, it is one the most efficient algorithm in the field of combinatorial optimization. Since the development of HSA, it attracted many researchers from various fields. Consequently, this algorithm guided researchers to improve on its performance to be in line with the requirements of the applications being developed. These improvements primarily cover

J. Del Ser (ed.), *Harmony Search Algorithm*, Advances in Intelligent Systems and Computing 514, DOI 10.1007/978-981-10-3728-3_2

three aspects: (1) improvements in terms of parameters setting, (2) improvements in terms of hybridizing HSA components with other meta-heuristic algorithms, (3) improvement in terms of structural characteristics. Most of the previous researches have sought to improve the performance of the HSA by changing parameters or hybridizing with other methods. However, there has been a lack of studies to improve the performance of the HSA by the formation of structures. For the improvement in terms of structural characteristics, Im et al. [3] suggested smallest-small world cellular harmony search algorithm (SSWCHS), cellular harmony search algorithm (CHS) using network theory, and Al-Betar et al. [4] suggested another cellular harmony search algorithm (cHS). Therefore, in this study, a new HSA that has structural characteristics, named multi-layered harmony search algorithm (MLHSA) is proposed. In addition, the MLHSA was applied to benchmark problems to verify the efficiency of proposed algorithm.

2 Multi-layered Harmony Search Algorithm

HSA can be conceptualized from a musical performance process involving searching for a best harmony [2]. The HSA uses harmony memory size (HMS), harmony memory considering rate (HMCR), pitch adjusting rate (PAR), and band width (BW) as optimization parameters. However, HSA has no structural characteristic in harmony memory (HM) and operators. Therefore, in this study, the structural characteristics with multi-layer concept are added to HSA for improvement of its optimization capability. The concept of the MLHSA is shown in Fig. 1.

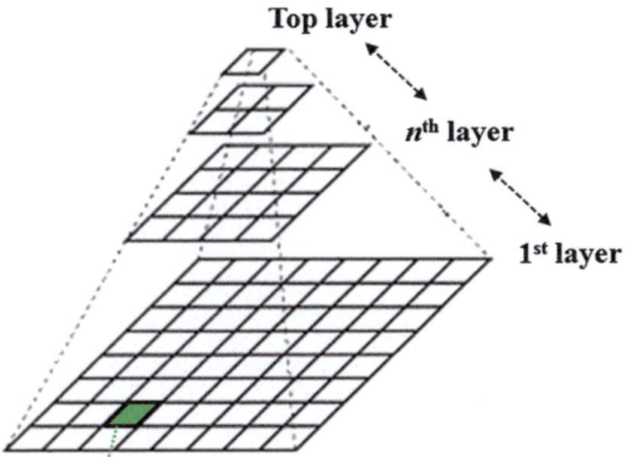

$$\text{Sub-memory} = \begin{bmatrix} x_1^1 & x_2^1 & \cdots & x_{nd}^1 & f(x^1) \\ x_1^2 & x_1^2 & \cdots & x_{nd}^2 & f(x^2) \\ \vdots & \vdots & \vdots & \vdots & \vdots \\ x_1^{SMS} & x_1^{SMS} & \cdots & x_{nd}^{SMS} & f(x^{SMS}) \end{bmatrix}$$

Fig. 1. Memory structure of MLHSA.

MLHSA has n layers and each layer has sub-memories. In this proposed algorithm, 7 of parameter are used for optimization process. The parameters sub-memory size of bottom layer (SMS_{Bottom}), sub-memory size of upper layers (SMS_{Upper}), and the number of layers (NOL) are used for structural setting of MLHSA. The parameters initial HMCR ($HMCR_{Initial}$), PAR of top layer (PAR_{Top}), PAR of bottom layers ($PAR_{Bottoms}$), and initial BW ($BW_{Initial}$) are also used for composition of characteristic for optimization respectively. In addition, the maximum threshold of HMCR ($HMCR_{Max} = 0.99$), minimum threshold of BW ($BW_{Min} = 10^{-6}$), and control parameter (CP = 0.999) for updating HMCR and BW are set as constant. The parameters HMCR and BW at t^{th} iteration are updated with Eqs. (1) and (2) respectively.

$$HMCR_t = Min.(1 - (1 - HMCR_{t-1}) \times CP, HMCR_{Max}) \quad (1)$$

$$BW_t = Max.(BW_{t-1} \times CP, BW_{Min}) \quad (2)$$

Figure 2 shows the operational steps of the MLHSA. In the step 1, the problem and parameters are composed. In the steps 2.1–2.3 the sub-memories in the 1st layer are generated. Steps 3.1–3.3 are setting up and improvising the memories of upper layers. Step 4 is replacement of the best solution of each sub-memory by using information exchange between layers. In the step 5, the stopping criterion is checked and the final solution is founded.

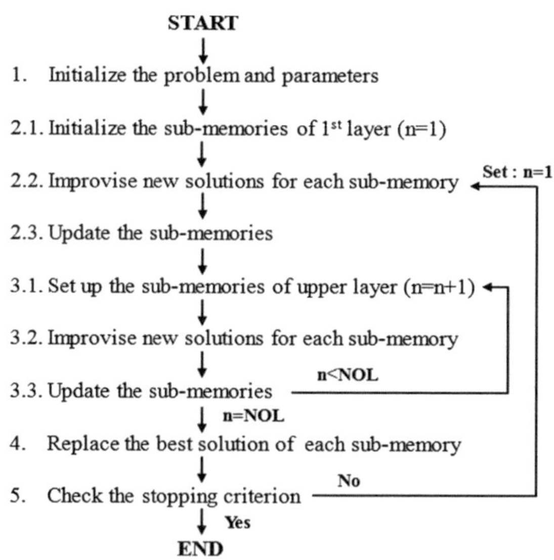

Fig. 2. Flowchart of MLHSA.

In this model, structural characteristics are added to HSA for improvement of exploration and exploitation capability (composition of sub-memories for exploitation, and information exchanges between layers for exploration). HSA, CHS, and SSWCHS can be included in the MLHSA category has specific parameter sets for structural characteristic. Therefore the MLHSA is one of generalized HSA.

3 Applications and Results

3.1 Mathematical Benchmark Problems

In this study, the proposed MLHSA is applied for solving mathematical benchmark functions, in particular the Griewank function and the Ackley function widely examined in the literature. The SSWCHS, CHS, and original HSA performed in Im et al. [3], are selected to have comparisons among optimizers. Benchmark functions have 30 design variables. The optimization task is carried out using 100 individual runs. MLHSA parameters are set equally for all benchmark functions ($SMS_{Bottom} = 50$, $SMS_{Uppers} = 1$, NOL = 2, $HMCR_{Initial} = 0.80$, $BW_{Initial} = 10^{-2}$, $PAR_{Top} = 0.08$, $PAR_{Bottoms} = 0.015$). Total number of iterations is fixed value 250,000 and the margin of error for feasible solution is set as 10^{-2} for all benchmark problems comparison with SSWCHS in same condition.

Table 1. Optimization results comparisons (The Griewank function)

Algorithm	MLHSA	SSWCHS	CHS	HSA
Mean	4.46E−03	1.03E−01	1.43E−01	6.94E−01
Best	4.22E−10	1.46E−06	2.71E−02	5.47E−06
Worst	2.22E−02	4.86E−01	4.85E−01	1.25E+00
SD	5.16E−03	9.80E−02	8.06E−02	4.47E−01
No. FS	93	10	0	3
Mean Iter	52,355	73,642	-	208,726
Iter SD	2.50E+04	4.39E+04	-	2.69E+04

Tables 1 and 2 show the obtained results for benchmark problems. The MLHSA shows best results for problems, compared to the other algorithms. The MLHSA finds 93 and 100 feasible solutions in benchmark problems. The MLHSA finds the feasible solution with smaller number of iterations than other algorithms. These results can conclude that the MLHSA converged faster than the others, and it can estimate a reliable optimal solution. In addition in this case study, MLHSA, SSWCHS and CHS include 2, 26 and 25 number of function evaluations for each iteration, respectively. Therefore, the MLHSA has advantage of time efficiency in optimization process.

3.2 Water Distribution Network Design

The Balerma network, an irrigation WDN located in the province of Almeria (Spain), was first investigated by Reca and Martinez [5]. It consists of 4 reservoirs, 8 loops, 454 pipes and 443 demand modes as shown in Fig. 3.

Table 2. Optimization results comparisons (The Ackley function)

Algorithm	MLHSA	SSWCHS	CHS	HSA
Mean	6.13E−06	4.62E−02	1.31E+00	1.58E+00
Best	3.42E−06	3.51E−03	6.10E−03	5.97E−03
Worst	1.07E−05	1.50E+00	1.65E+00	3.46E+00
SD	1.40E−06	2.11E−01	2.78E−01	7.17E−01
No. FS	100	96	3	8
Mean Iter	37,860	79,263	141,376	160,133
Iter SD	6.73E+03	6.48E+04	3.54E+04	5.86E+04

Fig. 3. The layout of Balerma Network.

A total of 10 commercial pipes with internal diameters from 113.0 mm to 581.8 mm are to be selected for this network. The objective function is mathematically assumed to be a cost function of pipe diameters and lengths. The minimum required pressure at each node is 20 m. For the cost of infeasible solutions with pressure deficient nodes, the penalty cost is added. The final objective function can be written as follows:

$$Min.Cost = \sum_{i=1}^{N} C_c(D_i) \cdot L_i + \sum_{j=1}^{M} PP_j, \qquad (3)$$

where $C_c(D_i)$ – construction cost according to pipe diameter per unit length (€/m); L_i – pipe length (m); D_i – pipe diameter (mm); PP_j – penalty function for ensuring pressure constraints are satisfied; N – number of nodes; M – number of pipes.

The parameters in the MLHSA, SMS_{Bottom} of 50, SMS_{Upper} of 1, NOL of 2, $HMCR_{Initial}$ of 0.95, PAR_{Top} of 0.05, $PAR_{Bottoms}$ of 0.01 are applied respectively. Total number of function evaluations is fixed value as 45,400 for comparison with other algorithms in the previous studies in same condition and 50 individual optimization runs are tested. Table 3 shows the optimal design results comparison of the Balerma network.

Table 3. Comparison of optimization results using different approaches.

Applied algorithms	$Min. Cost$ (€ M)	No. of function evaluations
Genetic algorithm [6]	3.738	45,400
Simulated annealing [6]	3.476	45,400
Mixed simulated annealing & Tabu search [6]	3.298	45,400
Harmony search algorithm [7]	2.601	45,400
Particle swarm harmony search [7]	2.633	45,400
Genetic heritage evolution by stochastic transmission [8]	2.178	45,400
Mine blast algorithm [9]	2.211	45,400
Improved mine blast algorithm [9]	2.064	45,400
MLHSA (Present study)	2.141	45,400

The MLHSA shows the second best results among applied algorithms with same number of function evaluations. Geem [7] found € 2.601 million from using HSA, while the MLHSA finds € 2.381 million with same condition. This means the structural characteristic is effectively combined with simulation based meta-heuristic algorithm and it can enhance the capability of optimization. The result shows the requirement of composition of structural improvement to enhance the efficient information exchange in optimization algorithms for real world sized engineering optimization problems.

4 Conclusion

Harmony search algorithm (HSA) has been successfully applied to many complex problems. The increasing complexity of problems has resulted in enormous challenges for current technique, and improved techniques are required. For the improvement in terms of structural characteristics, in this study, we proposed a new HSA that has structural characteristics, named multi-layered harmony search algorithm (MLHSA). In this new model, structural characteristics are added to HSA for improvement of exploration and exploitation capability. In addition, the MLHSA is applied to benchmark optimization problems to verify the efficiency of proposed algorithm. The results show the strength of MLHSA and it has competitiveness. This multi-layer concept can be added to other meta-heuristic algorithms, and it will be a helpful tool for improvement of the existing algorithms. In addition, the MLHSA can be applied to multi-objective optimization problems in the future research.

Acknowledgements. This research was supported by a grant from The National Research Foundation (NRF) of Korea, funded by the Korean government (MSIP) under no. 2016R1A2A1A05005306.

References

1. Blum, C., Roli, A.: Metaheuristics in combinatorial optimization: overview and conceptual comparison. ACM Comput. Surv. (CSUR) **35**(3), 268–308 (2003)
2. Geem, Z.W., Kim, J.H., Loganathan, G.V.: A new heuristic optimization algorithm: harmony search. Simulation **76**(2), 60–68 (2001)
3. Im, S.S., Yoo, D.G., Kim, J.H.: Smallest-small-world cellular harmony search for optimization of unconstrained benchmark problems. J. Appl. Math. **2013** (2013). Article ID: 635608
4. Al-Betar, M.A., Khader, A.T., Awadallah, M.A., Alawan, M.H., Zaqaibeh, B.: Cellular harmony search for optimization problems. J. Appl. Math. **2013** (2013). Article ID: 139464
5. Reca, J., Martínez, J.: Genetic algorithms for the design of looped irrigation water distribution networks. Water Resour. Res. **42**(5) (2006). Article ID: W05416
6. Reca, J., Martínez, J., Gil, C., BaÃśos, R.: Application of several meta-heuristic techniques to the optimization of real looped water distribution networks. Water Resour. Manage. **22**(10), 1367–1379 (2007)
7. Geem, Z.W.: Particle-swarm harmony search for water network design. Eng. Optim. **41**(4), 297–311 (2009)
8. Bolognesi, A., Bragalli, C., Marchi, A., Artina, S.: Genetic heritage evolution by stochastic transmission in the optimal design of water distribution networks. Adv. Eng. Softw. **41**(5), 792–801 (2010)
9. Sadollah, A., Yoo, D.G., Kim, J.H.: Improved mine blast algorithm for optimal cost design of water distribution systems. Eng. Optim. **47**(12), 1602–1618 (2015)

Application of Self-adaptive Method in Multi-objective Harmony Search Algorithm

Young Hwan Choi[1], Ho Min Lee[1], Do Guen Yoo[2], and Joong Hoon Kim[1(✉)]

[1] Department of Civil, Environmental and Architectural Engineering, Korea University, Seoul 02841, South Korea
{Younghwan87,dlgh86,jaykim}@korea.ac.kr

[2] K-water Research Institute, Korea Water Resources Corporation, 200 Sintangin-Ro, Daedeok-Gu, Daejeon 34350, South Korea
dgyoo411@kwater.or.kr

Abstract. The Harmony Search algorithm (HSA) is inspired by musical improvisation process searching for a perfect state of harmony. Although many variants have been released and increasing number of applications have appeared, selecting suitable parameter values for the optimization algorithm is not an easy task. To overcome the difficulty, many researchers developed skillful parameter-setting methods for the algorithm parameters such as Parameter-setting-Free methods and Self-adaptive methods. These methods have been applied in various research areas (e.g., mathematics, civil engineering, mechanic engineering, and economics) and considered different formulations for solving their problems. This study applies Self-adaptive methods in the multi-objective HSA framework to solve an engineering problem (i.e., water distribution network design). It can be efficiently applied to the search for Pareto optimal solutions in the multi-objective solution space.

Keywords: Harmony search algorithm · Parameter-setting-free method · Self-adaptive method

1 Introduction

In multi-objective optimization computing, it is important to assign suitable parameters to each optimization problem in order to obtain better solutions. Since the first attempt to apply multi-objective optimization concepts using the genetic algorithm [1], multi-objective optimization algorithms have been researched deeply, and are widely applied for various engineering applications. In particular, for water distribution network (WDN) design, Gessler and Walski [2] suggested the application of a multi-objective optimal design of the WDN optimization model. Halhal [3] adopted multi-objective optimal design in order to solve the maintenance problem of WDN, in which the maximization of pressure head allowance at each node and minimization of cost were considered. Ostfeld solved the multi-objective optimal design problem of WDNs using the

J. Del Ser (ed.), *Harmony Search Algorithm*, Advances in Intelligent Systems and Computing 514, DOI 10.1007/978-981-10-3728-3_3

split pipe method, where the minimal cost and maximal reliability are considered simultaneously [4]. Likewise, multi-objective optimization algorithms have the benefit of being able to handle various objectives that have a trade-off relationship, with each objective simultaneously satisfying its own constraints.

However, in the case of meta-heuristic algorithms, the importance of parameter settings is increasing because the settings for each optimization algorithm have a strong influence on the optimal solution. For this reason, some researchers have recently investigated a self-adaptive methods in order to solve the challenge of parameter setting in meta-heuristic algorithms. Mahdavi proposed an improved harmony search (IHS) algorithm, which is the first variant since the advent of harmony search (HS) [5]. This proposed algorithm has an adjustable PAR parameter setting and dynamic bandwidth. It is obviously different from the basic HS, where the PAR parameter and bandwidth have fixed values. The PAR value is updated and increases linearly in each iteration and the bandwidth value exponentially decreases. However, IHS is limited in that it only improves local search ability. Geem and Sim proposed the parameter-setting-free (PSF) method, which is an example of a parameter self-adaptive technique [6]. This enables the parameters to be automatically set (i.e. the harmony memory considering rate (HMCR) and pitch adjustment rate (PAR)) during the simulation. To overcome this limitation, the global-best HS algorithm (GHS) was developed [7]. They introduced a new improvisation mechanism by borrowing the concept of particle swarm optimization (PSO) and added a new operator, in which each particle represents a candidate solution to the optimization problem. Although recent studies emphasize finding optimal parameters, these studies are limited in that they do not apply the multi-objective concept in WDN.

Similarly to HS, various optimization algorithms have been developed to improve optimal design performance, applying self-adaptive techniques and automatic parameter tuning. The parameter self-adaptive algorithm to the Pareto differential evolution (PDE) developed by the self-adaptive PDE method [8]. This allows the inadequate range of the crossover rate and mutation rate to be automatically set. Huang applied PDE to multi-objective optimization [9]. Through such applications, the self-adaptive algorithm has been widely used in many optimization studies on parameter setting and constraint adjustment [10–12]. However, the majority of previous studies related to self-adaptive methods consider a single objective function and apply either a genetic algorithm, differential evolution, or PSO. Therefore, in this study, the self-adaptive MOHS is proposed. This algorithm uses not only the multi-objective harmony search algorithm, but also a self-adaptive method to solve multi-objective WDN design problems without requiring the burdensome tuning of adaptable parameters. In order to verify the self-adaptive MOHS, this method is applied to well-known benchmark networks in WDN design. The results are compared with those obtained using MOHS.

2 Self-adaptive Multi-objective Harmony Search

Self-adaptive MOHS is an integrated optimization model that employs the HS algorithm [13,14], the PSF method, which is an example of a self-adaptive method that adopts non-dominated sorting [15], and the crowding-distance concept [16] to solve multi-objective problems. The proposed algorithm is allowed the parameters of the optimization algorithm to be set up automatically and various objective functions to be considered simultaneously.

2.1 Self-Adaptive Technique

Initially, to apply to self-adaptive technique, the operation type memory (OTM) is constituted depending on the HM generation type. The equations for HM and OTM are presented in Eq. (1) as follows.

$$\text{HM} = \begin{pmatrix} x_1^1 & x_2^1 & \ldots & x_n^1 \\ x_1^2 & x_2^2 & \ldots & x_n^2 \\ \vdots & \ddots & \ddots & \vdots \\ x_1^{HMS} & x_2^{HMS} & \ldots & x_n^{HMS} \end{pmatrix}, \text{OTM} = \begin{pmatrix} y_1^1 & y_2^1 & \ldots & y_n^1 \\ y_1^2 & y_2^2 & \ldots & y_n^2 \\ \vdots & \ddots & \ddots & \vdots \\ y_1^{HMS} & y_2^{HMS} & \ldots & y_n^{HMS} \end{pmatrix}, \quad (1)$$

where x_n denotes a decision variable and y_n denotes a generation method for the decision variable. Each y_n represents one of three cases: (i) random selection, (ii) harmony memory consideration, and (iii) pitch adjustment. These three cases are chosen using the values of HMCR and PAR which are user parameters in original harmony search algorithm. And in the step of generating new solution, the HMCR and PAR are calculated by generation method for the new solution.

New HM elements can be generated through random selection, HM consideration, or pitch adjustment. Random selection is used to generate HM elements using a randomly selected value in the allowed HM range $(x_i^{Lower}, x_i^{Upper})$. When generating a new element of memory (x_1^{New}), it is selected from a specific HM $[x_1^1, x_1^{HMS}]$ as follows.

$$x_i^{new} = \begin{cases} x_i \in [x_i^{Lower}, x_i^{Upper}] & \text{(i) } 1 - \text{HMCR} \\ x_i \in HM = [x_i^1, x_i^2, \ldots, x_i^{HMS}] & \text{(ii) HMCR} \end{cases} \quad (2)$$

Pitch adjustment is a method that generates a new memory element by adjusting the according to Eq. (3), where new memory element generated by considering HM as

$$x_i^{new} = \begin{cases} x_i^{new} + b_w & \text{(i) PAR} \\ x_i^{new} & \text{(ii) } 1 - \text{PAR} \end{cases} \quad (3)$$

where new values are calculated for HMCR and PAR at every iteration. Because the solutions are generated by random selection with a high rate during the early iterations $(1, 2, \ldots, 100)$, HMCR and PAR are calculated after more than 100 iterations. In this study, 0.5 is selected for the initial HMCR and PAR values $(HMCR_1 = PAR_1 = 0.5)$.

After generating new HM, the new HMCR and PAR are calculated using the OTM. The individual HMCR and PAR are determined as follows:

$$\text{HMCR}_i = \frac{\#(y_i^j = \text{Memory and Pitch})}{HMS}, \text{PAR}_i = \frac{\#(y_i^j = \text{Pitch})}{HMS}, \quad (4)$$

where $\#(\cdot)$ is a count function representing the number of specific operations (e.g. harmony memory considering or pitch adjusting) in HM. HMCR_i and PAR_i represent the HMCR and PAR of the i-th decision variable, respectively.

As the number of iterations increases, the HMCR generally increases, but the PAR decreases. This trend enables us to increase the HMCR to 1 and decrease the PAR to 0. In order to prevent such a problem, noise value Φ is set between 0–1 in the calculation of HMCR_i and PAR_i. The equations governing this are as follows:

$$\text{HMCR}_i = \begin{cases} \text{HMCR}_i + \Phi \times U(-1,1) & \text{if } \text{HMCR}_i \in [0,1] \\ \text{HMCR}_i & \text{if } \text{HMCR}_i \notin [0,1] \end{cases} \tag{5}$$

$$\text{PAR}_i = \begin{cases} \text{PAR}_i + \Phi \times U(-1,1) & \text{if } \text{PAR}_i \in [0,1] \\ \text{PAR}_i & \text{if } \text{PAR}_i \notin [0,1] \end{cases} \tag{6}$$

where $U(-1,1)$ is a random value between $[-1,1]$.

3 Objective Functions and Constraints

In this study, to evaluate the proposed algorithm, the minimize construction cost [17] and the maximize system resilience [18] used as the objective functions.

$$\text{Min. Cost} = \sum_{i=1}^{N} C(D_i)L_i, \tag{7}$$

$$\text{Max. Resilience} = \frac{\sum_{j=1}^{N} q_j(h_j - h_j^*)}{\sum_{k=1}^{NR} Q_k H_k + \sum_{i=1}^{NP} P_i - \sum_{j=1}^{N} q_j h_j^*} \tag{8}$$

where, $C(D_i)$ is the cost function of the i-th pipe per unit length (m) of each pipe diameter, L_i is the length (m) of the i-th pipe, D_i is the pipe diameter (mm) of the i-th pipe, and N is the total number of pipes. N, NR, and NP are the number of nodes, reservoirs, and pumps, respectively, h_j is a pressure head at node j, h_j^* is the head requirement at node j, H_K is the water level of reservoir K, Q_K is the water flow of reservoir K, and P_i is the power of pump i. The system resilience is in the range of 0–1, where a higher value indicates better resilience.

The constraints of this study is used the minimum pressure. If the pressure at each node does not meet the minimum pressure constraint during the hydraulic analysis of the optimal design, a large penalty point value is added to the penalty constant. This penalty value affects the value of the objective function.

4 Application Results

The purpose of this study is to confirm the search efficiency and improvement in convenience provided by the application of the proposed self-adaptive method to the MOOA. The proposed algorithm and MOHS are applied and fairly compared using same HMS depending on the test problem. This minimizes the effects of the parameter tuning and initial solution quality on the final solution quality. The initial solutions were generated using different numbers of HMS (i.e. Two-loop network: 30, Hanoi network: 50) by random search and this initial solution set was used in all iterative simulation runs. And the convergence and diversity performance index are used to quantitatively evaluate two algorithms (i.e., self-adaptive multi-objective harmony search and MOHS).

In this section, the results of the optimal WDN design using two algorithm (i.e., Self-adaptive MOHS, MOHS). In Table 1, self-adaptive MOHS obtains better convergence and diversity for almost all cases to compare MOHS.

Table 1. Performance index of benchmark network.

Index	Algorithms	Two-loop network	Hanoi network
Convergence [21]	MOHS	0.667	0.4633
	Self-adaptive MOHS	0.811	0.7011
Diversity [22]	MOHS	0.8238	0.9242
	Self-adaptive MOHS	0.8696	0.9094

In the diversity aspect, the self-adaptive MOHS generated wider Pareto fronts than MOHS. This is demonstrated visually in Table 1. The value of diversity index of self-adaptive MOHS shows 0.811 in Two-loop network. It can evaluate the diversity of optimal MOOA solutions to use the difference of minimum and maximum values of the objective function. And the convergence of self-adaptive MOHS also show the better convergence. In Table 1, self-adaptive MOHS achieves remarkably high convergence/non-domination compared with MOHS in the two-loop and Hanoi networks.

5 Conclusions

In this study, the self-adaptive MOHS was developed in order to consider various design factors and enhance the efficiency of the optimal design for a WDN. The self-adaptive MOHS model proposed in this study was applied to three well known benchmark networks for optimal WDN design. Finally, performance indices were compared to evaluate the convergence and diversity of the solutions as well as the homogeneous spatial distributions of consecutive solutions. The evaluation of the self-adaptive MOHS Pareto fronts for three different WDN design problems indicates that the proposed method generally performed better in terms of both convergence and diversity than MOHS. Furthermore, self-adaptive MOHS achieved similar results in terms of convergence and diversity even if no parameter sensitivity analysis was performed. Such a characteristic enables the inconvenience of setting parameters to be eliminated. Moreover, it increases search efficiency by automatically setting up the optimal parameters during the iterations of the solution search and considering various design factors simultaneously. Therefore, this method can be regarded as a useful algorithm for solving more complex and practical optimization problems in the future.

Acknowledgments. This subject is supported by Korea Ministry of Environment as Global Top Project (2016002120004).

References

1. Schaffer, J.D.: Multiple objective optimization with vector evaluated genetic algorithms. In: Grefenstette, J.J. (ed.) Proceedings of International Conference on Genetic Algorithms and Their Applications, PA, Pittsburgh, pp. 93–100 (1985)
2. Gessler, J., Walski, T.M.: Water Distribution System Optimization (No. WES/TR/EL-85-11). Army Engineer Waterways Experiment Station Vicksburg MS Environmental Lab (1985)
3. Halhal, D., Walters, G., Savic, D., Ouazar, D.: Scheduling of water distribution system rehabilitation using structured messy genetic algorithms. Evol. Comput. **7**(3), 311–329 (1999)
4. Ostfeld, A., Oliker, N., Salomons, E.: Multiobjective optimization for least cost design and resiliency of water distribution systems. J. Water Resour. Plann. Manage. **140**(12), 1943–5452 (2013)
5. Mahdavi, M., Fesanghary, M., Damangir, E.: An improved harmony search algorithm for solving optimization problems. Appl. Math. Comput. **188**(2), 1567–1579 (2007)
6. Geem, Z.W., Sim, K.B.: Parameter-setting-free harmony search algorithm. Appl. Math. Comput. **217**(8), 3881–3889 (2010)
7. Omran, M.G., Mahdavi, M.: Global-best harmony search. Appl. Math. Comput. **198**(2), 643–656 (2008)
8. Abbass, H., Sarker, R., Newton, C.: PDE: a pareto-frontier differential evolution approach for multi-objective optimization problems. In: Congress on Evolutionary Computation, vol. 2, pp. 971–978 (2001)
9. Huang, V.L., Zhao, S.Z., Mallipeddi, R., Suganthan, P.N.: Multi-objective optimization using self-adaptive differential evolution algorithm. In: IEEE Congress on Evolutionary Computation, pp. 190–194 (2009)
10. Deb, K.: Multi-Objective Optimization Using Evolutionary Algorithms, vol. 16, pp. 315–338. Wiley, Chichester (2001)
11. Tessema, B., Yen, G.G.: A self-adaptive penalty function based algorithm for constrained optimization. In: IEEE Congress on Evolutionary Computation, pp. 246–253 (2006)
12. Chen, H.P., Gu, F., Lu, B.Y., Gu, C.S.: Application of self-adaptive multi-objective genetic algorithm in flexible job shop scheduling. J. Syst. Simul. **8**, 053 (2006)
13. Geem, Z.W., Kim, J.-H., Loganathan, G.V.: A new heuristic optimization algorithm: harmony search. Simulation **76**(2), 60–68 (2001)
14. Kim, J.H., Geem, Z.W., Kim, E.S.: Parameter estimation of the nonlinear muskingum model using harmony search. J. Am. Water Resour. Assoc. **37**(5), 1131–1138 (2001)
15. Fonseca, C.M., Fleming, P.J.: Genetic algorithms for multiobjective optimization: formulation discussion and generalization. In: Proceedings of ICGA 1993, pp. 416–423 (1993)

16. Deb, K., Agrawal, S., Pratap, A., Meyarivan, T.: A fast elitist non-dominated sorting genetic algorithm for multi-objective optimization: NSGA-II. In: Schoenauer, M., Deb, K., Rudolph, G., Yao, X., Lutton, E., Merelo, J.J., Schwefel, H.-P. (eds.) PPSN 2000. LNCS, vol. 1917, pp. 849–858. Springer, Heidelberg (2000). doi:10.1007/3-540-45356-3_83
17. Shamir, U., Howard, C.D.: Water distribution systems analysis (1968)
18. Todini, E.: Looped water distribution networks design using a resilience index based heuristic approach. Urban Water **2**(2), 115–122 (2000)
19. Alperovits, E., Shamir, U.: Design of optimal water distribution systems. Water Resour. Res. **13**(6), 885–900 (1977)
20. Fujiwara, O., Khang, D.B.: A two phase decomposition method for optimal design of looped water distribution networks. Water Resour. Res. **26**(4), 539–549 (1990)
21. Zitzler, E., Deb, K., Thiele, L.: Comparison of multiobjective evolutionary algorithms: empirical results. Evol. Comput. **8**(2), 173–195 (2000)
22. Zitzler, E.: Evolutionary Algorithms for Multiobjective Optimization: Methods and Applications. Shaker, Ithaca (1999)

A Comparative Study of Exploration Ability of Harmony Search Algorithms

Anupam Yadav[1], Neha Yadav[2], and Joong Hoon Kim[2(✉)]

[1] Department of Sciences and Humanities,
National Institute of Technology Uttarakhand, Srinagar (Garhwal), India
anupam@nituk.ac.in

[2] Department of Civil, Environmental and Architectural Engineering,
Korea University, Seoul 02841, South Korea
nehayadav441@gmail.com, jaykim@korea.ac.kr

Abstract. Harmony Search Algorithm (HSA) is one of the efficient optimization algorithms among various optimization algorithms proposed over the years. In literature, many variants of the Harmony search algorithms are proposed based on the various ideas. In this article the exploration ability of the four HS algorithms are compared. An experimental comparative study of exploration ability of different versions of HS algorithms is done. The article provides a detailed explorative abilities of various HS algorithms. The comparison is based on the theoretical studies performed in [1]. The theoretical conclusions of the exploration analysis are justified with experimental studies. This article concludes with the searching ability of various HS algorithms along with their ranking with most explorative HS algorithm.

Keywords: Harmony search · Exploration · Variance

1 Introduction

Harmony Search Algorithm is one of the efficient optimization algorithm developed by Geem et al. [2]. It is inspired from the music improvisation process. The analogy between music improvisation and optimization can be established by creating correspondence between music player to the decision variable. In order to execute the technique in real time optimization each decision variable chooses and possible range together to make a solution. This solution is then improved by creating harmony memory, and pitch adjusting. Over the year various optimization algorithms [6–8] are proposed but HSA remains one of the best choice for function optimization. In order to introduce the HS algorithm for engineering optimization.

Let $X_i = (x_i^1, x_i^2, ..., x_i^{HMS})$ be the initially generated harmony where i varies from $1 : Dim$ and let $X'(x_1, x_2, ...x_{Dim})$ be a new harmony. If $X_i = (x_i^1, x_i^2, ..., x_i^{HMS})$ is the ith harmony vector from all the initialized harmony vectors then the improvised harmony is generated based on the following rule.

J. Del Ser (ed.), *Harmony Search Algorithm*, Advances in Intelligent Systems and Computing 514, DOI 10.1007/978-981-10-3728-3_4

In order to understand the process of harmony search a brief detail of each step is presented here:

- Definition of Optimization problem.
  It is defined as

$$\text{Max or Min} F(x_i) \qquad \text{subject to} \qquad x_i \in X_i, i = 1, 2, 3, ..., N \tag{1}$$

  where $F()$ is objective function to be optimized and x_i is the solution vector of decision variables and $L_i \leq X_i \leq U_i$, L_i and U_i are the lower and upper bounds of the decision variables. At this stage we define two parameters HM size (HMS) (population size), HMCR which is harmony considering rate and PAR: pitch adjusting rate.
- In this step, the HM is generated using the formula:

$$x_i = L_i + rand(0,1)(U_i - L_i) \tag{2}$$

- New Harmony Improvisation: a new harmony vector $x'(x'_1, \ldots, x'_{HMS})$ is generate based on memory consideration, pitch adjustment and random selection. The selection follows the following criterion.

$$x'_i = \begin{cases} x_i \in x_i^1, x_i^2, x_i^{HMS} \text{ if probability HMCR} \\ \text{new } x_i \text{ if probability (1-HMCR)} \end{cases} \tag{3}$$

  The pitch adjustment is done using the following equation

$$x'_i = \begin{cases} x'_i \pm rand(0,1).bw \text{ with probability PAR} \\ x'_i \text{ with probability (1 - PAR)} \end{cases} \tag{4}$$

  where bw is an arbitrary bandwidth number and $rand(0,1)$ is the random number uniformly distributed between 0 and 1.
- Harmony Memory Update: In this step new harmony vector $x'(x'_1, x'_2, \ldots x'_{HMS})$ is compared with the worst harmony available in the harmony memory, if new harmony is better than worst then it will be replaced with the worst one.
- Stopping criterion: The algorithms will stop if it will meet with the predefined stopping criterion.

A brief working procedure of the HS algorithm is given in Algorithm 1.

Till today, a number of harmony search algorithms are designed based on various ideas such as improved harmony search [3], global best harmony search [4] and self adaptive global best harmony search [5]. Recently Yadav et al. [9] proposed a comparative studies of convergence of HSA. These algorithms have been applied to many optimization problems. The exploration ability of any optimization algorithm always remains an important aspect. Das et al. [1] have proposed a theoretical concept for measuring the searching capability of the harmony search algorithm. In this article the exploration ability of a number of harmony search

Algorithm 1. Harmony Search Algorithm

Initialization of Harmony Vector
for $i = 1 : N$ **do**
  if $U(0,1) \leq HMCR$ then /* memory consideration */ **then**
    begin
    $x_i' = x_j'$, where j $U(1,2,,HMS)$
    if $U(0,1) \leq PAR(t)$ **then** /* pitch adjustment */
      begin
      $x_i' = x_k^{best}$ where best is the index of the best harmony in the HM and $k \in U(1,N)$
    end if
  else/* random selection */
    $x_i' = LB_i + rand(0,1)(UB_i - LB_i)$
  end if
end for

algorithms is analyzed and their comparative study is presented. The theoretical comparisons of searching abilities are then justified with experimental results. The theoretical studies proposed by Das et al. [1] provides a detailed theoretical background for the exploration ability of the Harmony Search Algorithm. The expected variance of the population of harmony search is given by the following equation:

$$Exp(Var(Y)) = \frac{(m-1)}{m} \cdot [HMCR.Var(x) + HMCR \cdot (1 - HMCR) \cdot \overline{x}^2 + \frac{1}{3} HMCR \cdot PAR \cdot bw^2 + \frac{a^2}{3} \cdot (1 - HMCR)].$$

2 Comparative Studies of HSA

In this article based on the formula for expected population variance, a comparative exploration ability of five variants of HSA is compared with each other. In the next section a brief idea of the variants of harmony search algorithms are presented.

2.1 Harmony Search Algorithm

The original harmony search was proposed by Geem et al. [2] and this exactly follows the scheme presented in Introduction section.

2.2 Improved Harmony Search

It was proposed by Mahadavi et al. [3]. The major enhancement in this variant of HSA was to provide HSA with dynamic PAR and bw values. The following formulae are proposed for the dynamic calculation of parameter values:

$$PAR_{(gen)} = PAR_{min} + ((PAR_{max} - PAR_{min}))/NI_{gen}, \tag{5}$$

$$bw(gen) = bw_{max} exp(c.gen), \tag{6}$$

where $c = \ln(\frac{bw_{min}}{bw_{max}})/NI$, PAR_{min} : minimum pitch adjusting rate, PAR : pitch adjusting rate, NI: number of solution vector generations, PAR_{max} : maximum pitch adjusting rate, gen: generation number.

2.3 Global Best Harmony Search

It was proposed by Mahamed and Madhavi [4]. The major advancement in this variant was the introduction of global best harmony on the line of particle swarm optimization. Rest of the procedure is same as the original HSA.

2.4 Self Adaptive Global Best HS Algorithm

It was proposed by Kuan-ke et al. [5]. In this variant of HSA the a self adaptive strategy for the parameter values is coined. The PAR and $HMCR$ values are chosen statistically. The following formula is proposed for the calculation of bw values:

$$bw = \begin{cases} (bw_{min} + (\frac{(bw_{max} - bw_{min})}{NI}) \text{ if } t < NI/2 \\ bw_{min} \text{ if } t \geq NI/2 \end{cases} \tag{7}$$

where bw_{min} and bw_{max} are the lower and upper bounds for bw value, t is the generation number and NI is the total number of generations.

3 Experimental Studies and Analysis

As proposed by Das et al. [1], the expected variance formula:

$$Exp(Var(Y)) = \frac{(m-1)}{m} \cdot [HMCR \cdot Var(x) + HMCR \cdot (1 - HMCR) \cdot \overline{x}^2 + \frac{1}{3} HMCR \cdot PAR \cdot bw^2 + \frac{a^2}{3} \cdot (1 - HMCR)]$$

A generation wise expected variance of the HSA, Improved Harmony Search, self adaptive global best harmony search and global best harmony search are compared. All the parameter values for each scheme is kept same as proposed in the respective articles. To compare the results each algorithm is run for 1000 iterations over the benchmark functions Sphere, Rastrigin, Schewfel and Greiwank with 50 dimension and 100 HMS. The plots of the expected variance against the iterations are plotted in Figs. 1, 2, 3, 4, 5, 6, 7, 8, 9, 10, 11, 12, 13, 14, 15 and 16. The comparative plots suggests that the exploration ability of Improved harmony search is better than the other HS algorithms.

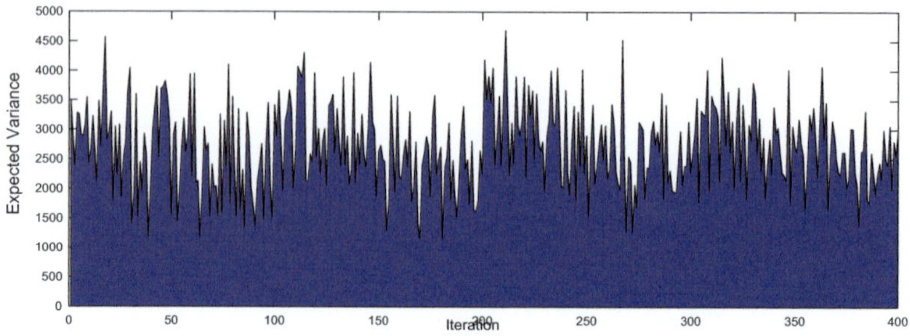

Fig. 1. Iterative variation of expected variance of Harmony search over sphere function

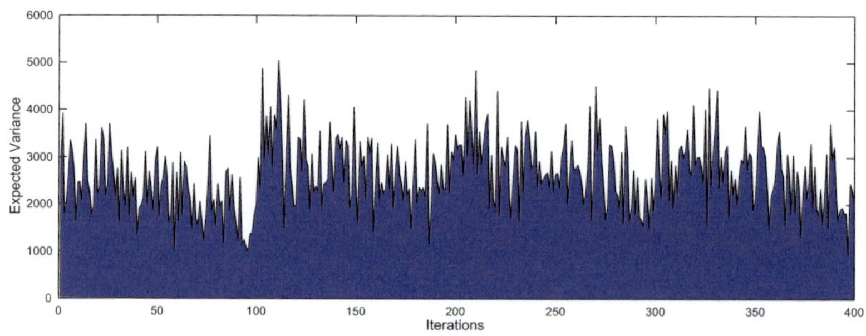

Fig. 2. Iterative variation of expected variance of Improved Harmony search over sphere function

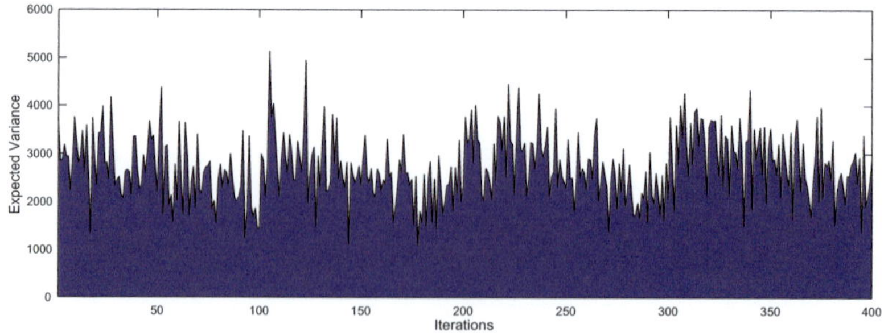

Fig. 3. Iterative variation of expected variance of global best harmony search over sphere function

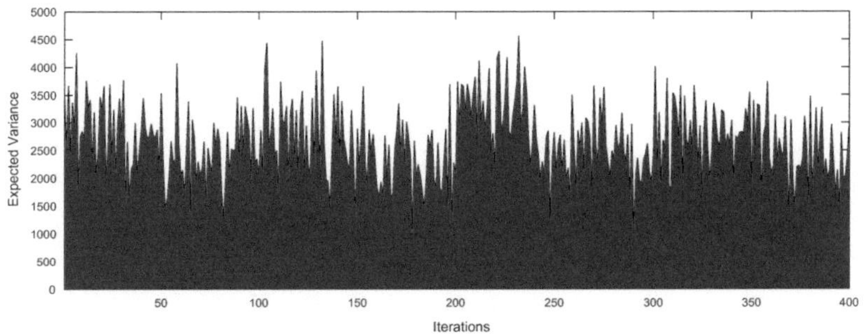

Fig. 4. Iterative variation of expected variance of self adaptive global best HS algorithm over sphere function

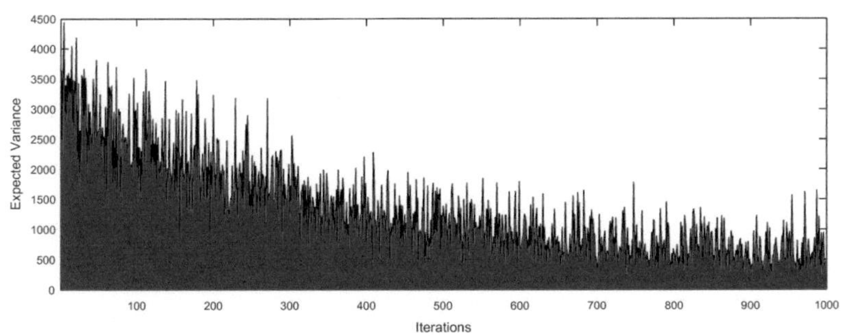

Fig. 5. Iterative variation of expected variance of Harmony search over Rastrigin function

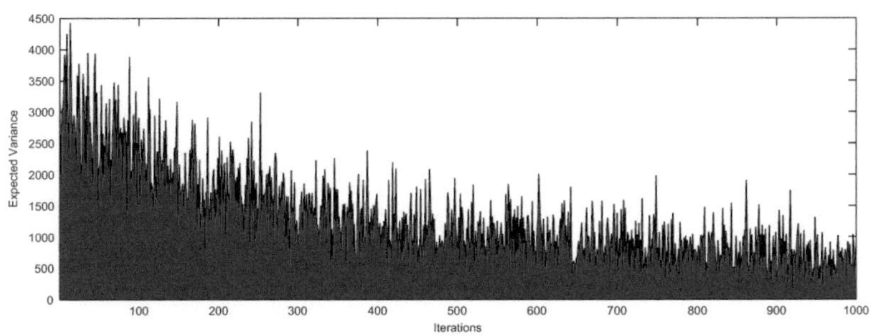

Fig. 6. Iterative variation of expected variance of Improved Harmony search over Rastrigin function

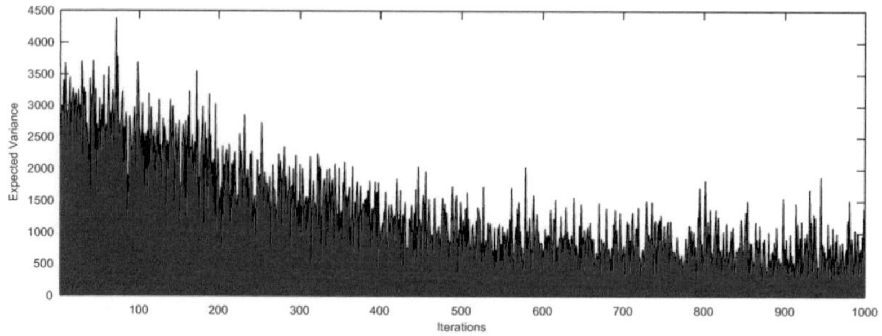

Fig. 7. Iterative variation of expected variance of global best harmony search over Rastrigin function

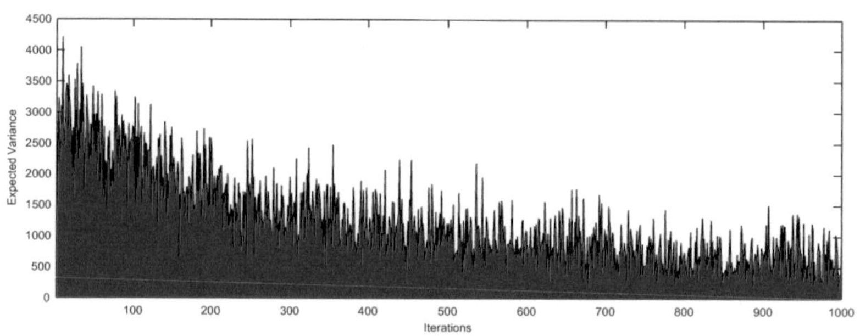

Fig. 8. Iterative variation of expected variance of self adaptive global best HS algorithm over Rastrigin function

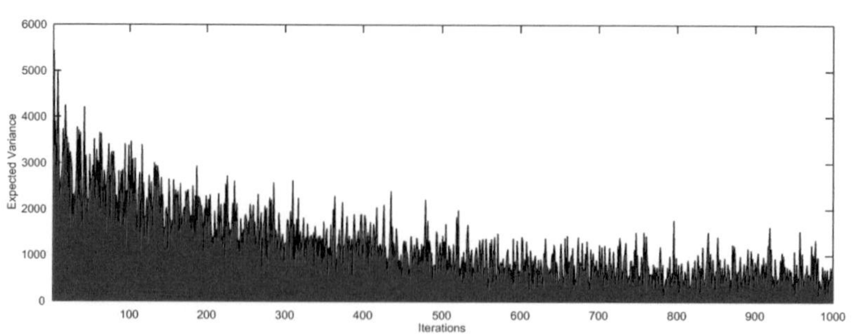

Fig. 9. Iterative variation of expected variance of Harmony search over Greiwank function

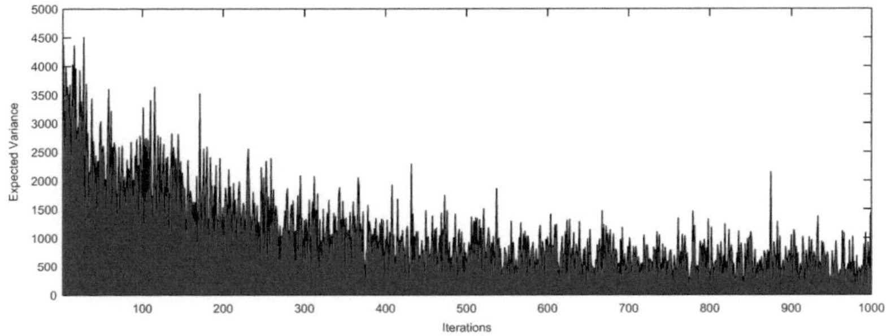

Fig. 10. Iterative variation of expected variance of Improved Harmony search over Greiwank function

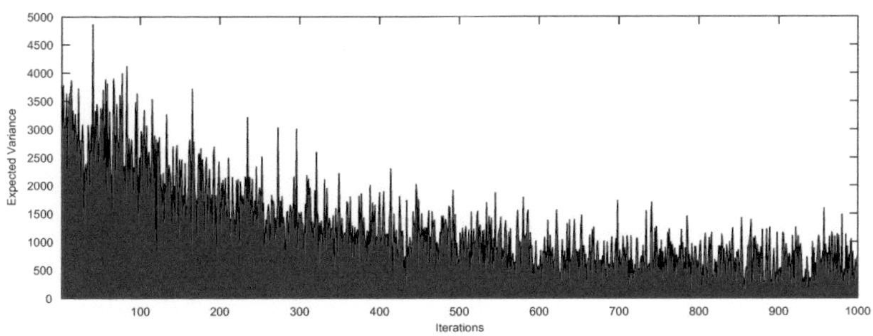

Fig. 11. Iterative variation of expected variance of global best harmony search over Greiwank function

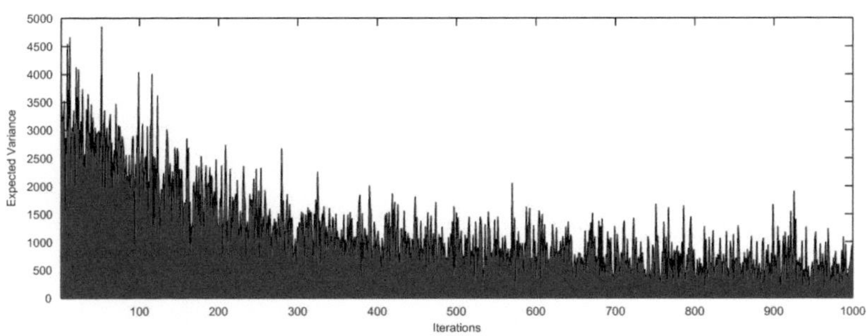

Fig. 12. Iterative variation of expected variance of self adaptive global best HS algorithm over Greiwank function

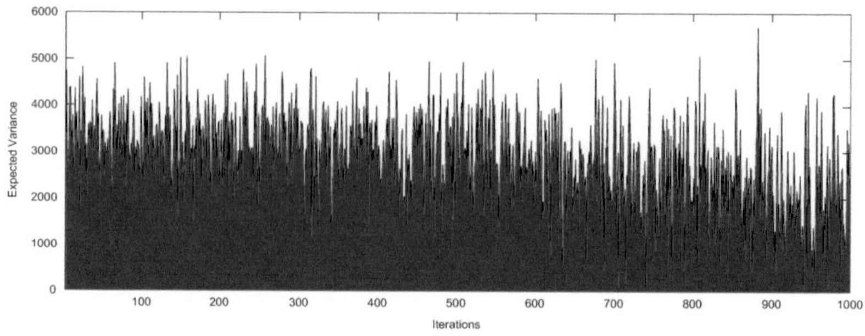

Fig. 13. Iterative variation of expected variance of Harmony search over Shwefel function

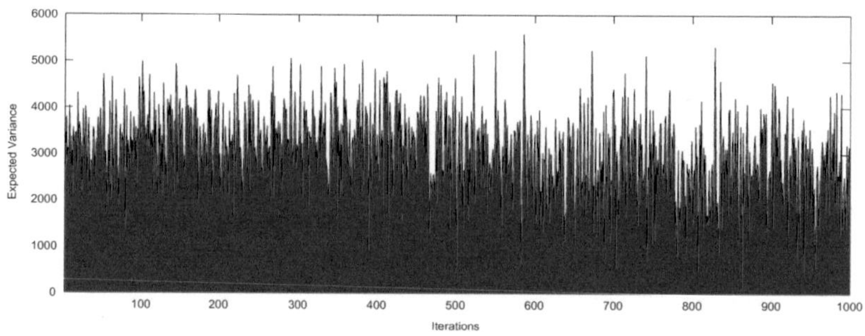

Fig. 14. Iterative variation of expected variance of Improved Harmony search over Shwefel function

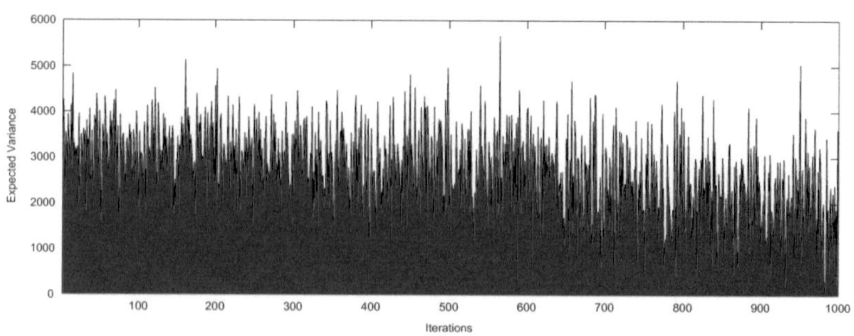

Fig. 15. Iterative variation of expected variance of global best harmony search over Shwefel function

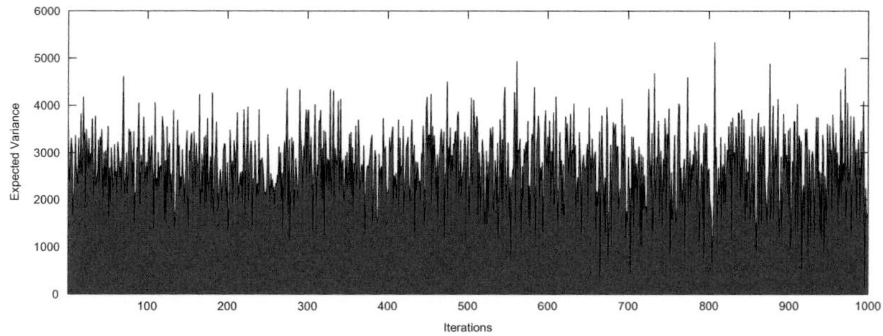

Fig. 16. Iterative variation of expected variance of self adaptive global best HS algorithm over Shwefel

4 Conclusion

In this article four variants of HSA are compared on the basis of expected variance, the expected variance is plotted against the iterations. The plots are created for four functions and the comparative results suggest that the exploration capacity of Improved harmony search is better than other algorithms.

References

1. Das, S., Mukhopadhyay, A., Roy, A., Abraham, A., Panigrahi, B.K.: Exploratory power of the harmony search algorithm: analysis and improvements for global numerical optimization. IEEE Trans. Syst. Man Cybern. Part B Cybern. **41**(1), 89–106 (2011)
2. Geem, Z.W., Kim, J.H., Loganathan, G.V.: A new heuristic optimization algorithm: harmony search. Simulation **76**(2), 60–68 (2001)
3. Mahdavi, M., Fesanghary, M., Damangir, E.: An improved harmony search algorithm for solving optimization problems. Appl. Math. Comput. **188**(2), 1567–1579 (2007)
4. Omran, M.G.H., Mahdavi, M.: Global-best harmony search. Appl. Math. Comput. **198**(2), 643–656 (2008)
5. Pan, Q.K., Suganthan, P.N., Tasgetiren, M.F., Liang, J.J.: A self-adaptive global best harmony search algorithm for continuous optimization problems. Appl. Math. Comput. **216**(3), 830–848 (2010)
6. Yadav, A., Deep, K.: Shrinking hypersphere based trajectory of particles in PSO. Appl. Math. Comput. **220**, 246–267 (2013)
7. Yadav, A., Deep, K.: An efficient co-swarm particle swarm optimization for non-linear constrained optimization. J. Comput. Sci. **5**(2), 258–268 (2014)
8. Yadav, A., Deep, K., Kim, J.H., Nagar, A.K.: Gravitational swarm optimizer for global optimization. Swarm Evol. Comput. **31**, 64–89 (2016)
9. Yadav, A., Yadav, N., Kim, J.H.: A study of harmony search algorithms: exploration and convergence ability. In: Kim, J.H., Geem, Z.W. (eds.) ICHSA2015. AISC, vol. 382, pp. 53–62. Springer, Heidelberg (2016). doi:10.1007/978-3-662-47926-1_6

New Bio-inspired Heuristics

The Extraordinary Particle Swarm Optimization and Its Application in Constrained Engineering Problems

Thi Thuy Ngo[1], Ali Sadollah[1], Do Guen Yoo[2], Yeon Moon Choo[1], Sang Hoon Jun[1], and Joong Hoon Kim[1(✉)]

[1] Department of Civil, Environmental and Architectural Engineering, Korea University, Seoul 136-713, South Korea
tide4586@yahoo.com, jaykim@korea.ac.kr

[2] K-water Research Institute, Korea Water Resources Corporation, 200 Sintangin-Ro, Daedeok-Gu, Daejeon 34350, South Korea
dgyoo411@kwater.or.kr

Abstract. The particle swarm optimization (PSO) is a natural-inspire optimization algorithm mimicking the movement behavior of animal flocks for food searching. Although the algorithm presents some advantages and widely application, however, there are several drawbacks such as trapping in local optima and immature convergence rate. To overcome these disadvantages, many improved versions of PSO have been proposed. One of the latest variants is the extraordinary particle swarm optimization (EPSO). The particles in the EPSO are assigned to move toward their own determined target through the search space. The applicability of EPSO is verified by several experiments in engineering optimization problems. The application results show the outperformance of the EPSO than the other PSO variants in terms of solution searching and as well as convergence rate.

Keywords: Extraordinary particle swarm optimization · Engineering optimization problem · Urban drainage system

1 Introduction

Nowadays, metaheuristics algorithms have been widely applied in many engineering fields and for solving different complex problems. Among them, bio-inspired algorithms are based on successful characteristics of biological systems and natural phenomena such as the genetic algorithms (GAs) [1], the ant system [2], the particle swarm optimization (PSO) [3]. The PSO is one of the most common optimizer originated by Kennedy and Eberhart [3] for solving various problems arising in sciences and engineering. The metaheuristic global optimization technique inspired the movement behavior of bird and fish flock when searching flood.

In the PSO, each individual namely particle share information with others. The movement of particles is impacted by the best particle among them

J. Del Ser (ed.), *Harmony Search Algorithm*, Advances in Intelligent Systems and Computing 514, DOI 10.1007/978-981-10-3728-3_5

(global best, *gbest*) and their best experience (personal best, *pbest*) as well. The trade-off between exploration and exploitation processes is involved by combining local and global searches. The fast convergence, easy implementation, and simple computation are considered as advantages of PSO. However, the PSO exhibits some disadvantages in trapping in local optima and slow convergence rate in the later iterations when particles are in the adjacent area of the optimal solution. Therefore, several variants of PSO are proposed to overcome aforementioned drawbacks such as the cooperative PSO (CPSO) [4], the comprehensive learning PSO (CLPSO) [5], the fully informed particle swarm (FIPS) [6], the Frankenstein's PSO (F-PSO) [7], and the adaptive inertia weight PSO (AIW-PSO) [8].

In this paper, a new collaborative approach is proposed to adjust the movement of particles. Particles in the new movement strategy have interaction with their target that can be global best, local bests, or even the worst particle. This approach may help the PSO to escape from the local optima by a probability of breaking the rule of following the best locations. In the other word, particles may not fly toward its best and global best and may break the swarm rule used in the PSO and become extraordinary particles. The improved variant of PSO, therefore, is called as extraordinary particle swarm optimization (EPSO). The performance of EPSO is verified by three constrained optimization problems in mechanical design and drainage system.

2 Extraordinary Particle Swarm Optimization

The EPSO is a new variant of PSO in which each particle can be seen as an extraordinary particle which could fly through the search space by its own direction [9]. In standard PSO, each particle ith flies toward current best (global best and personal best) and is indicated by position $X_i^t = (x_i^1, x_i^2, \ldots, x_i^D)$ and velocity $V_i^t = (v_i^1, x_i^2, \ldots, x_i^D)$ in the D-dimensional search space. The movement of particles toward the current bests is updated by iteration using cognitive and social coefficients and inertia weight.

In the EPSO, in addition to moving toward global and personal bests, particles may fly to their determined target. Each particle has its own target at each iteration. The target could be the global best, personal best or any particle with different cost/fitness. Particles choose their target stochastically and change their targets through iterations. The particles in the EPSO may not follow the current bests (i.e., personal and global bests) and tend to interrupt the movement rules used in the standard PSO. That is the reason we call them as extraordinary particles. At every iteration, each extraordinary particle randomly chooses its own target to move toward. The updating equations for the particle velocity and its new location are given as follows:

$$\overrightarrow{V_i}(t+1) = C(\overrightarrow{X_{Ti}}(t) - \overrightarrow{X_i}(t)) \tag{1}$$

$$\overrightarrow{X_i}(t+1) = \overrightarrow{X_i}(t) + \overrightarrow{V_i}(t+1) \tag{2}$$

where $\overrightarrow{X_{Ti}}(t)$ is the determined target T of particle i^{th} at iteration t; C is combined component including cognitive and social factors.

The target particle could be the global best or the personal best that extraordinary particle has experienced or just any particle in the swarm population. The uniform random search process is also taken part in the EPSO if the index of determined target (T) beyond the range of the potential target [0, *Tup*] where Tup can be calculated as round ($\alpha \times Npop$) with Îś is a user-defined parameter so introduced to control the range of target solution. The detailed procedure of the EPSO is described in the following steps:

Step 1: Setting initial parameters: combined coefficient C ranging from zero to two, and upper bound α of a potential target in the entire population, ranging from zero to one.

Step 2: Initializing the swarm population.

Step 3: Sorting particles based on their cost/fitness.

Step 4: Selecting the target: Select the target for each particle at each iteration among all particles using the determined index (T) of particles.

$$T = round(rand \times N_{pop}) \tag{3}$$

where N_{pop} is the population size and rand is uniformly distributed random number [0,1]. The index of the target (T) ranges from zero to Npop.

Step 5: Updating particles to their new locations using the following equation:

$$\overrightarrow{X_i(t+1)} = \begin{cases} \overrightarrow{X_i}(t) + \overrightarrow{V_i}(t+1) & \text{if } T \in Tup \\ \overrightarrow{LB_i} + rand(\overrightarrow{UB_i} - \overrightarrow{LB_i}) & \text{otherwise} \end{cases} \tag{4}$$

where LB and UB are lower and upper bounds of the search space, respectively.

Step 6: Check the stopping criterion. If the stopping condition is met, stop. Otherwise, return to Step 3.

3 Engineering Constrained Benchmarks

This section provides applications of EPSO for some practical engineering design problems. The EPSO accompanied with other optimizers including the standard PSO, cooperative PSO (CPSO) [4], hybrid real-binary particle swarm optimization (HPSO) [10], particle swarm optimization with differential evolution (PSO-DE) [11], hybrid nelder-mead simplex search and particle swarm optimization (NM–PSO) [12] have been applied and compared in terms of statistical optimization results. The optimal solutions of EPSO are obtained after 30 independent runs with different numbers of function evaluations (NFEs) while those of other PSO variants are extracted from literature [13,14].

3.1 The Pressure Vessel Design Problem

The pressure vessel design problem aims to minimize the total cost of material, forming and welding subject to constraints of design variables [15]. There are

Table 1. Comparison of statistical optimization results gained by the EPSO and other optimizers for the pressure vessel design problem

Optimizer	Worst	Average	Best	SD	NFEs
PSO	14076.324	8756.6803	6693.7212	1.49E+03	8,000
CPSO	6363.8041	6147.1332	6061.0777	8.64E+01	240,000
HPSO	6288.6770	6099.9323	6059.7143	8.62E+01	81,000
PSO-DE	N/A	6059.7140	6059.7140	N/A	42,100
NM-PSO	5960.0557	5946.7901	5930.3137	9.16E+00	80,000
EPSO	7315.6752	6254.1804	5885.3383	4.24E+02	10,000
	6076.6205	5920.8442	5885.3328	5.21E+01	100,000

four decision variables in this problem including the thickness of the shell and the head, the inner radius, and the length of the cylindrical section. Table 1 presents obtained results from EPSO and other PSO variants.

The comparison results show that EPSO could find the better solution by less NFEs than other PSO variants. After 30 independent runs, EPSO derives the best, worst and average solutions far lower than those of other optimizers. The standard deviation (SD) of multiple runs of EPSO is also smaller than other showing the stability of the algorithm.

3.2 The Tension/Compression Spring Design Problem

The objective of tension/compression spring design problem is to minimize the weight of a tension/compression spring subject to several constraints on the minimum deflection, shear stress, surge frequency, limits on outside diameter, and design variables [16]. The optimal results obtained from PSO variants are shown in Table 2.

Table 2. Comparison of statistical optimization results gained by the EPSO and other optimizers for the tension/compression spring design problem

Optimizer	Worst	Average	Best	SD	NFEs
PSO	0.071802	0.019555	0.012857	1.16E−02	2,000
CPSO	0.012924	0.012730	0.012675	5.20E−04	240,000
HPSO	0.012719	0.012707	0.012665	1.58E−05	81,000
PSO-DE	0.012665	0.012665	0.012665	1.20E−08	24,950
NM-PSO	0.012633	0.012631	0.012630	8.47E−07	80,000
EPSO	0.016911	0.014056	0.012670	1.27E−03	2,000
	0.014218	0.013030	0.012669	3.64E−04	10,000

In this problem, with the same NFEs, the obtained results of EPSO surpass other variants of PSO in terms of the best, worst and average solutions. The lower standard deviation in many runs of EPSO shows that the algorithm is more stable than others. The smaller NFEs and better optimal solutions illustrate better convergence rate of EPSO than other variants.

4 Drainage System Design and Operation Problem

The design and operation optimization problem are used to verify the ability of EPSO. The study area is a detention basin in Daerim 3 catchment (Seoul, Korea) including side-weir B drains storm water from conduit A to store in storage unit C and slowly releases to the main conduit E by a pump station D (Fig. 1). The optimal detention facilities aim to minimize flooding damages in the neighbor area of detention reservoir C and at control node F as well. The detailed description of the case study and problem formulation can be extracted from literature [17]. In this study, the decision variables are operating level of two pump units and crest height of side-weir. The optimization techniques are integrated with flow routing model (EPA-SWMM) to simulate historical flood events in 2010 and 2011 to get optimal solutions. The comparison between optimal solutions derived from EPSO with the current facilities is also presented in Table 3.

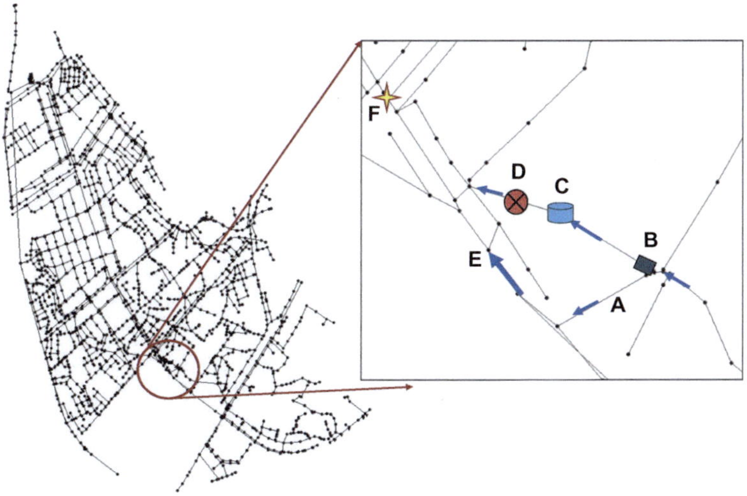

Fig. 1. Daerim3 detention reservoir facilities and the system layout

In general, optimal solution dramatically reduces flooding situation in both the adjacent area of detention basin and the entire watershed. The flooded duration in the detention reservoir neighbor area is totally cut off. It means that flooding does not occur in this area and water level in detention reservoir is retained lower than the maximum possible depth. The water level at control node is maintained with little fluctuation. In the whole watershed, flooded volume and number of flooded nodes are partially decreased under optimal design and operation (79 % in flood event 2010 and 19 % in flood 2011). The number of flooded nodes in the whole watershed is also reduced in optimal facilities. The result demonstrates the high efficiency of EPSO to search optimal design and operation rule for flooding reduction purpose.

Table 3. Comparison of the present and optimal detention reservoir facilities regarding flood damage in flood events 2010 and 2011

Criteria	Flood event in 2010		Flood event in 2011	
	Present	Optimal	Present	Optimal
Max depth at detention basin (m)	3.20	3.01	3.20	1.17
Flooded duration at detention basin (10 min)	15	0	1	0
Max depth at control node (m)	1.90	1.89	1.89	1.91
Total flooded volume ($10^3 m^3$)	6.729	1.426	0.607	0.491
Number flooded nodes	18	17	14	14

5 Conclusion

The new cooperative approach for particle swarm optimization using extraordinary particles, namely EPSO is introduced in this paper. The two constrained constructional problems are described to verify algorithm's efficiency. The algorithm surpasses other PSO variants in terms of solutions and convergence rates. A real optimization problem of the drainage system is also introduced to present ability of the EPSO when combining with a simulation model. The comparison results demonstrate the promising applicability of EPSO in complex engineering optimization problems.

Acknowledgments. This research was supported by a grant (13AWMP-B066744-01) from the Advanced Water Management Research Program funded by the Ministry of Land, Infrastructure, and Transport of the Korean government.

References

1. Holland, J.H.: Adaptation in Natural and Artificial Systems: An Introductory Analysis with Applications to Biology, Control, and Artificial Intelligence. University of Michigan Press, Michigan (1975)
2. Dorigo, M., Maniezzo, V., Colorni, A.: Ant system: optimization by a colony of cooperating agents. IEEE Trans. Syst. Man Cyb. **26**(1), 29–41 (1996)
3. Kennedy, J., Eberhart, R.C.: Particle swarm optimization. In: IEEE International Conference on Neural Networks, pp. 1942–1948. Piscataway, NJ (1995)

4. Bergh, F.V.D., Engelbrecht, A.P.: A Cooperative approach to particle swarm optimization. IEEE T. Evolut. Comput. **8**(3), 225–239 (2004)
5. Liang, J.J., Qin, A.K.: Comprehensive learning particle swarm optimizer for global optimization of multimodal functions. IEEE Trans. Evolut. Comput. **10**(3), 281–295 (2006)
6. Mendes, R., Kennedy, J., Neves, J.: The fully informed particle swarm: simpler, maybe better. IEEE Trans. Evolut. Comput. **8**(3), 204–210 (2004)
7. Oca, M.A., Stutzle, T.: Frankenstein's pso: a composite particle swarm optimization algorithm. IEEE Trans. Evolut. Comput. **13**(5), 1120–1132 (2009)
8. Nickabadi, A., Ebadzadeh, M.M., Safabakhsh, R.: A novel particle swarm optimization algorithm with adaptive inertia weight. Appl. Soft Comput. **11**(4), 3658–3670 (2011)
9. Ngo, T.T., Sadollah, A., Kim, J.H.: A cooperative particle swarm optimizer with stochastic movements for computationally expensive numerical optimization problems. J. Comput. Sci. **13**, 68–82 (2016)
10. Jin, N., Rahmat-Samii, Y.: Hybrid real-binary particle swarm optimization (HPSO) in engineering electromagnetics. IEEE T. Antenn. Propag. **58**(12), 3786–3794 (2010)
11. Liu, H., Cai, Z., Wang, Y.: Hybridizing particle swarm optimization with differential evolution for constrained numerical and engineering optimization. Appl. Soft Comput. **10**, 629–640 (2010)
12. Zahara, E., Kao, Y.T.: Hybrid Nelder-Mead simplex search and particle swarm optimization for constrained engineering design problems. Expert Sys. Appl. **36**, 3880–3886 (2009)
13. Sadollah, A., Bahreininejad, A., Eskandar, H., Hamdi, M.: Mine Blast Algorithm: a new population based algorithm for solving constrained engineering optimization problems. Appl. Soft Comput. **13**, 2592–2612 (2013)
14. Eskandar, H., Sadollah, A., Bahreininejad, A., Hamdi, M.: Water cycle algorithm - a novel metaheuristic optimization method for solving constrained engineering optimization problems. Comput. Struct. **110–111**, 151–166 (2012)
15. Kannan, B.K., Kramer, S.N.: An augmented lagrange multiplier based method for mixed integer discrete continuous optimization and its applications to mechanical design. J. Mech. Design **116**, 405–411 (1994)
16. Arora, J.S.: Introduction to Optimum Design. McGraw-Hill, New York (1989)
17. Ngo, T.T., Yoo, D.G., Lee, Y.S., Kim, J.H.: Optimization of upstream detention reservoir facilities for downstream flood mitigation in Urban Areas. Water **8**(7), 290 (2016)

A Multi Dynamic Binary Black Hole Algorithm Applied to Set Covering Problem

José García[1,2(✉)], Broderick Crawford[2], Ricardo Soto[2], and Pablo García[1]

[1] Telefónica Investigación y Desarrollo, Santiago, Chile
{joseantonio.garcia,pablo.garciab}@telefonica.com
[2] Pontificia Universidad Católica de Valparaíso, Valparaíso, Chile
{broderick.crawford,ricardo.soto}@pucv.cl

Abstract. The set covering problem seeks for minimum cost family of subsets from n given subsets, which together covers the complete set. In this article, we present multi dynamic binary black hole algorithm for resolving the set covering problem. This algorithm has the particularity to propose a generic dynamic binarization method to manage the exploration and exploitation properties. Furthermore we explore the implementation of the algorithm on Apache Spark distributed framework.

Keywords: Metaheuristics · Binarization · Set covering problem · Black hole · Spark big data framework

1 Introduction

Global optimization problems arise in many areas of science. Several of these problems are NP-hard and discrete or binary type. Examples of discrete and binary optimization in the industry are transport aircraft wing, task assignment problem, scheduling jobs on grid computing, smart grid. Moreover, information brings an explosion of multiple sources [22]. Several uses for this information comes from NP-hard optimization problems. Some of these problems need time resolution near real time, with solution not necessarily perfect but with suitable quality.

Metaheuristic techniques are suitable for many optimization problems. The metaheuristic algorithms are flexible, simple and scalable. However many metaheuristics algorithms working properly in continuous spaces [14]. It is not obvious how the continuous version can be applied to a discrete or binary problems. Exist different approach to convert a continuous metaheuristic in a discrete or binary algorithm. Examples of discrete adaptation are found using rounding off discretization method in Ant colony [1], PSO optimization [19], Firefly [3], Artificial bee colony [3]. In binary problems we found approach in Magnetic Optimization Algorithm [2], Gravitational Search Algorithm [15], Firefly Algorithm [9], Shuffled Frog Leaping Algorithms [10].

The motivation of this article is to propose a generic way to translate the properties of exploration and exploitation of continuous algorithms getting

J. Del Ser (ed.), *Harmony Search Algorithm*, Advances in Intelligent Systems and Computing 514, DOI 10.1007/978-981-10-3728-3_6

binary algorithms. We propose a multi dynamic binary black hole (MDBBH) algorithm based on ranking of the particles to estimate the transition probability in a binary space of each feasible solution.

We test our algorithm with large-scale of set covering real problems. We exploit the distributed spark framework (frequently used in industry). This framework is designed to provide efficient support to iterative algorithms.

The tests of quality and convergence time were verified on three classes of known set covering problems. OR-Library benchmarks, Airline and bus scheduling problems and Railway scheduling problems. In the Sect. 2 we describe a set covering problem. In Sect. 3 we describe the distributed framework under which we implement the distributed algorithm. Section 4 describes the proposed algorithm, Sect. 5 shows our result and finally we conclude in Sect. 6 suggesting some future research.

2 Set Covering Problem

Set covering problem (SCP) is a classic problem which belongs to the class of NP-complete [11]. Several efficient algorithms have been developed to solve SCP instances. There are exact algorithms who generally rely on the branch-and-bound and brach-and-cut method to obtain optimal solutions. These methods have the problem that can solve instances of limited sizes. To address this concern, it has recently developed new heuristics able to find optimal solutions in reasonable time [4].

For example [13], presented a number of greedy algorithms based on a Lagrangian relaxation, Caprara et al. [6], introduced relaxation-based Lagrangian heuristics applied to the set covering problem. Metaheuristics also has been applied to solve SCP in present days. For example we have classical approach such as genetic algorithm [20], simulated annealing [5], colony optimization [8], and more recent metaheuristics as cat swarm [7] and Cuckoo Search [17].

Le $A = (a_{ij})$ a n $\times$ m zero-one matrix, then a column j cover a row i if $a_{ij} = 1$. Besides a column j is associated with a non-negative real cost c_j. Let $I = \{1, ..., n\}$ and $J = \{1, ...m\}$, the set of rows and columns respectively. The SPC search a minimum cost subset $S \subset J$ where each row $i \in I$ is covered by at least one column $j \in J$ i.e.:

$$\text{Minimize} f(x) = \sum_{j=1}^{m} c_j x_j \tag{1}$$

$$\text{Subject to } \sum_{j=1}^{m} a_{ij} x_j \geq 1, \forall i \in I, \text{and } x_j \in \{0, 1\}, \forall j \in J \tag{2}$$

The SCP is a problem whose domain can be represented by binary values. Let $x_j \in \{0,1\}, \forall j \in \{1,..,m\}$. In this binary representation $x_j = 1$ if column j belongs to the feasible solution. In the case of our algorithm, each particle represent a binary potential solution.

3 Spark

Spark is a state of art distributed computing framework for big data. It is designed to deal with iterative procedures that recursively perform operations over the same data [21]. This has been widely used in machine learning algorithms [16]. Spark born as in-memory cluster computing framework for processing and analyzing large amounts of data. It provides a simple programming interface, which enables an application developer to easily use the CPU, memory, and storage resources across a cluster of servers for processing large datasets in memory [21].

Spark is positioned quickly as a general purpose platform. it provides a unified integrated platform for different types of data processing jobs. It can be used for batch processing, iterative process, interactive analysis, stream processing, machine learning and graph computing.

Resilient Distributed Datasets (RDDs) are the core data units in Spark. These units are distributed, immutable i.e. the transformation of RDDs are RDDs and fault-tolerant memory abstraction. Exist two types of operations: transformations, who take RDDs and produce RDDs and actions who take RDDs and produce values.

4 Multi Dynamics Binary Black Hole Algorithm

In our approach we introduce a new dynamic method that translate a particle ranking in exploration and exploitation properties. Let $X = (x^1, x^2, ...x^d, .., x^n)$ a vector. Then x^d correspond to the value of X in the d position of the vector. We use the black hole algorithm [12], as an example of binarization in which the method ranking particle is applied. Let BH a black hole. Then for each particle X approaching to a BH, we calculate $BH_{\backslash X}$, $X_{\backslash BH}$ and $BH \cap X$ vectors where each position is defined as:

$$BH^d_{\backslash X} := \begin{cases} x^d = 1, \text{ if } x^d = 1 \text{ in BH and } x^d = 0 \text{ in X} \\ x^d = 0, \qquad \text{otherwise} \end{cases} \tag{3}$$

$$X^d_{\backslash BH} := \begin{cases} x^d = 1, \text{ if } x^d = 0 \text{ in BH and } x^d = 1 \text{ in X} \\ x^d = 0, \qquad \text{otherwise} \end{cases} \tag{4}$$

$$(BH \cap X)^d := \begin{cases} x^d = 1, \text{ if } x^d = 1 \text{ in BH and } x^d = 1 \text{ in X} \\ x^d = 0, \qquad \text{otherwise} \end{cases} \tag{5}$$

The Algorithm 1, describes the method used to perform the transition in the different particles. In each iteration we perform a particle ranking using the objective function. The set of particles are classified using the ranking in n groups, (for our results n = 5), we define for each group a transition probability. The transition probability let us manage exploitation properties and exploration properties. A higher transition probability, the jumps are bigger therefore favour exploration. Let $\mathfrak{M}$ a map from $X \in \mathbb{R}^n$ to the [0,1]. $\mathfrak{M}(X)$ represent a transition probability of group. For our case the transition probability was $\mathfrak{M}(X) = k * \alpha$, where $\alpha = 0.1$ and $k = \{1, .., 5\}$ the number of the group X.

Then we start with X a feasible solution in the search space, then we apply the Eq. 7 to elements x^d of X that meet the condition $BH^d_{\setminus X} = 1$ in $BH_{\setminus X}$. Let X_i the result of apply the Eq. 7. Then consider the elements of x^d_i in X_i satisfying $X^d_{\setminus BH} = 1$, then apply the condition 8 to these elements. The results we call X_j. The rest of algorithm applies a repair operator, the list of black hole is updated and replaced with new solution when some distance rule is satisfied.

$$\mathfrak{M} : X \to [0, 1], \mathfrak{M}(X) := TransitionProb(Group(X)) \tag{6}$$

Let $BH_{\setminus X}$ as before, $BH^d_{\setminus X} = 1$ and X a particle Then:

$$x^d = \begin{cases} 1 \text{ if } random() \geq \mathfrak{M}(X) \\ 0, \qquad \text{otherwise} \end{cases} \tag{7}$$

Let $X_{\setminus BH}$ as before, $X^d_{\setminus BH} = 1$ and X a particle Then:

$$x^d = \begin{cases} 0 \text{ if } random() \geq \mathfrak{M}(X) \\ 1, \qquad \text{otherwise} \end{cases} \tag{8}$$

Some solutions can violate the constraints of the problem. For example a new SCP solution has uncovered rows. In order to provide viable solutions we apply in step seven of particle update algorithm, a repair solution operator. This operator is described in Algorithm 2. Our repair operator is inspired in the generic greedy heuristic [18]. We generalize this algorithm using partitions in order to work with large-scale real problem in good runtime. Let S the set of columns in the actual solution, for the case of new solution $S = \emptyset$. Let $\hat{S}$ the complement of S, we partition $\hat{S}$ in different groups. U_R are the uncovered rows in the feasible solution also we partitioning U_R. Then while the set of columns S not cover all rows, we select candidates for columns in one partition of the complement of S and one partition in U_R. For these restricted candidates we randomly selected a measure between the seven measures proposed by Vasko [18] obtaining our column candidate. Finally update S and U_R sets. For the case of medium problems, we used full set of columns and rows candidates getting the original Vasko algorithm.

Algorithm 1. Particle update Algorithm

Initialize population of stars
Select black hole BH
while $Iteration < MaxIteration$ **do**
  for *all* $X \in particles$ **do**
    X_i = *result to apply equation 7 to* X
    X_j = *result to apply equation 8 to* X_i
    X = *result to Apply Repair Solution Operator to* X_j
    if $f_X < f_{BH}$ **then**
      Update BH
    end if
    if $R > (len(X) - len(X \cup BH))$ **then**
      Replace with new star
    end if
  end for
end while

Our algorithm use a multiple instances of black hole process and we allow communication between the different instances. For each instance, in each iteration, we get a black hole candidate, using the fitness function. If the candidate exists in the list of black holes, then it is not considered. If the candidate, not exist, we compared it with the BH of its island. In the case that candidate has a better value than the island black hole, we replace it. Otherwise, we compared randomly with one of others islands black holes in the list. If the candidate has a better value, it replaces the black hole.

We implement our MDBBH algorithm in Spark framework. We want to address big problems thus we need to scale in computing capacity. Moreover, we want to be close to enterprise development capacity. Spark framework fulfills both expectations. It scale well for iterative algorithms and it is a framework widely used in the enterprise. This map distributes the calculation for each island.

First of all, we use a specific partitioner by range to decrease the data shuffling. Also we put the SCP binary matrix representation and its weight vector as a distributed shared variables making visible in all worker nodes. The first mapping is used to generate the initial population This mapping allow us to distribute the particle generation in different workers. After that we do another map. This map distributes the calculation by island. Each island of particles are assigned to the same worker. Once the islands iterations are finished, we collect and update the black hole list. Then we iterate the process until the threshold is complete.

5 Results

In this section, we present computational experiments with the proposed MDBBH algorithm. We test MDBBH on three classes of the known set covering problems.

Algorithm 2. Repair Solution Operator

$S =$ *The set of columns in the solution*
$U_R =$ *The set of uncovered row*
while S is infeasible solution **do**
    $S_f =$ *GetColumnPartitionCandidates(S)*
    $R_f =$ *GetRowPartitionCandidates(R)*
    Tmeasure $=$ *SelectRandomMeasure()*
    for $i \in R_f$, $j \in S_f$ *(i in ascending order)* **do**
        $measure_{ij} = getMeasure(R_f(i), S_f(j))$
    end for
    $s_j = getMinimumMeasure$ – *the best column*
    $S = updateS(s_j)$
    $U_R = updateR(s_j)$
end while

1. OR-Library benchmarks: This class includes 65 small and medium size randomly generated problems that were frequently used in the literature. Most metaheuristic approaches for the SCP have been tested on these problems. They are available in OR-Library and are described in Table 1. We used the medium size E, F, G and H class of problems.
2. Airline and bus scheduling problems: this class includes fourteen real-world airline scheduling problems (AA instances) and two bus driver scheduling problems (bus instances). These problems were obtained from [20].
3. Railway scheduling problems: this class includes seven large-scale railway crew scheduling problems from Italian railways and are available in OR-Library.

The implementation of the MDBBH algorithm was done in python 2.7 using spark library. The experiments were been performed in Azure platform using spark 1.5.2 and hadoop 2.4.1 versions, with 16 cores. The total number of particles were 100 and 10 the number of islands. Best, average and worst results were achieved by running 30 times every problem where each execution have 2000 iterations. We also develop testing scalability of our algorithm using 4,8,12 and 16 cores.

In Table 1 we see the results of the OR library problems. For E.x instances, our MDBBH algorithm obtained four minimum values. The average was very close to the best known. The iterations were between 119 and 157 having a runtime of approximately two minutes. For the case of F.x problems, MDBBH obtained 80% of maximum values. The average also was very close to the best, except to the instance F.3 and F.4. The iterations and execution time was very similar to E.x instances. The group G.x, was the worst performer, only got one maximum. The average execution time was 3.5 min approximately. We emphasize that these instances have the lowest density of all the OR-library. MDBBH got two maximum in H.3 and H.5. The average also was very close to the best known. The execution times were about 3 min on average, higher than the previous instances. In Fig. 1(b) we include the boxplots with confidence intervals, medians and IQR. The plot uses as measure %-Gap defined in the Eq. 9. The results of execution times shown in Table 1 were performed using a configuration of 16 cores. Additionally we developed an scalability experiment. This experiment run the same setting algorithm with 4, 8, 12, 16 cores. The results are shown in Fig. 1(a). All instances scale properly, however the larger instances G and H scale a little bit better.

$$\% - Gap = 100 \times \frac{BestKnow - FoundSolution}{BestKnown} \tag{9}$$

In Table 1 we show the results of MDBBH applied to airline and bus driver crew scheduling medium size problems. The MDBBH algorithm obtained two maximum for the instances AA03 and AA05. The difference between the best known and our best value was 0,78% on average. The iterations and execution times increased compared with the OR instances. The average execution was a little less than 8 min. The iterations were 550 on average.

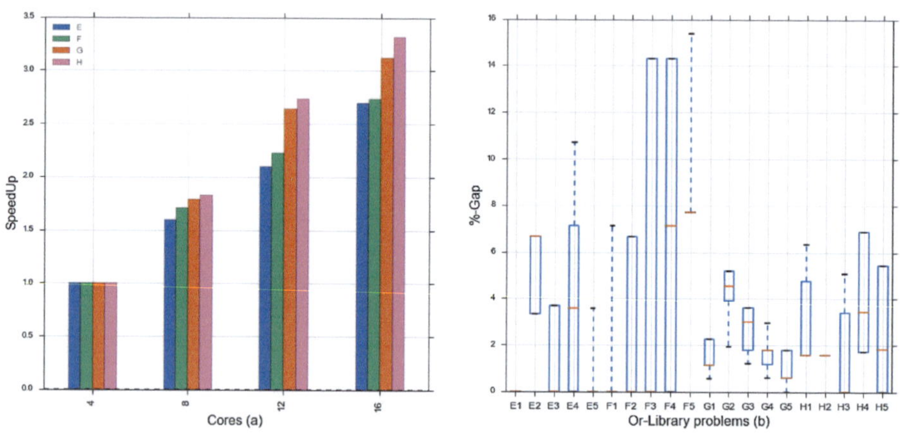

Fig. 1. The figure on the left (a), shows the results of the execution of the MDBBH algorithm in Spark using different cores. The figure on the right (b) displays the behaviour of the results obtained in 30 iterations for instances of the OR library.

Finally our third group of problems belongs to large size railway crew scheduling problems. We note that Rail507, 516 and 582 there are no so large than Rail2586 and 2536. To address these problems, we had to change the repair operator. It allows the operator uses subsets of columns. This have consequence in the quality of our repair algorithm, but we won at execution time. For the case of problems 507, 516 and 582 we use 5 partitions. For the case of 2586 and 2536 we use 50 partitions. The results are very different. When we use 5 partitions we get regular results. The difference between the best known and our best result was five percent in average. For the case of 2586 and 2536 we use 50 partitions. The results were very bad.

Table 1. Results for OR-Library benchmarks E, F, G, H problems. Airline and bus driver crew scheduling and Rail Way problems

Instance	row	col	Density	Best Known	mdbbh (best)	mdbbh (avg)	mdbbh (worst)	Iterations	Time
OR-Library benchmarks E, F, G, H									
E.1	500	5000	10%	29	29	29	29	126	87
E.2	500	5000	10%	30	31	31.6	32	157	127
E.3	500	5000	10%	27	27	27.4	28	141	118
E.4	500	5000	10%	28	28	29.1	31	119	127
E.5	500	5000	10%	28	28	28	29	144	114
F.1	500	5000	20%	14	14	14.1	15	137	116
F.2	500	5000	20%	15	15	15.3	16	112	93
F.3	500	5000	20%	14	14	14.8	16	149	125
F.4	500	5000	20%	14	14	14.9	15	132	101
F.5	500	5000	20%	13	14	14.1	15	139	125
G.1	1000	10000	2%	176	177	178.5	180	224	187
G.2	1000	10000	2%	154	157	160.6	162	247	206
G.3	1000	10000	2%	166	168	170.4	172	235	209
G.4	1000	10000	2%	168	169	170.9	173	278	213
G.5	1000	10000	2%	168	168	169.8	171	218	182
H.1	1000	10000	5%	63	64	64.9	67	231	186
H.2	1000	10000	5%	63	64	64	64	183	154
H.3	1000	10000	5%	59	59	60	62	212	179
H.4	1000	10000	5%	58	59	60.4	62	245	201
H.5	1000	10000	5%	55	55	56.4	58	172	159
Airline and bus driver crew scheduling problems									
AA03	106	8661	4.05%	33155	33155	33507.3	33852	273	165
AA04	106	8002	4.05%	34573	34967	35150.9	35498	409	435
AA05	105	7435	4.05%	31623	31623	31745.8	32103	142	150
AA06	105	6951	4.11%	37464	37490	37681.2	38409	244	240
AA11	271	4413	2.53%	35478	35494	35938.6	36471	486	540
AA12	272	4208	2.52%	30815	31257	31560.8	31838	456	270
AA13	265	4025	2.60%	33211	33670	34015.7	34481	836	960
AA14	266	3868	2.50%	33219	33388	33532.5	33782	1069	1020
AA15	267	3701	2.58%	34409	34619	34847.9	35407	727	690
AA16	265	3558	2.63%	32752	33045	33334.0	33755	642	558
AA17	264	3425	2.61%	31612	32188	32448.2	32901	345	376
AA18	271	3314	2.55%	36782	36952	37217.6	37540	741	409
AA19	263	3202	2.63%	32317	32737	33196.6	33684	639	375
AA20	269	3095	2.58%	34912	35399	35792.1	36173	683	432
BUS1	454	2241	1.88%	27947	28176	28283.5	28352	821	505
BUS2	681	9524	0.51%	67760	68462	68502.8	68549	502	324
Rail Way problems									
Rail507	507	63009	1.2%	174	187	191.4	195	1438	3432
Rail516	516	47311	1.3%	182	186	187.3	190	1267	2520
Rail582	582	55515	1.2%	211	223	225.8	228	1468	3060
Rail2586	2586	920683	0.4%	948	1152	1156	1158	1650	25920
Rail2536	2536	1081841	0.4%	691	836	842	844	1750	13680

6 Conclusion

In this work we presented MDBBH algorithm. This algorithm use ranking of the particle given by the value of the objective function at each iteration as a general principle to manage the transition of the particles in a binary search space. This algorithm was applied to three class of medium and large set covering problems. Also we exploited the possibility of multi black hole in a islands configuration. For the islands communication, we proposed a random interchange algorithm. It algorithm select the best candidate of each island compared with its black hole. In the case of the candidate don't replace the black hole, we let compared random with one of the other islands black holes. The MDBBH was implemented in Apache spark distributed framework, with the intention of improving the execution times and take advantage of the islands implementation to address medium and large problems. The performance of our MDBBH was very good. In the case of OR-library the best known deviation was 0.89% on average. In Airline and bus driver crew scheduling problems, the average performance was 0.78%. In the case of Railway crew scheduling large problems. The result was not really satisfactory. The three first problems our deviation was 5.11%, quite more distanced from our previous cases. But in the two last problems the result was very bad. The deviation was 21.1%.

As a line of work we can improve the algorithm incorporating local search and Perturbation operators. The latter algorithm for the case to be trapped in a local optimum. As future work, it's interesting to quantify the contribution of the different operators in obtaining the final solution. Moreover it is interesting to study the behaviour of the distributed framework under different conditions such as changing the number of particles and using other types of partitioners. Finally with respect to the binarization it is interesting to study the mechanism with other continuous metaheuristics and other problems. Also we want to improve or change the approach of our weaker repair operator together with grow of our cluster capacity to improve the results in the Rail way problems.

Acknowledgments. Broderick Crawford is supported by Grant CONICYT/FONDECYT/REGULAR/1140897 and Ricardo Soto is supported by Grant CONICYT/FONDECYT/REGULAR/1160455.

References

1. Abdelaziz, A.Y., Osama, R.A., El-Khodary, S.M., Panigrahi, B.K.: Distribution systems reconfiguration using the hyper-cube ant colony optimization algorithm. In: Panigrahi, B.K., Suganthan, P.N., Das, S., Satapathy, S.C. (eds.) SEMCCO 2011. LNCS, vol. 7077, pp. 257–266. Springer, Heidelberg (2011). doi:10.1007/978-3-642-27242-4_30
2. Tayarani-N, M.H., Akbarzadeh-T, M.R.: Magnetic optimization algorithms, a new synthesis. In: IEEE International Conference on Evolutionary Computation, pp. 2659–2664 (2008)
3. Bacanin, N., Brajevic, I., Tuba, M.: Firefly algorithm applied to integer programming problems. Recent Adv. Math. 143–148 (2013)

4. Beasley, J.E., Chu, P.C.: A genetic algorithm for the set covering problem. Eur. J. Oper. Res. **94**(2), 392–404 (1996)
5. Brusco, M.J., Jacobs, L.W., Thompson, G.M.: A morphing procedure to supplement a simulated annealing heuristic for cost-and coverage-correlated set-covering problems. Ann. Oper. Res. **86**, 611–627 (1999)
6. Caprara, A., Fischetti, M., Toth, P.: A heuristic method for the set covering problem. Oper. Res. **47**(5), 730–743 (1999)
7. Crawford, B., Soto, R., Berríos, N., Johnson, F., Paredes, F., Castro, C., Norero, E.: A binary cat swarm optimization algorithm for the non-unicost set covering problem. In: Mathematical Problems in Engineering (2015)
8. Crawford, B., Soto, R., Monfroy, E., Paredes, F., Palma, W.: A hybrid ant algorithm for the set covering problem. Int. J. Phys. Sci. **6**(19), 4667–4673 (2014)
9. Crawford, B., Soto, R., Olivares-Suárez, M., Paredes, F.: A binary firefly algorithm for the set covering problem. In: Silhavy, R., Senkerik, R., Oplatkova, Z.K., Silhavy, P., Prokopova, Z. (eds.) Modern Trends and Techniques in Computer Science. AISC, vol. 285, pp. 65–73. Springer, Heidelberg (2014). doi:10.1007/978-3-319-06740-7_6
10. Crawford, B., Soto, R., Peña, C., Riquelme-Leiva, M., Torres-Rojas, C., Johnson, F., Paredes, F.: Binarization methods for shuffled frog leaping algorithms that solve set covering problems. In: Silhavy, R., Senkerik, R., Oplatkova, Z.K., Prokopova, Z., Silhavy, P. (eds.) Software Engineering in Intelligent Systems. AISC, vol. 349, pp. 317–326. Springer, Heidelberg (2015). doi:10.1007/978-3-319-18473-9_31
11. Gary, M.R., Johnson, D.S.: Computers and intractability. A Guide to the Theory of NP-Completeness (1979)
12. Hatamlou, A.: Black hole: a new heuristic optimization approach for data clustering. Inf. Sci. **222**, 175–184 (2013)
13. Beasley, J.: A lagrangian heuristic for set-covering problems. Naval Res. Logistics (NRL) **37**(1), 151–164 (1990)
14. Lozano, M., Molina, D., García-Martínez, C., Herrera, F.: Evolutionary algorithms and other metaheuristics for continuous optimization problems (2010)
15. Rashedi, E., Nezamabadi-Pour, H., Saryazdi, S.: GSA: a gravitational search algorithm. Inf. Sci. **179**(13), 2232–2248 (2009)
16. Shyam, R., Kumar, S., Poornachandran, P., Soman, K.P.: Apache Spark a big data analytics platform for smart grid. Procedia Technol. **21**, 171–178 (2015)
17. Soto, R., Crawford, B., Olivares, R., Barraza, J., Johnson, F., Paredes, F.: A binary cuckoo search algorithm for solving the set covering problem. In: Ferrández Vicente, J.M., Álvarez-Sánchez, J.R., de la Paz López, F., Toledo-Moreo, F.J., Adeli, H. (eds.) IWINAC 2015. LNCS, vol. 9108, pp. 88–97. Springer, Heidelberg (2015). doi:10.1007/978-3-319-18833-1_10
18. Vasko, F.J., Knolle, P.J., Spiegel, D.S.: An empirical study of hybrid genetic algorithms for the set covering problem. J. Oper. Res. Soc. **56**(10), 1213–1223 (2005)
19. Venter, G., Sobieszczanski-Sobieski, J.: Multidisciplinary optimization of a transport aircraft wing using particle swarm optimization. Struct. Multi. Optim. **26**(1–2), 121–131 (2004)
20. Yelbay, B., Birbil, Şİ., Bülbül, K.: The set covering problem revisited: an empirical study of the value of dual information. J. Ind. Manage. Optim. **11**(2), 575–594 (2015)
21. Zaharia, M., Chowdhury, M., Franklin, M.J., Shenker, S., Stoica, I.: Spark: cluster computing with working sets. HotCloud **10**, 10 (2010)
22. Zhai, Y., Ong, Y.-S., Tsang, I.W.: The emerging big dimensionality. IEEE Comput. Intell. Mag. **9**(3), 14–26 (2014)

ACO Based Model Checking Extended by Smell-Like Pheromone with Hop Counts

Keiichiro Takada[1(✉)], Munehiro Takimoto[1], Tsutomu Kumazawa[2], and Yasushi Kambayashi[3]

[1] Tokyo University of Science, 2641 Yamazaki, Noda-shi, Chiba 278-8510, Japan
{k-takada,mune}@cs.is.noda.tus.ac.jp
[2] Software Research Associates, Inc., 2-32-8 Minami-Ikebukuro, Toshima-ku, Tokyo 171-8513, Japan
tkumazawa@acm.org
[3] Nippon Institute of Technology, 4-1 Gakuendai, Miyashiro-machi, Minamisaitama-gun, Saitama 345-8501, Japan
yasushi@nit.ac.jp

Abstract. Model checking is a technique of verifying the model that represents the design of software or hardware. A model checker checks whether a model satisfies a given specification. Since general model checkers work based on a deterministic algorithm, checking may require too much execution time and memory resource. In order to address this problem, an approach based on ACO was proposed. In this paper, we extend the ACO approach with smell-like pheromone that behaves like smell diffusing from food. In our approach, the smell-like pheromone records the number of hops from an acceptance state at each edge. Since our model checker searches the final states in the direction where the number of hops decreases, it can complete the checking more efficiently. We have implemented a model checker based on our approach. The experimental results show that our approach practically decreases the execution time without sacrificing the length of counter examples.

Keywords: Model checking · Ant colony optimization · Swarm intelligence · State explosion

1 Introduction

Model Checking is a technique that checks whether the design of software or hardware system described as a state transition model satisfies the property specified by the user, which is called a specification. Tools that automatically perform the model checking are called model checkers. General model checkers use a deterministic algorithm for the checking, which often consumes too much machine resources because they exhaustively check their search spaces. In order to mitigate this problem, techniques based on Ant Colony Optimization (ACO) have been proposed [1–4], which are nondeterministic algorithms. ACO is a swarm intelligence-based method inspired by the behaviors of ants

J. Del Ser (ed.), *Harmony Search Algorithm*, Advances in Intelligent Systems and Computing 514, DOI 10.1007/978-981-10-3728-3_7

collecting food, which can be regarded as a multi-agent system that exploits artificial stigmergy (artificial pheromone) for the solution of combinatorial optimization problems. In other words, ACO is a kind of statistic algorithms that is categorized into meta-heuristics [5]. Since ACO explores a search space based on the property in which pheromone attracts ants, it helps the model checker focus on a small part of the entire search space, so that the ACO based model checker can suppress the use of the computational resources.

One of the major features of the model checker is to present an error trace to its users (called a counter example), which helps the users understand why the model violates the specification [6]. Thus, the shorter the presented counter example is, the more comprehensible it is for the users. ACO based approach enables model checkers to generate the shorter counter examples.

Most model checking techniques deal with properties roughly categorized into safety and liveness [7,8]. Safety properties state that undesirable things never hold, while liveness properties assert that the desirable things finally hold. Our approach handles the safety and focuses on the fact that finding the violation of the safety is reduced to the reachability problem on directed graphs. In other words, the model checking for the safety just searches final (i.e. error) states starting with a specific start state. For safety checking, a path from the start state to a final state is a counter example.

In the ACO based model checking, ants move in the directed graph to look for a final state. Ants probabilistically choose the move direction according to the amounts of pheromone deposited by preceding ants on their trails. This behavior of ants helps other ants to find relatively short paths to the final states, because the paths that have much pheromone tend to be selected by many ants. Thus, pheromone trails represent the quality of paths to the final state in terms of path length. In order to further assist ants to search final states, in our previous work [9], we have introduced another special pheromone that attracts ants to the final states such as smell emitted from food. The attraction of the pheromone efficiently guides ants, and localizes the search space, so that our approach has achieved the less consumption of the resources, and reduce the execution time without sacrificing the quality(length) of counter examples, compared with the conventional ACO-based approach.

In this paper, we give a further improvement of the smell-like pheromone based approach. In the original approach, the smell-like pheromone guides ants to the final states. However, when the pheromone diffuses from the final states to the surroundings, it keeps the initial strength. Therefore it tends to become difficult to correctly guide ants to the final states as the pheromone diffuses in the wider area. In order to mitigate this problem, we propose smell-like pheromones with hop counts from the final states. The hop counts, which are increased each time the smell-like pheromones diffuse along an edge, are recorded to states along with the pheromones. If a smell-like pheromone visits a state that another smell-like pheromone has already visited, their hop counts are compared, and then, the smaller count is adopted. The recording manner of hop counts contributes to exposing the direction to the final states by the decreasing directions of the hop counts.

The rest of the paper is organized as follows. The next section presents preliminaries of our approach. In the third section, we give a brief explanation of a traditional ACO based approach for safety. Our approach of new ACO algorithm is contained in the fourth section. The fifth section describes the experimental results and discuss of its efficiency. Conclusion and future work are described in the sixth section.

2 Model Checking

In this section, we give an overview of model checking, and explain the problems that we address. Model Checking is a formal and fully automatic verification technique that verifies the correctness of software or hardware [10]. Model Checking is one of the most successful research topics in software engineering, and it has been applied for the verification of communication protocols [11].

Model Checking consists of the following steps: (1) representing the model of a system as a state transition system such as an automaton, (2) describing a specification that represents the software's required properties with temporal logic such as Linear Temporal Logic (LTL) [12], and (3) checking whether the state transition system satisfies the specification. For example, SPIN [8,13] is one of the most popular model checkers. It checks whether a model described in Promela satisfies the specification described in LTL. Promela is the description language specifically designed for SPIN. At this time, SPIN converts the model and the negation of the specification into Büchi automata, then builds their intersection to exhaustively search for its accepting paths on it. If some accepting paths are found, it means the model does not hold the specification, and the corresponding paths are shown as the counter examples. Otherwise the fact of satisfaction is notified.

In this paper, we tackle with two problems of model checking: *state explosion problem* and *incomprehensive counter examples*. In model checking, the automata tend to be very large. Actually, the size of an automaton increases exponentially in the size of the model. The exponential increase is called state explosion, which may cause exhaustion of resources and may lead to system failure. In order to mitigate the state explosion, techniques such as Partial Order Reduction that decreases the number of states [14], and Bitstate Hashing that reduces the memory area occupied by each state [8] have been proposed. The other problem is related to the length of counter examples. Because a counter example is presented to the user as a diagnosis of specification violation, it is highly desirable to make it comprehensible for the human user. As mentioned above, a counter example is represented as the path over the directed graph. It means that the counter example is more understandable if it is small enough in lengthwise. That is to say, we need to aim at the model checking methods that generate as short counter examples as possible.

Recent researches have shown that various meta-heuristic approaches provide more favorable results than traditional deterministic algorithms do in terms of efficiency. Alba et al. showed the effectiveness of Ant Colony Optimization (ACO) for the model checking [1]. Their approach is extended to the checking of safety with Partial Order Reduction [3] and that of liveness [4]. As another work, Francesca et al. proposed a deadlock detection technique using ACO [15]. They utilize the heuristics based on the structures of models to tell ants estimated directions to food.

3 ACO Based Approach

In this section, we give the basics of ACO, and then, describe ACOhg that is a variant of ACO proposed by Alba et al. [1–3] for model checking with large state spaces.

3.1 ACO

ACO is one of the meta-heuristic search algorithms inspired by the behaviors of ants that discover paths from their nest to food. It is known that ACO is effective for some optimization problems such as routing, assigning and scheduling problems. When ACO is applied to model checking problems, a problem is a model expressed in a directed graph, where the nest and food are the start and error states respectively. The paths between these nodes represent candidate solutions of the model checking problem, i.e. counter examples.

In ACO, agents corresponding to ants cooperatively search the shortest paths through indirect communications using pheromones. The optimal paths found by ants correspond to the optimal solution in optimization problems. The effect of the pheromones is represented as weight on edges, which are assigned to the paths that ants have visited. The pheromones play the role of guides that induce ants to select shorter paths to error states.

ACO is the repetition of *ConstructAntSolutions* step for searching paths, and *UpdatePheromones* step for updating the distribution of pheromone. Also, *terminationCondition* is satisfied when some solutions are found or the number of the repetition exceeds a fixed number. In the following, we discuss each step in more detail.

Initialization step initializes pheromones on edges for preparation of searching by ants. While regarding the start state and final states as the nest and food respectively, pheromones with random strength are randomly located between these states.

ConstructAntSolutions step makes each ant probabilistically transit states starting from the start state to find candidate solutions. At this time, a transition to the next state is selected using the following equation:

$$p_{ij}^{k} = \frac{[\tau_{ij}]^{\alpha}[\eta_{ij}]^{\beta}}{\Sigma_{l \in N_i}[\tau_{il}]^{\alpha}[\eta_{il}]^{\beta}}, \quad if\ j \in N_i \tag{1}$$

In the equation, N_i is the set of states to which an ant can move from state i; j is the destination state of one step movement of the ant; τ_{ij} is the value of pheromone on the edge (i, j); and η is the heuristic value representing the number of transitions to a final state, which is set to a smaller value than actual one. The heuristic value is estimated based on the locations of final states, or the length and property of a specification that are described in LTL. Also, α and β are empirically determined values in order to adjust the effects of the heuristic value and pheromone value, respectively. When ants have moved on edges in the fixed number of times, or some candidates of solutions are found, *UpdatePheromones* step follows.

UpdatePheromones step updates the pheromone value on each edge. The pheromone strengths on selected edges are increased according to the appropriateness of the edges. At the same time, the pheromones on the other edges are updated as follows:

$$\tau_{ij} \leftarrow (1 - x_i)\tau_{ij} \tag{2}$$

In the equation, x_i is set to a value between 0 and 1. This is the value representing the degree of evaporation of pheromones. Through the update process, the priorities of previously selected edges are decreased step by step until some ants select the edges again.

3.2 ACOhg

It is difficult for traditional ACOs to handle state transition systems for model checking that have thousands of nodes. Because there can be billions of edges in such systems, they require a number of megabytes in a memory to record pheromone values. Especially, the size of a model of a concurrent system is known to be huge. For example, the size of the model of Dining Philosophers with n philosophers is 3^n, which increases exponentially. The simple solutions such as to prohibit revisiting to the same states are not effective, because some ants may run into brick walls, or may wander from state to state for a long time even if they finally reach the final states. Thus, the behaviors of the traditional ACO in model checking may result in the state explosion without finding any candidates of solutions. In addition, as a more fatal problem, the traditional ACO has to initially set pheromone values to all the transitions. This may also consume too much memory.

In order to mitigate the problems of the traditional ACO, Alba et al. proposed *ACO for huge graphs* (ACOhg), which can perform searching with less memory than the traditional ones [1,2]. The basic idea of ACOhg is to give the upper limit λ_{ant} to the number of move steps of ants at one stage. This search manner suppresses the time and memory consumption, but limiting of move steps may prevent ants from reaching the final states. That is, λ_{ant} has to be decided so as to find the final states. ACOhg gives two kinds of techniques for determining λ_{ant}; they are *expansion technique* and *missionary technique*.

In the expansion technique, once the system cannot find a final state in the search for the current λ_{ant}, λ_{ant} is increased by the value given as a parameter, and then the system searches the wider area defined by the new λ_{ant} again. The process starts with small λ_{ant} and repeats until it finds some final states.

The expansion technique increases λ_{ant} just enough, so that the memory consumption can be suppressed. Also, it is easy to implement because it is a simple extension of the traditional ACO. On the other hand, its behavior becomes closer to the traditional ACO's as λ_{ant} increases.

The missionary technique is similar to the expansion technique, but searches are performed from not the start state but some states on the edge of the previously searched area, i.e. ignoring old pheromone. The new start state on the edge is selected from the states that ants reach at the previous stage. This search manner enables ACO to gradually extend the search area without changing λ_{ant}, so that it only requires constant time and memory consumption at each stage.

Both approaches decide the strength of pheromones to be assigned based on a *fitness function*. The fitness function returns the degree of penalty for the trail of each ant, which becomes very large in the case where the trail includes some cycles, or no final state. Based on the fitness function f, a^{best} with the lowest penalty $f(a^{best})$ is determined, and then, the pheromone of its trace is strengthened as follows:

$$\tau_{ij} \leftarrow \tau_{ij} + \frac{1}{f(a^{best})}, \forall(i,j) \in a^{best} \tag{3}$$

ACOhg, which is based on MMAS, also uses the fitness function for calculating the limit values τ_{max} and τ_{min} of the pheromone value for each trail as follows:

$$\tau_{max} = \frac{1}{\rho f(a^{best})}, \tag{4}$$

$$\tau_{min} = \frac{\tau_{max}}{a} \tag{5}$$

In the equation, ρ and a are constants that control the range of the pheromone values. Also, the missionary approach uses the fitness function to decide the next start states at each stage.

3.3 EACOhg

In ACOhg, ants have to randomly search for the final states until some ants reach them for the first time. On the other hand, natural ants can search for food with its smell, even if there is no pheromone.

We have proposed *EACOhg* [9] where we have introduced another kind of pheromone similar to smell, which we call goal pheromone, to ACOhg. Goal pheromone is stronger than the normal pheromone. Once the goal pheromones are put on transition edges, the edges are selected in preference to the edges with the normal pheromone as shown in Fig. 1. The search manner localizes the search area further, contributing to finding the final states more quickly and generating shorter counter examples, although memory consumption may increase a little to hold the goal pheromones.

The existence of the goal pheromone shows that there are some paths from the current state to the final states. Therefore, it is enhanced in order to attract

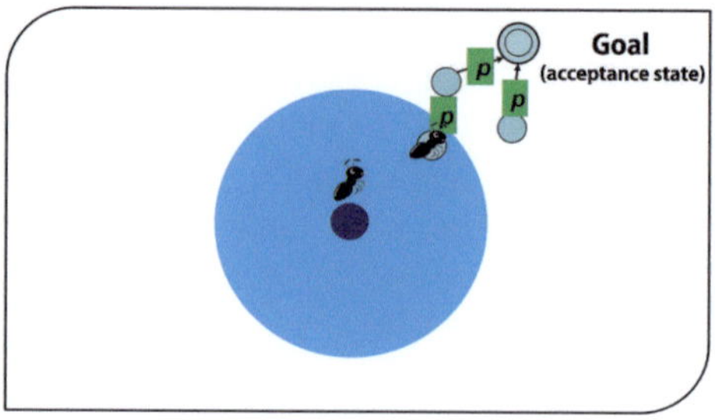

Fig. 1. Diffusing of smell

ants more strongly through the following properties: (1) attracting ants more strongly than normal pheromone, (2) non-volatility, and (3) defusing from the final states to their peripheries.

Note that the above properties induce ants to select relatively short paths to the final states, which leads to short counter examples.

The normal pheromones behave along with MMAS and hence, have the value less than or equal to τ_{max}. The goal pheromone is set to the value more than τ_{max} in order to attract ants more strongly than normal pheromones. Moreover, while the normal pheromone evaporates, the goal pheromone does not evaporate and has a constant value. It is derived from the fact that the smell is always supplied by the source, i.e. food. Thus, once an ant finds some goal pheromones, it moves to the edges with them with high probability.

The goal pheromone is assigned on in-edges of the current states in order to make them diffuse. At that time, the pheromone is scattered on only some edges that are randomly selected. The scatter manner contributes to suppressing time and memory consumption.

4 The Extension with Hop Counts

The EACOhg works well for shortening counter examples in most models, but there are some cases where the execution time and memory consumption are hardly reduced in large models. We have found that it is because of emitting too much goal pheromone to the wide area in the large models. That is, in such a model, because most goal pheromones diffuse the area that are far from paths connecting the start state and final states, it takes much time until some ants meet the goal pheromone. In such case the goal pheromone does not provide enough guidance to the ants.

In order to mitigate the problem, we make each goal pheromone have hop count from the final state that generates the pheromone. The goal pheromone records the hop count in the reached state while increasing the hop count each time it reversely traverses a transition edge. Simultaneously, the goal pheromone only marks its value on the edges whose source state has greater hop count than destination state has.

In more details, the diffusion process through an edge, which is represented as (*src*, *dst*), of the goal pheromone consists of the following two steps.

In the first step, the goal pheromone tries to record a new hop count at *src*, which is one bigger than the hop count at *dst*. At this time, if *src* has already had a count smaller than the new hop count, or no hop count, the new hop count is recorded at *src*; otherwise, the hop count at *src* is not modified, as shown in Fig. 4. Notice that the final states have zero as a hop count. The manner calculating the hop count at each state allows the hop count to be close to the smallest hop count from the final states.

In the second step, the pheromone value is selectively deposited on the edges whose source states have the smallest hop count. As shown in Fig. 4, the diffusion manner of the goal pheromone suppresses depositing pheromone value on unnecessary edges, and contributes to making sequences of states with pheromone close to the shortest paths to the final states.

5 Numerical Experiments

5.1 Overview of Experimental Strategy

In order to demonstrate the effectiveness of our approach, we have implemented our algorithm, on a practical model checker called LTSA (Labeled Transition System Analyzer) [7,16]. LTSA is one of the model checkers that can be customized easily, and can handle models that are suitable for our purpose. LTSA supports the checking of Fluent Linear Temporal Logic [16], which is a kind of LTL specialized for event-based systems described as Labeled Transition Systems (Fig. 2).

We used values as shown in Table 1 for the coefficients and constants that appear in equations and algorithms. P_p and P_c are weights of penalties used by the fitness function for no final state and the path with some cycles. *phspeed* is the diffusion speed of goal pheromone, which is used in addition to the traditional coefficients on this experiment. We use the same settings as EACOhg for the traditional coefficients because our algorithm is a simple extension of EACOhg (Fig. 3).

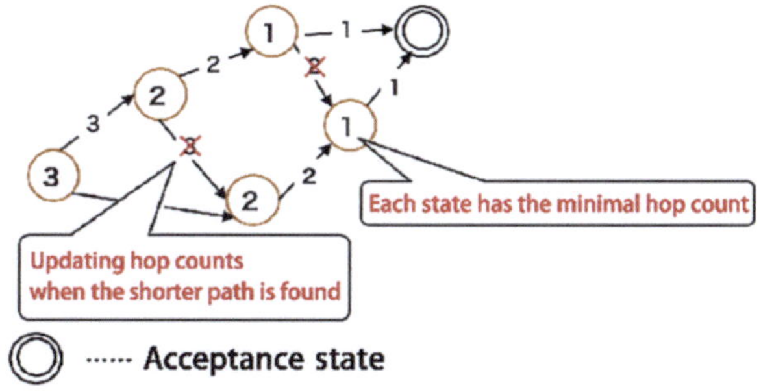

Fig. 2. Updating minimal hop counts

Table 1. Settings of coefficients

Coefficients	Values
mstep	100
colsize	10
α	1.0
β	2.0
η	1.0
ρ	0.2
a	5
P_p	70
P_c	70
phspeed	10

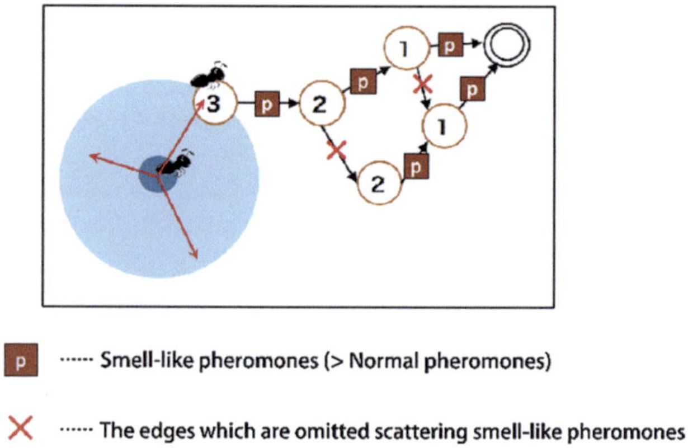

Fig. 3. Scattering smell-like pheromones

5.2 Results

In the experiments, we prepared three models: model A with 5884 states, model B with 39274 states, and model C with 266,218 states by adjusting the parameters of three kinds of examples *Mutex_fluent*, *DeadlockFreePhilosophers* and *DatabaseRing*.

We conducted experiments of them for the four metrics: execution time, the length of counter examples, the amount of memory consumption and the number of goal pheromone scattered. In each experiment, we compared our approach with EACOhg for models A, B and C, where we show the average of the results of one hundred times applications for each model. Then we applied t-test to check whether the differences of each average values are statistically significant. We consider they are meaningful if their p-values in the t-test are less than 0.05.

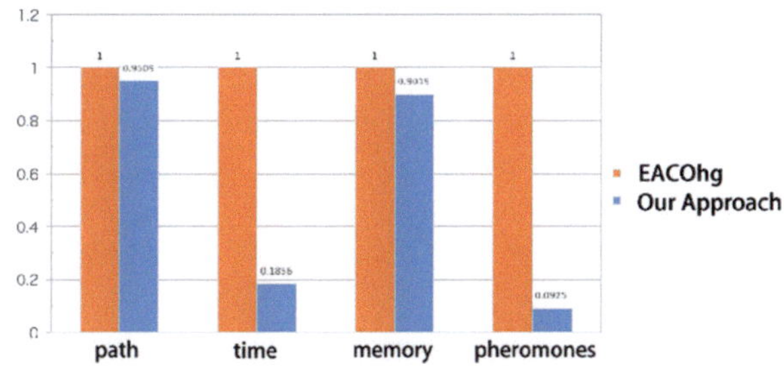

Fig. 4. Result of model A

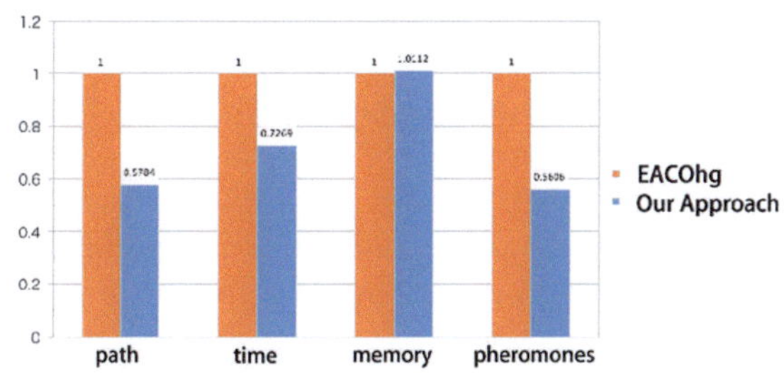

Fig. 5. Result of model B

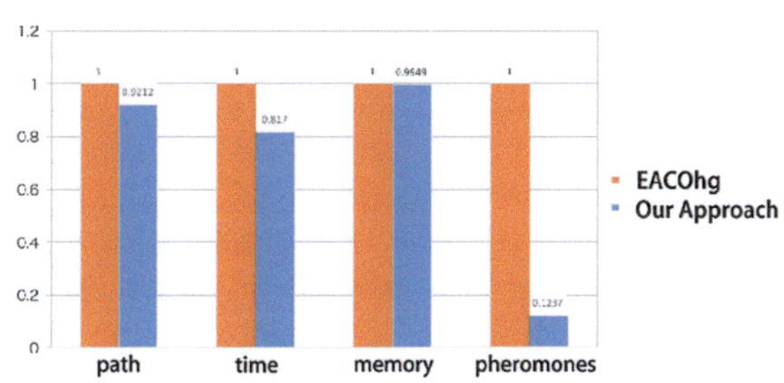

Fig. 6. Result of model C

As shown in Fig. 4, in model A, the values of our approach were better than EACOhg for all the evaluation metrics. The p-values of execution time, the length of counter examples, the amount of memory consumption and the number of scattered goal pheromones were 0.055, 0.092, 0.899, and 0.115 respectively. The test values show that the results of EACOhg and our approach hardly have any difference, although the length of counter examples seems to have a little meaningful difference. In model B, our approach produces much better results than EACOhg does for the evaluation metrics other than memory consumption, as shown in Fig. 5. These p-values were 1.675×10^{-14}, 7.214×10^{-4}, 0.164, and 6.794×10^{-4} respectively, which show that these results are significant. On the other hand, the amounts of memory consumption, of both approaches are approximately equal. In model C, our approach produces better results than EACOhg does for all the metrics, as shown in Fig. 6. The p-values of execution time and the number of scattered goal pheromones are 6.313×10^{-5} and 9.122×10^{-17} respectively, which show that the results are significant. On the other hand, with regard to the length of counter examples and the amount of memory consumption, it is difficult to say that our approach has advantages, because of their p-values, 0.320 and 0.326.

5.3 Discussion

As shown in the previous section, we did not have significant advantages for some metrics in some models. It is worth discussing their reasons.

In model A, we suppose that the size of the model affects the results. In fact, model A is the smallest of all the models, and therefore, ants could find some goal pheromones before goal pheromones diffuse in the wide area. As mentioned above, if the diffusion of the goal pheromone is limited to the area surrounding the final states, EACOhg works well. Therefore, in model A, EACOhg would be useful enough. In model B, our approach works better than EACOhg does in the metrics other than the amount of memory consumption. Especially, our approach has a remarkable advantage for the length of counter examples. In fact, our approach mostly generated counter examples of length 20, while most counter examples generated by EACOhg has the length from 30 to 50. This is because our diffusion manner of goal pheromone suppresses unnecessary markings, so that the goal pheromone more effectively contributes to guiding ants to the final states. Also, shortened traces of ants would contribute to reducing execution time. In model C, we could not have remarkable effectiveness for the length of counter examples. This would be because the original ACO occupied the effectiveness of shortening the counter examples instead of the goal pheromone, which would be derived from the structure of the model. However, considering the decrease of the number of scattered goal pheromones, we can say that the similar results show an advantage of our approach.

6 Conclusion and Future Work

We proposed a new diffusion technique of the smell-like pheromone that suppresses unnecessary markings of edges based on hop counts from the final states. Since the smell-like pheromone tend to mark the shortest path to the final states, the number of scattered pheromone can be decreased without sacrificing the quality of the path.

We have implemented our approach on a practical model checker, and have conducted experiments. The results of the experiments have shown that our approach can improve ACO based model checking further in some models.

As a future direction, we will investigate a technique for adjusting the emission speed of the smell-like pheromone depending on the size of models.

References

1. Alba, E., Chicano, F.: Finding safety errors with ACO. In: Proceedings of the 9th Annual Conference on Genetic and Evolutionary Computation, pp. 1066–1073 (2007)
2. Alba, E., Chicano, F.: ACOhg: dealing with huge graphs. In: Proceedings of the 9th Annual Conference on Genetic and Evolutionary Computation, pp. 10–17 (2007)
3. Chicano, F., Alba, E.: Ant colony optimization with partial order reduction for discovering safety property violations in concurrent models. Inf. Process. Lett. **106**(6), 221–231 (2008)
4. Chicano, F., Alba, E.: Finding liveness errors with ACO. In: IEEE Congress on Evolutionary Computation, pp. 2997–3004 (2008)
5. Boussaïd, I., Lepagnot, J., Siarry, P.: A survey on optimization metaheuristics. Inf. Sci. **237**, 82–117 (2013)
6. Clarke, E.M.: The birth of model checking. In: Grumberg, O., Veith, H. (eds.) 25MC Festschrift. LNCS, vol. 5000, pp. 1–26. Springer, Heidelberg (2008). doi:10.1007/978-3-540-69850-0_1
7. Magee, J., Kramer, J.: Concurrency: State Models and Java Programs. Wiley, New York (1999)
8. Holzmann, G.J.: The Spin Model Checker Primer and Reference Manual. Addison-Wesley, Reading (2004)
9. Kumazawa, T., Yokoyama, C., Takimoto, M., Kambayashi, Y.: Ant colony optimization based model checking extended by smell-like pheromone. EAI Endorsed Trans. Ind. Netw. Intell. Syst. **16**(7), e1 (2016)
10. Clarke, E.M., Emerson, E.A.: Design and synthesis of synchronization skeletons using branching time temporal logic. In: Kozen, D. (ed.) Logic of Programs 1981. LNCS, vol. 131, pp. 52–71. Springer, Heidelberg (1982). doi:10.1007/BFb0025774
11. Edelkamp, S., Leue, S., Lluch-Lafuente, A.: Directed explicit-state model checking in the validation of communication protocols. Int. J. Softw. Tools Technol. Transf. **5**(2–3), 247–267 (2004)
12. Manna, Z., Pnueli, A.: The Temporal Logic of Reactive and Concurrent Systems: Specification. Springer, New York (1992)
13. Holzmann, G.J.: The model checker spin. IEEE Trans. Softw. Eng. **23**(5), 279–295 (1997)
14. Lluch-Lafuente, A., Edelkamp, S., Leue, S.: Partial order reduction in directed model checking. In: Bošnački, D., Leue, S. (eds.) SPIN 2002. LNCS, vol. 2318, pp. 112–127. Springer, Heidelberg (2002). doi:10.1007/3-540-46017-9_10
15. Francesca, G., Santone, A., Vaglini, G., Villani, M.L.: Ant colony optimization for deadlock detection in concurrent systems. In: IEEE Annual Computer Software and Applications Conference, pp. 108–117 (2011)
16. Giannakopoulou, D., Magee, J.: Fluent model checking for event-based systems. In: European Software Engineering Conference, pp. 257–266 (2003)

Harmony Search and Data Mining/Big Data

On the Creation of Diverse Ensembles for Nonstationary Environments Using Bio-inspired Heuristics

Jesus L. Lobo[1(✉)], Javier Del Ser[1,2,3], Esther Villar-Rodriguez[1], Miren Nekane Bilbao[2], and Sancho Salcedo-Sanz[4]

[1] TECNALIA, 48160 Derio, Spain
{jesus.lopez,javier.delser,esther.villar}@tecnalia.com
[2] University of the Basque Country UPV/EHU, 48013 Bilbao, Spain
{javier.delser,nekane.bilbao}@ehu.eus
[3] Basque Center for Applied Mathematics (BCAM), 48009 Bilbao, Spain
[4] Universidad de Alcalá, 28871 Alcalá de Henares, Spain
sancho.salcedo@uah.es

Abstract. Recently the relevance of adaptive models for dynamic data environments has turned into a hot topic due to the vast number of scenarios generating nonstationary data streams. When a change (concept drift) in data distribution occurs, the ensembles of models trained over these data sources are obsolete and do not adapt suitably to the new distribution of the data. Although most of the research on the field is focused on the detection of this drift to re-train the ensemble, it is widely known the importance of the diversity in the ensemble shortly after the drift in order to reduce the initial drop in accuracy. In a Big Data scenario in which data can be huge (and also the number of past models), achieving the most diverse ensemble implies the calculus of all possible combinations of models, which is not an easy task to carry out quickly in the long term. This challenge can be formulated as an optimization problem, for which bio-inspired algorithms can play one of the key roles in these adaptive algorithms. Precisely this is the goal of this manuscript: to validate the relevance of the diversity right after drifts, and to unveil how to achieve a highly diverse ensemble by using a self-learning optimization technique.

Keywords: Concept drift · Diversity · Bioinspired optimization

1 Introduction and Related Work

The increasing number of applications favoring the generation of data streams – such as mobile phones, sensor networks and in general all scenarios under the so-called *Internet of Things* paradigm [1] – has led the research community to the necessity for new approaches capable of dealing with fast incoming information flows. In these practical situations it is often assumed that the process behind

J. Del Ser (ed.), *Harmony Search Algorithm*, Advances in Intelligent Systems and Computing 514, DOI 10.1007/978-981-10-3728-3_8

the generation of such data streams is stationary, i.e. the statistical properties of the underlying phenomena that produce the information to be processed do not vary along time. Unfortunately, in many real scenarios this assumption does not hold since the data generation process becomes affected by a nonstationary event (such as eventual changes in the users' habits, seasonality, periodicity, sensor errors, etc.). Under these circumstances the statistical distribution of the data may change (drift), which ultimately causes that models trained over these data sources are obsolete and do not adapt suitably to the new distribution of the data. Therefore, in the context of data mining in such nonstationary environments the construction of learning models requires adaptive approaches to ease the adjustment of such model to drifts, either from an active (i.e. drift detection, which triggers a subsequent model adaptation) or a passive perspective (the adaptation of the model whenever new data arrive).

Ensembles are one of the most useful approaches to deal with concept drift, and have been successfully used to improve the accuracy of single classifiers in incremental learning. Diversity among the constituent learners in ensemble models has been empirically proven to be crucial when dealing with concept drift [2]. Specifically this study evinces that the diversity plays an important role *before* and *after* a concept drift, importance that is also subject to the severity of the drift: before the drift, ensembles with less diversity obtain better test errors, while shortly after the drift more diverse ensembles use to score lower test error rates. Their difference in terms of test error performance when compared to lower diversity ensembles is usually more significant when the severity is higher. Therefore, it is a good strategy to maintain highly diverse ensembles and utilize them shortly after the drift (independently from the type of drift) to obtain good performance scores. The so-called Diversity for Dealing with Drifts (DDD) approach published in [3] leverages this empirically validated conclusion, and is one of the most recognized methods to manage diverse ensemble in the presence of concept drift from an active perspective.

Due to the above noted importance of achieving a good balance between adaptability (diversity) and performance along time, there is a latent need for novel mechanisms to optimally balance the diversity in ensemble learning. This work falls within this research trend and formulates the diversity balance as an optimization problem. We explore the benefits of a bio-inspired solver for the construction of ensembles with different levels of diversity, in particular the Harmony Search (HS) algorithm [4]. HS has demonstrated to be competitive respect to other evolutionary heuristics for optimization paradigms in diverse fields such as energy [5,6], bio-informatics [7], telecommunications [8,9], data mining and concept drift [12], and logistics [13], among many others [10,11]. However, to the knowledge of the authors no previous contribution has gravitated on the diversity-accuracy trade-off in ensemble learning over nonstationary data.

The idea behind this research work is to use the HS algorithm to build ensembles of models with maximum and minimum diversity, and then utilize them shortly after the drift to show that ensembles with high diversity yield better classification performance those with low diversity. In this regard it has been widely acknowledged in the literature (see e.g. [14–16] and references therein) that the Area Under the ROC Curve (AUC) is the most strongly recommended score due to the fact that the naive accuracy metric is not a reliable indicator in severely imbalanced data sets. This work will embrace this recommendation in what follows, specially in Sect. 4 for comparing results among different ensembles.

The rest of the paper is organized as follows: Sect. 2 introduces the analyzed scenario. Section 3 delves into the proposed approach based on HS, whereas Sect. 4 presents and discusses the simulation results obtained over the SEA data set [20]. Finally, Sect. 5 ends the paper and sketches future research lines.

2 Analyzed Scenario

In batch learning [15] the level of diversity among base learners in the ensemble is a relevant topic that has grasped notable attention in the literature. The success of ensemble learning algorithms is based, to a certain point, on the accuracy and diversity among the base learners [17]; some studies have revealed that it exists a positive correlation between accuracy of the ensemble and diversity among its members [18]. In [2] it was concluded that it is a good strategy to maintain highly diverse ensembles to obtain good responses shortly after the drift, independently of the type of the drift. However, at this point it is necessary to choose a metric to measure the diversity of an ensemble so as to build ensembles with different levels of diversity depending on the instant at which it is applied (before or shortly after the drift).

Following the recommendations of [18], in which a thorough analysis of 10 measures was discussed, the Yule's Q statistic [19] is selected for the purpose of minimizing the error of ensembles. The advantages of this measure are its simplicity and ease of interpretation. Considering two classifiers C_i and C_j, the Yule's Q statistic metric can be calculated as

$$Q_{i,j} \doteq \frac{N^{11}N^{00} - N^{01}N^{10}}{N^{11}N^{00} + N^{01}N^{10}}, \tag{1}$$

where N^{ab} is the number of training samples for which the classification given by C_i is a and the classification given by C_j is b. We further assume that 1 represents a correct classification and 0 is a misclassification. Q varies between -1 and 1. Classifiers that tend to recognize the same objects correctly will have positive values of Q, and those which commit errors on different objects will render Q negative. For an ensemble E of M classifiers, the $\bar{Q}$ statistic averaged over all pairs of classifiers is given by

$$Q_{averaged} \doteq \frac{2}{M(M-1)} \sum_{i=1}^{M-1} \sum_{j=i+1}^{M} Q_{i,j}. \tag{2}$$

where, as mentioned above, higher/lower $\bar{Q}$ values are associated with lower/higher diversity, establishing an inversely proportional relation.

Bearing these definitions in mind a batch learning technique based on an ensemble composed of several base learners has to deal with applications that provide fast incoming information flows in the form of data batches. In this scenario one different model can be trained with each newly incoming batch, hence the total number of combinations that may yield possible diverse ensembles at a concrete time step can be too high in the long run for its exhaustive evaluation. Due to the time constraints or computation costs in certain cases of these applications, the task of finding diverse ensembles may not be affordable in practice. For this reason this challenge can be dealt with by formulating the choice of diverse base learners as an optimization problem, for which a bio-inspired technique can find an optimal diverse ensemble at each moment on time.

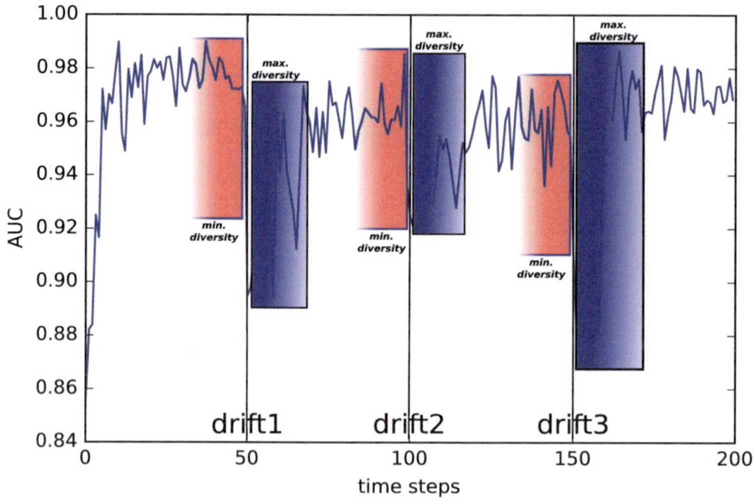

Fig. 1. Diversity importance before and after the drift for the SEA data set.

In the case of the online learning approach proposed in [3], the learning process of the ensemble is carried out for a fixed number of times (defined by the rate parameter λ characterizing the $Poisson(\lambda)$ distribution) with the same current training data. As it is not possible to store past data, so the ensemble can not learn from past information. In this way, higher/lower λ values are associated with higher/lower $\bar{Q}$ values (lower/higher diversity). In the case of a batch learning approach, a model can be trained with incoming training examples, and be part of the ensemble if it is considered, being it possible to have an ensemble formed by members trained in the past. This work follows a batch learning approach, and uses an HS solver to maximize and minimize the diversity of the ensemble. Figure 1 represents a batch learning process during 200 time steps with the AUC score, and it shows how the importance of the diversity is at each moment before and after the drift for the SEA dataset [20].

In order to test the feasibility of HS to achieve different levels of diversity for the ensemble, this work has been evaluated when applied over one of the most widely used synthetic data sets for assessing new concept drift developments: the SEA data set. Following the original data set generation procedure posed in this work, a total of 10000 3-dimensional samples have been generated at random within the range $\mathbb{R}[0, 10)$. Only the first two dimensions (features) are set informative for the class to be predicted, whereas the remaining dimension is irrelevant and acts as a noisy component for the target label. Points have been split in 200 batches of length 250 samples, which have been further divided into 4 main groups or blocks characterized by different concepts: a data sample belongs to class 1 if $x_1 + x_2 \leq \Theta$, where x_1 and x_2 represent the first two features of the sample and Θ is a threshold value that sets the frontier between the two classes. A recurrent series of values (i.e. $\Theta = \{4, 7, 4, 7\}$) has been used to generate the four concept blocks. An additional class noise has been also inserted within each block by randomly changing the class of 5% of the total instances.

3 Proposed HS-Based Approach

HS works by imitating the activity of musicians while improvising new music pieces. The choice of which note to play next is something which takes years to learn to do effectively. Each musician in the band (ensemble) is often faced with the problem of picking the next note. To do so they can resort to their knowledge of the notes in the key they are playing in (which notes sound aesthetically pleasant in the context of the song), as well as the notes they have played previously (what notes sounded good in the recent context). The notes they played recently are most likely to sound pleasantly. Also, it is wise to select a particular note that the audience might expect and adjust the pitch ignoring the expected note to create an artistic effect and a new, potentially better, harmony. HS seeks an optimal combination of inputs, just as a musician seeks a good harmony. HS generates "harmonies" of inputs which are evaluated for quality, and iterates this process until it finds the best one possible. The aesthetics of a musical harmony are analogous to the fitness of a particular solution, so following this simile HS attempts to achieve a good combination of inputs, just like musicians optimize their note selection using their own heuristics. Each input to the problem is considered as a different instrument in an ensemble, each potential note corresponds to each potential value of the inputs that the function might adopt. The musical harmony of notes is modeled as a programmatic harmony of values. Each iteration a new harmony is generated its quality is calculated: if it makes the cut it is included in the musician's memory. This way, iteration by iteration, old poor quality harmonies are discarded and replaced by better ones. The average quality of the set of harmonies in this memory as a whole gradually increases as these new harmonies replace poor ones.

This being said, notes in the HS solver particularized to the problem tackled in this paper represent the members of the ensemble. At each time step t_i, HS optimizes the diversity of the ensemble formed by 10 members, combining all past models trained from t_0 to t_{i-1}. Special attention deserves the fact that the more past time steps are handled at the time where the ensemble is to be built, the more necessary an efficient optimization technique is, because there are more candidates (models) to form the ensemble. In the last time step there are 199 different models, thus considering that the order of the selected models does not matter and that a model can not be selected more than once, there are 2.13×10^{16} possible combinations (being $n \doteq 199$ and $m \doteq 10$) to form an ensemble of 10 base learners at this time. Taking into account that a Big Data application may have millions of time steps (and models), the need for an optimization technique can be solidly argued.

Table 1. HS similarities in the proposed approach

Element	HS original definition	Proposed approach
Instrument	One of the inputs to the quality function	Each ensemble at time step t_i composed of 10 possible models from t_0 to t_{i-1}
Note	One of the possible values of an input	$Q_{averaged}$ value for each ensemble
Harmony	A combination of each instrument playing a particular note	The formation of an ensemble composed of trained models
Quality	A quantitative measure of a harmony's desirability	The Yule's Q diversity metric
Harmony memory	The collection of good harmonies stored in memory	The collection of ensembles
HMCR	The process of generating a new harmony using random notes from the memory	The probability of choosing a model (note) of the former ensemble (harmony) for the new one
PAR	The process of moving a particular note of an instrument up or down	The probability of choosing a "similar" model (note) to the current one from the new ensemble (new harmony)

The superior performance of HS over other solutions finds its roots in their operators; the search process of HS is controlled by three different operators iteratively applied to a set of candidate solutions [4]. In a nutshell, the Harmony Memory Considerate Rate (HMCR) operator generates a new harmony using random notes from the rest of harmonies in the memory, whereas the Pitch Adjustment Rate (PAR) mutates a particular note of an instrument to a value of the vicinity of its previous value. Table 1 shows the similarities between the original definition of HS and the proposed approach.

The HS approach is applied over a total of 200 data batches, with $\mathbf{X}_{tr}(t)$ and $\mathbf{X}_{tst}(t)$ being composed by 250 samples. Every 50 batches a concept drift occurs, with 3 drift events in total. All base learners are Decision Trees and the ensemble is of size $M = 10$. This work has followed an active approach (using "perfect" drift detection) that triggers the selection of those ensembles that are more appropriate for each time slot. The study will show how to achieve an optimal level of diversity for the ensemble in each moment by the use of HS, evidencing that shortly after the drift an ensemble with high diversity obtains a better classification performance (AUC score) than an ensemble with low diversity. HS has been used to minimize the $\bar{Q}$ metric during 10 time steps after the drift.

4 Experiments and Results

The main purpose of this work is to demonstrate the feasibility of using HS to build ensembles with maximum and minimum diversity specifically when the learning process requires it due to detected drift statistics by an external detector. In this case, Fig. 1 shows three drift moments at time steps 50, 100, and 150; as already explained in Sect. 2, it should be a good strategy to maintain highly diverse ensembles to obtain good classification performance (AUC score) shortly after the drift. The experiments discussed in what follows aim at corroborating this recommendation first posed in [2] by means of an HS-based selection of learners for the ensemble.

In order to avoid using complex algorithm for adaptive learning (which falls out of the scope of this study), we have built ensembles of size $M = 10$ by following two different perspectives. The first refers to the ensemble formed by the best M past learners (hereafter labeled as `BEST`), i.e. those M past learners that obtain the best AUC score, being trained with their training batch data at their time but testing the current batch training data. In the second approach (corr. `LAST`) the ensemble is built with the M last learners again trained with their training batch data at that time, but testing the current batch training data. It is assumed that diversity can help mainly to reduce the initial increase in the error caused by drifts at time steps 50, 100, and 150.

Table 2. Mean AUC and mean $\bar{Q}$ scores over 15 Monte Carlos for the `BEST` approach.

		High diversity ensemble	Low diversity ensemble
After drift 1	AUC	0.911 ± 0.020	0.910 ± 0.025
	$\bar{Q}$	0.507 ± 0.038	0.839 ± 0.021
After drift 2	AUC	0.932 ± 0.008	0.923 ± 0.009
	$\bar{Q}$	0.495 ± 0.029	0.835 ± 0.019
After drift 3	AUC	0.894 ± 0.022	0.848 ± 0.036
	$\bar{Q}$	0.469 ± 0.036	0.837 ± 0.024

This experiment maximizes or minimizes the diversity of the ensemble after drifts 1, 2 and 3 by using HS during the period of 50–59 time steps, 100–109

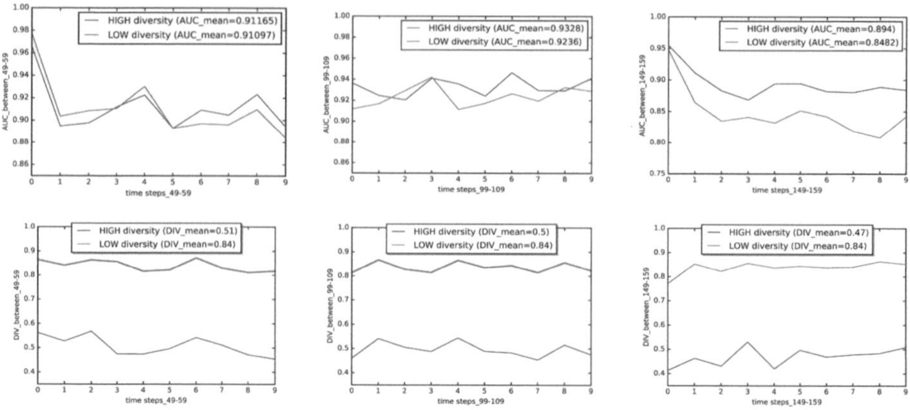

Fig. 2. Mean AUC and $\bar{Q}$ over 15 Monte Carlos after drifts 1, 2, and 3 respectively for the `BEST` approach.

time steps, and 150–159 time steps, respectively. After that, the AUC scores in these periods averaged over 15 Monte Carlo iterations are compared to confirm that it is a good strategy to have an ensemble with high diversity shortly after the drift. The HS algorithm has been configured as follows: 500 improvisations, a harmony memory size of 50 candidate solutions, a HMCR of 0.5 and a PAR of 0.1. As the role of a high diversity ensemble becomes progressively less important after the drift, it is assumed in this work that after 10 time steps high diversity is no longer recommended to have a positive impact in the AUC score. Next, the results of the experiment are discussed.

As it is shown in Table 2, for the "10 best past learners" perspective, the best AUC scores are achieved when the diversity of the ensemble is maximized,

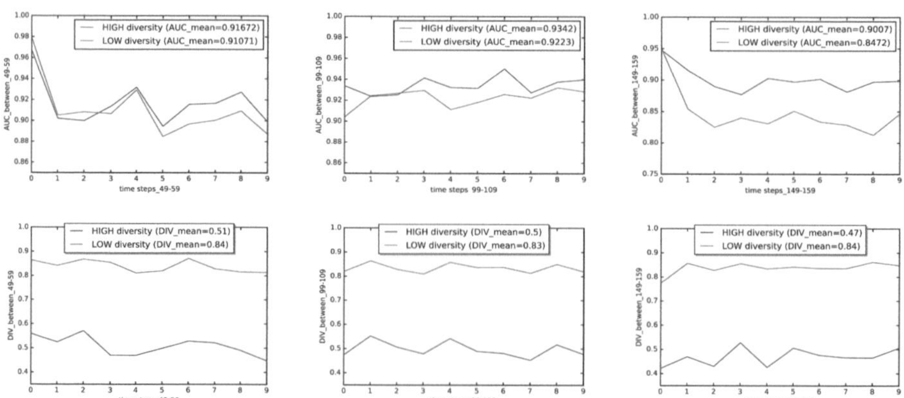

Fig. 3. Mean AUC and $\bar{Q}$ over 15 Monte Carlos after drifts 1, 2, and 3 for the `LAST` approach.

Table 3. Mean AUC and mean $\bar{Q}$ scores over 15 Monte Carlos for the `LAST` approach.

		High diversity ensemble	Low diversity ensemble
After drift 1	AUC	0.916 ± 0.020	0.910 ± 0.025
	$\bar{Q}$	0.508 ± 0.037	0.838 ± 0.022
After drift 2	AUC	0.922 ± 0.008	0.934 ± 0.007
	$\bar{Q}$	0.497 ± 0.029	0.833 ± 0.017
After drift 3	AUC	0.847 ± 0.036	0.900 ± 0.018
	$\bar{Q}$	0.469 ± 0.034	0.837 ± 0.024

in contrast with the version in which the diversity is minimized. Considering the results of the corresponding period of time, the Fig. 2 shows the AUC scores of the high and low diversity ensembles, displaying a better behavior after the drifts in the case of the high diversity one. This also makes sense when checking the level of diversity at each time step.

In the `LAST` case the same results are shown in Table 3, and Fig. 3 shows the best performance for the high diversity ensemble.

5 Conclusions and Future Research Lines

It has been confirmed in Sect. 4 that it is indeed a good strategy to maintain highly diverse ensembles to obtain good classification performance shortly after the drift. Furthermore, the use of a bio-inspired solver such as HS is an proper way of building high diversity ensembles for batch learning scenarios where the evaluation of all possible ensembles of past learners at each time cannot be performed by an exhaustive method. When the time requirements or computational cost are stringent constraints, the HS algorithm allows reducing the number of improvisations and the size of the harmony memory, achieving a solution suitably balancing optimality and computational complexity under these conditions.

After the drift is detected, it is very critical to determine the time range over which a high diversity ensemble is convenient. This work has considered 10 time steps as a relevant interval for high diversity just to show the importance of a high diversity ensemble after the drift. However, this time period might change depending on the type of drift, its severity, the reliability of the drift detection and the statistics of the data considered in the problem. In general it is widely accepted that after a *large* number of time steps since the beginning of the drift, maintaining a high diversity becomes less important and even counterproductive with respect to low diversity ensembles. However, the exact quantification of this *large* number of time steps remains an open problem.

Diversity by itself is helpful to reduce the initial drop in accuracy that happens right after a drift, but not to provide convergence to the new concept. Although high diversity ensembles may help to cushion the initial increase in the error soon after the drift, they do not quickly adapt to the new concept (recovery from drifts). A practical workaround is to create a new ensemble after a drift detection. In this way, the technique would achieve the required equilibrium between *stability* and *plasticity* [21] to reduce the initial drop of accuracy after the drift while, at the same time, to be able to adapt to the new concept.

Also as a future research line it might be of interest to delve into the influence of the size of the ensemble in order to establish a mechanism to find the proper size at each point while simultaneously maintaining a certain level of diversity in the ensemble. There is a trade-off between the severity degree of the disagreement among the members of the ensemble and the number of base learners within it. On the other hand, also the number of samples in the batch may affect this equilibrium, which will also be subject of further investigation in the future.

Acknowledgments. This work has been supported by the Basque Government through the ELKARTEK program (ref. KK-2015/0000080, BID3A project) and BID3ABI project.

References

1. Atzori, L., Iera, A., Morabito, G.: The internet of things: a survey. Comput. Netw. **54**(15), 2787–2805 (2010)
2. Minku, L.L., White, A.P., Yao, X.: The impact of diversity on online ensemble learning in the presence of concept drift. IEEE Trans. Knowl. Data Eng. **22**(5), 730–742 (2010)
3. Minku, L.L., Yao, X.: DDD: a new ensemble approach for dealing with concept drift. IEEE Trans. Knowl. Data Eng. **24**(4), 619–633 (2012)
4. Geem, Z.W., Kim, J.H., Loganathan, G.: A new heuristic optimization algorithm: harmony search. Simulation **76**(2), 60–68 (2001)
5. Pandi, V.R., Panigrahi, B.K., Das, S., Cui, Z.: Dynamic economic load dispatch with wind energy using modified harmony search. Int. J. Bio-inspired Comput. **2**(3/4), 282–289 (2010)
6. Salcedo-Sanz, S., Pastor-Sánchez, A., Del Ser, J., Prieto, L., Geem, Z.W.: A Coral reefs optimization algorithm with harmony search operators for accurate wind speed prediction. Renew. Energy **75**, 93–101 (2015)
7. Scalabrin, M.H., Parpinelli, R.S., Benítez, C.M., Lopes, H.S.: Population-based harmony search using GPU applied to protein structure prediction. Int. J. Comput. Sci. Eng. **9**(1/2), 106 (2014)
8. Zhang, R., Hanzo, L.: Iterative multiuser detection and channel decoding for DS-CDMA using harmony search. IEEE Signal Process. Lett. **16**(10), 917–920 (2009)
9. Manjarres, D., Del Ser, J., Gil-Lopez, S., Vecchio, M., Landa-Torres, I., Lopez-Valcarce, R.: A novel heuristic approach for distance- and connectivity-based multihop node localization in wireless sensor networks. Soft. Comput. **17**(1), 17–28 (2013)
10. Manjarres, D., Landa-Torres, I., Gil-Lopez, S., Del Ser, J., Bilbao, M.N., Salcedo-Sanz, S., Geem, Z.W.: A survey on applications of the harmony search algorithm. Eng. Appl. Artif. Intell. **26**(8), 1818–1831 (2013)
11. Geem, Z.W., Tseng, C.L., Williams, J.C.: Harmony search algorithms for water and environmental systems. In: Geem, Z.W. (ed.) Music-Inspired Harmony Search Algorithm, vol. 191, pp. 113–127. Springer, Heidelberg (2009)

12. Karimi, Z., Abolhassani, H., Beigy, H.: A new method of mining data streams using harmony search. J. Intell. Inf. Syst. **39**(2), 491–511 (2012)
13. Bilbao, M.N., Del Ser, J., Salcedo-Sanz, S., Casanova-Mateo, C.: On the application of multi-objective harmony search heuristics to the predictive deployment of firefighting aircrafts: a realistic case study. Int. J. Bioinspired Comput. **7**(5), 270–284 (2015)
14. Žliobaitė, J., Pechenizkiy, M., Gama, J.: An overview of concept drift applications. In: Japkowicz, N., Stefanowski, J. (eds.) Big Data Analysis: New Algorithms for a New Society, vol. 16, pp. 91–114. Springer, Cham (2016)
15. Gama, J., Zliobaite, I., Bifet, A., Pechenizkiy, M., Bouchachia, A.: A survey on concept drift adaptation. ACM Comput. Surv. (CSUR) **46**(4), 1–37 (2014)
16. Ditzler, G., Polikar, R., Chawla, N.: An incremental learning algorithm for nonstationary environments and class imbalance. In: International Conference on Pattern Recognition (ICPR), pp. 2997–3000 (2010)
17. Ditterrich, T.G.: Machine learning research: four current directions. Artif. Intell. Mag. **4**, 97–136 (1997)
18. Kuncheva, L.I., Whitaker, C.J.: Measures of diversity in classifier ensembles. Mach. Learn. **51**(2), 181–207 (2003)
19. Yule, G.U.: On the association of attributes in statistics: with illustrations from the material of the childhood society, & c. In: Philosophical Transactions of the Royal Society of London. Series A, Containing Papers of a Mathematical or Physical Character, vol. 194, pp. 257–319 (1900)
20. Street, W.N., Kim, Y.: A streaming ensemble algorithm (SEA) for large-scale classification. In: ACM SIGKDD International Conference on Knowledge Discovery and Data Mining, pp. 377–382 (2001)
21. Grossberg, S.: Nonlinear neural networks: principles, mechanisms, and architectures. Neural Netw. **1**(1), 17–61 (1988)

A Novel Grouping Harmony Search Algorithm for Clustering Problems

Itziar Landa-Torres[1(✉)], Diana Manjarres[1], Sergio Gil-López[1], Javier Del Ser[1,2,3], and Sancho Salcedo-Sanz[4]

[1] TECNALIA, 48160 Derio, Spain
{itziar.landa,diana.manjarres,sergio.gil,javier.delser}@tecnalia.com
[2] University of the Basque Country UPV/EHU, 48013 Bilbao, Spain
javier.delser@ehu.eus
[3] Basque Center for Applied Mathematics (BCAM), 48009 Bilbao, Spain
[4] Universidad de Alcalá, 28871 Alcalá de Henares, Spain
sancho.salcedo@uah.es

Abstract. The problem of partitioning a data set into disjoint groups or clusters of related items plays a key role in data analytics, in particular when the information retrieval becomes crucial for further data analysis. In this context, clustering approaches aim at obtaining a good partition of the data based on multiple criteria. One of the most challenging aspects of clustering techniques is the inference of the optimal number of clusters. In this regard, a number of clustering methods from the literature assume that the number of clusters is known a priori and subsequently assign instances to clusters based on distance, density or any other criterion. This paper proposes to override any prior assumption on the number of clusters or groups in the data at hand by hybridizing the grouping encoding strategy and the Harmony Search (HS) algorithm. The resulting hybrid approach optimally infers the number of clusters by means of the tailored design of the HS operators, which estimates this important structural clustering parameter as an implicit byproduct of the instance-to-cluster mapping performed by the algorithm. Apart from inferring the optimal number of clusters, simulation results verify that the proposed scheme achieves a better performance than other naïve clustering techniques in synthetic scenarios and widely known data repositories.

Keywords: Clustering · Grouping encoding · Harmony Search

1 Introduction

Clustering is an important subgroup of unsupervised learning technique that involves grouping data objects into groups or clusters, which may be disjoint (*crisp* clustering) or overlap among each other (*fuzzy* clustering) [1]. A loose definition of clustering could be casted as the process of classifying unlabeled objects into groups in such a way that the members within a determinate cluster (or group) are similar to each other under a given measure of similarity.

J. Del Ser (ed.), *Harmony Search Algorithm*, Advances in Intelligent Systems and Computing 514, DOI 10.1007/978-981-10-3728-3_9

In essence, clustering aims at grouping an input set of *samples* into a finite number of clusters using only the information contained in such samples. The so-called *samples* (also denoted in the related literature as observations or instances) are normally modeled as numerical vectors whose items (features) represent numerical information to be used in the similarity measure. Mathematically, given a feature space $\mathcal{U}$, and if $\boldsymbol{\mathcal{X}} \doteq \{\mathbf{X}_1, \ldots, \mathbf{X}_N\}$ denotes a set of N samples in such an space, the challenge of clustering problems lies in finding an K-sized optimal partition of $\boldsymbol{\mathcal{X}}$, i.e. $\boldsymbol{\mathcal{X}}^* = \{\boldsymbol{\mathcal{X}}_1^*, \ldots, \boldsymbol{\mathcal{X}}_K^*\}$ (with $\boldsymbol{\mathcal{X}}_k^* \bigcap \boldsymbol{\mathcal{X}}_{k'}^* = \emptyset$ $\forall k \neq k'$, $\bigcup_{k=1}^{K} \boldsymbol{\mathcal{X}}_k^* = \boldsymbol{\mathcal{X}}$ and $\boldsymbol{\mathcal{X}}_k^*$ denoting the k-th partition of $\boldsymbol{\mathcal{X}}^*$). This clustering arrangement collects in the same cluster samples that are similar to each other as measured by a given objective function $f(\boldsymbol{\mathcal{X}}^*)$, which can defined under different similarity-based criteria.

Most of the clustering techniques in the literature can be divided into two general classes: hierarchical [2] and partitional [3]. The first class corresponds to those methods that create a hierarchical decomposition of the dataset under study. They can be agglomerative, when they start the clustering process with each sample on a separate cluster and successively combine clusters; or divisive, if they begin with all the patterns in a single cluster and perform this partition. By contrast, partitional algorithms obtain a single partition of the data instead of a hierarchy, i.e. they begin with a initial partition that is iteratively refined in order to obtain the final solution.

Interestingly under the scope of this manuscript, clustering algorithms can be also sorted depending on the deterministic or stochastic nature of the underlying algorithm: as such, deterministic approaches are not controlled by any probabilistic process, hence the instance-to-cluster mapping is fixed whenever the parameters of the algorithm and the dataset being clustered do not vary along time. On the other hand, stochastic clustering models are governed by probability-based processes so that this randomness may help the search process escape from local optima, at the cost of a certain degree of variability imposed on the instance-to-cluster mapping even when the input data does not vary. It is also worth mentioning other clustering techniques that rely on other criteria with stochastic and deterministic ingredients in their algorithmic core, such as Tabu Search [4], neural networks [5] and kernel spaces [6].

In this paper we focus on stochastic clustering models ruled by meta-heuristic solvers, in particular those incorporating the so-called Harmony Search (HS) algorithm as their constituent heuristic engine. The authors in [7] present a framework for simultaneous feature selection and clustering using the HS algorithm, whereas in [8] a centralized cluster-based protocol based on HS is proposed to minimize the intra-cluster distance and thus optimize the energy consumption of wireless networks. The performance of HS is compared to that of conventional clustering techniques in [9]. Despite the massive upsurge of different clustering techniques along the years, to the authors' knowledge grouping-encoded algorithms have not been tested in clustering problems. Intuitively, the grouping algorithm should perform well when applied to clustering problems, since it is originally conceived and well-adapted to manage groups of items [10].

This work takes a step further by adapting a grouping-encoded HS to accommodate a varying number of groups (clusters), which is estimated along the search process undertaken by the HS operators. This novel ingredient is deemed of paramount importance for practical clustering scenarios where traditional tools in this regard (e.g. the Elbow method) have been proven not to be efficient nor effective in most cases.

2 Proposed Approach

The Grouping Harmony Search Algorithm for Clustering (GHSC) proposed in this paper adopts the classical grouping encoding first contributed by Falkenauer in [10] and recently implemented in [12]. This variable-length grouping encoding is carried out by splitting each candidate vector $\mathbf{s}$ handled by the heuristic solver into two parts, i.e. $\mathbf{s} = [\mathbf{s_x}|\mathbf{s_y}]$: the first part, $\mathbf{s_x}$, is the *assignment part*, which consists of N integer indices with values drawn from the set $\{1, \ldots, k_y\}$ with k_y denoting the length of the second part of the solution. This part establishes to which cluster is assigned each sample. The second part $\mathbf{s_y}$ of the encoded solution (*group part*) maintains a list of tags associated to each of the clusters of the solution. It is composed by a k_y-length vector of integer indices from the set $\{1, \ldots, K\}$, which serves as a indexing reference for the assignment part. Therefore, the length of $\mathbf{s_x}$ is fixed and equal to N for a given problem, whereas the length of the group part is not fixed and may vary among the solutions handled by the search process. The underlying heuristic solver searches for the best length $1 \leq k_y \leq K$ of the group part in terms of an objective function. Following this notation, a solution for a clustering problem with $N = 10$ samples and $k_y = 4$ clusters could be represented as [3 2 2 4 1 3 1 2 3 4 | 1 2 3 4] in which, according to the notation introduced in Sect. 1, the partition $\boldsymbol{\mathcal{X}}^*$ would be $\boldsymbol{\mathcal{X}}_1^* = \{\mathbf{X}_5, \mathbf{X}_7\}$, $\boldsymbol{\mathcal{X}}_2^* = \{\mathbf{X}_2, \mathbf{X}_3, \mathbf{X}_8\}$, $\boldsymbol{\mathcal{X}}_3^* = \{\mathbf{X}_1, \mathbf{X}_6, \mathbf{X}_9\}$ and $\boldsymbol{\mathcal{X}}_3^* = \{\mathbf{X}_4, \mathbf{X}_{10}\}$.

2.1 Redundancy of the Clustering Encoding

As often stated in related contributions, most solution encoding strategies proposed to represent groups suffer from redundancies [13,14]. This section will elaborate on how to properly design a cluster encoding that mitigates this redundancy by ensuring that a clustering arrangement cannot be represented by more than one encoded individual. To this end, in the following set of encoded solutions (with colors compounding the group part for clarity),

$$\begin{array}{l}
[1\ 2\ 3\ 2\ 3\ 1\ 2\ 1\ 3 \mid \mathrm{G\ R\ Y}], \\
[1\ 3\ 2\ 3\ 2\ 1\ 3\ 1\ 2 \mid \mathrm{G\ Y\ R}], \\
[3\ 1\ 2\ 1\ 2\ 3\ 1\ 3\ 2 \mid \mathrm{R\ Y\ G}], \\
[2\ 1\ 3\ 1\ 3\ 2\ 1\ 2\ 3 \mid \mathrm{R\ G\ Y}], \\
[2\ 3\ 1\ 3\ 1\ 2\ 3\ 2\ 1 \mid \mathrm{Y\ G\ R}], \\
[3\ 2\ 1\ 2\ 1\ 3\ 2\ 3\ 1 \mid \mathrm{Y\ R\ G}],
\end{array} \tag{1}$$

it can be noted that all the represented clustering arrangements are equal: elements $\{\mathbf{X}_1, \mathbf{X}_6, \mathbf{X}_8\}$ always belong to cluster G (green), $\{\mathbf{X}_2, \mathbf{X}_4, \mathbf{X}_7\}$ to cluster

R (red) and $\{\mathbf{X}_3, \mathbf{X}_5, \mathbf{X}_9\}$ to Y (yellow). The proposal made recently by the authors in [15] proposes to sort the group part according to a pre-established criterion. In this particular case, for example, the colors can be listed according to its wavelength λ ($\lambda_R = 618$, $\lambda_G = 497$ and $\lambda_Y = 570\,\text{nm}$). As a result, the 6 individuals would be encoded as [1 3 2 3 2 1 3 1 2 | G Y R].

This being said, the proposal of this work is to sort the indices of the clusters in order of appearance along the assignment part, in such a way that the 6 equivalent solutions in (1) are encoded as [1 2 3 2 3 1 2 1 3 | 1 2 3]. This encoding reduces the solution space of the problem, as it implies that the first position in the solution vector will always be a 1 and higher indexes will be less likely to appear. Additionally, the group part of the harmony is not needed, as all the information from the solution can be extracted from the assignment part; it determines each of the samples to which cluster is assigned and the number of clusters k_y can be deduced from the maximum value in the assignment part, i.e. $k_y = \max \mathbf{s}_\mathbf{X}$. As a counterpoint the proposed solution encoding requires several modifications in the improvisation operators of the nominal Harmony Search algorithm, which is the heuristic engine selected to evolve this numerical representation of clustering arrangements.

2.2 Proposed Grouping Harmony Search Algorithm for Clustering

The heuristic solver that lies at the core of the proposed clustering scheme is Harmony Search (HS), a population-based meta-heuristic algorithm that since its invention in [11] has rendered excellent results in the field of combinatorial optimization [16]. It mimics the behavior of a music orchestra when aiming at composing the most harmonious melody, as measured by aesthetic standards. Just like jazz musicians improvise harmonies time after time searching for aesthetically pleasant melodies, the HS algorithm improves the fitness of the solution vector in an iterative fashion by applying several operators to a φ-sized set of solutions, stored in the so-called Harmony Memory (HM). The flow diagram of the HS algorithm can be summarized in four steps: (i) initialization of the HM; (ii) improvisation of a new harmony; (iii) update of the HM with the new generated harmony if its fitness improves that of the worst currently in the HM; and (iv) repeat termination criterion (e.g. maximum number of iterations or fitness stall) is satisfied. The improvisation procedure is controlled by two different probabilistic operators, which are sequentially applied to each variable so as to produce new improvised candidate solutions:

- The Harmony Memory Considering Rate, HMCR $\in [0, 1]$, establishes the probability that the new value for a certain variable is drawn uniformly from the values of this same note in all the remaining melodies. Otherwise (i.e. with a probability 1 − HMCR), the value is chosen uniformly at random from the alphabet of the variable. This latter case is commonly referred to as *random consideration*. However, some works (e.g. [17]) implement the random consideration as a third, separated probabilistic operator.

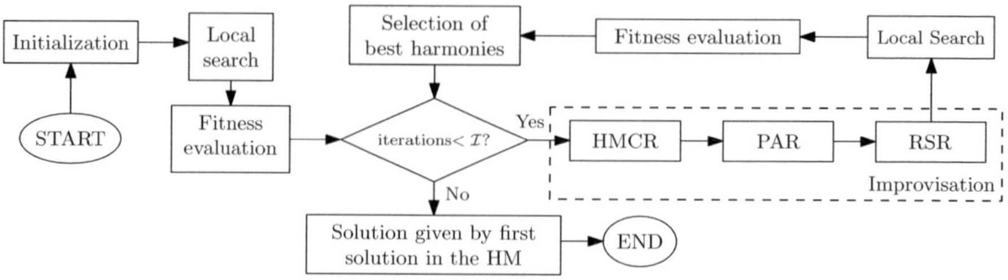

Fig. 1. General scheme of the proposed GHSC algorithm.

- The Pitch Adjusting Rate, PAR $\in [0, 1]$, sets the probability that the new value x_{new} for a given variable is drawn from its neighboring values in the alphabet. To yield a PAR operator well-suited to the problem at hand, the variable alphabet should be sorted according to the fitness to be optimized so that subtle pitch adjustments do not imprint large changes in the value of the fitness function.

The flow diagram of the proposed GHSC algorithm is schematically shown in Fig. 1, and comprises 5 different steps:

A. Initialization, only executed at the first iteration: each of the φ candidate solutions included in the HM is assigned a random value from the set $\{1, \ldots, K\}$ (number of clusters), from which the N entries of the $\mathbf{s_x}$ part of the harmony are drawn uniformly at random.

B. Improvisation, which generates new harmonies by operating on the s_x part of each solution. This process is sequentially applied to each entry of the s_x part of every harmony in the HM. Additionally, as done in [12,15,17], the proposed improvisation procedure differs from the original HS implementation by introducing a third parameter *Random Selection Rate*, RSR $\in [0, 1]$, which allows for an improved control of the tradeoff between the explorative and the exploitative behavior of the algorithm. Thus, the improvisation of the proposed GHSC is controlled by means of the HMCR, PAR and RSR operators. In light of the results obtained in previous works [17], the parameters of the CHS are modified so as to achieve a more effective information exchange and exploit information related to the distances among samples:

- HMCR: the proposed method does not exchange information between harmonies, but it leverages the information of the current partition $\boldsymbol{\mathcal{X}}^*$ and attempts to enhance it based on the proximity between samples. The HMCR applied to a certain sample establishes the probability that such an instance is assigned to another cluster by addressing the connections of the surrounding samples. These are the main steps:
  1. All the samples in the network are listed in increasing order of the distances to the sample to which the HMCR is applied (from smallest to largest). The first samples of this list are denoted as *nearby candidates*.

2. One out of the *nearby candidates* computed in the previous step is uniformly chosen at random, and the sample on which HMCR is applied is assigned to the cluster of the chosen *nearby candidate.*

- PAR: as explained before, this process executes subtle adjustments in the chosen harmony. In the proposed GHSC algorithm the PAR stands for the probability that the actual sample is assigned to the cluster of its roughly *nearest* sample is assigned to.
- RSR: as can be inferred from the above descriptions, none of the previous processes involves a variation in the number of clusters. The RSR will therefore modify this number and, in order to ensure that the variation in a harmony from one iteration to the next one is gradual, this parameter is applied to each harmony, not to each note. The RSR has two operation modes in which the possibility of increasing or decreasing the number of clusters is given with 50% of probability each; in the first case, a new cluster is created by splitting one of the actual solution (chosen randomly), and in the second case, one randomly selected cluster is eliminated and its samples are reassigned to the closest remaining cluster.

C. Local Search: this procedure is utilized to find local optima in the vicinity of a certain harmony. Specifically, this method measures the metric obtained when a determinate sample is assigned to all the other clusters in the solution and selects the best cluster for each sample. The process is applied to the harmonies under a certain probability and is sequentially repeated until all samples are assigned to their most optimal cluster.

D. Fitness evaluation: at every iteration the quality evaluation of newly improvised harmonies is made based on an objective function. To this end two widely known metrics will be utilized: the Davis-Bouldin (DB) index [18] and the Silhouette coefficient [19]. The Davis-Bouldin index is defined for a given cluster arrangement $\boldsymbol{\mathcal{X}}^*$ and a similarity metric $d(\cdot,\cdot)$ as

$$\mathrm{DB}(\boldsymbol{\mathcal{X}}^*) \doteq \frac{1}{K}\sum_{k=1}^{K}\max_{k\neq k'}\left\{\frac{\sum_{x\in\boldsymbol{\mathcal{X}}_k^*} d^2(x,\mu_k)+\sum_{x\in\boldsymbol{\mathcal{X}}_{k'}^*} d^2(x,\mu_{k'})}{d^2(\mu_k,\mu_{k'})}\right\}, \tag{2}$$

with μ_k denoting the centroid of cluster $\boldsymbol{\mathcal{X}}_k^*$. This index favors solutions with small distances between the samples assigned to the same cluster and large distances among different clusters. It is important to note that compact and well separated clusters entail low values of the DB index. Furthermore, this score does not present a monotonic behavior with the number of clusters, so it also allows validating the optimal number of clusters for a given data set. On the other hand, the Silhouette coefficient is a commonly utilized measure in clustering problems as it allows evaluating the quality not only of a single solution, but also that of each of the arranged clusters. For the n-th sample $\mathbf{X}_n$ assigned to cluster $k(n)$, the Silhouette coefficient Υ_n is defined as

$$\Upsilon_n \doteq \frac{\alpha_{k(n)}-\beta_{k(n)}}{\max\{\alpha_{k(n)},\beta_{k(n)}\}}, \tag{3}$$

where the parameter $\alpha_{k(n)}$ represents the average distance between samples in the cluster $\boldsymbol{\mathcal{X}}^*_{k(n)}$ (i.e. the intra-cluster distance) to which the n-th sample belongs, whereas $\beta_{k(n)}$ stands for the minimum distance between the samples in cluster $\boldsymbol{\mathcal{X}}^*_{k(n)}$ to the remaining samples assigned to different clusters $k' \neq k(n)$. For a cluster $\boldsymbol{\mathcal{X}}^*_k$, the silhouette coefficient is defined as the average of the Silhouette coefficients of its constituent samples. By using this coefficient good partitions featuring compact and well-defined clusters while separated from each other are obtained when its value gets close to its maximum value. The evaluation of these metric functions and their comparison to the fitness of harmonies from previous iterations permits to update the HM with the φ best harmonies.

E. Stop criterion: the search process stops when a fixed number of iterations I is reached. This criterion has been established in order to provide a fair comparison between the algorithms compared in the later discussed experiments.

Besides the novel encoding solution for avoiding redundancies presented in Sect. 2, two additional concepts are included in the proposed GHSC scheme. The first one is related with the differential characteristic of the defined operators, thus no population-based knowledge is used and a new harmony is improvised from its state at previous iteration. On the other hand, the improvisation operators are defined on the basis of the structural relationship of the feature space as provided by the Euclidean distances between samples and their closest neighbors. Last but not least, the fact that only feasible solutions are improvised during the iterative process guarantees that the computational complexity of the algorithm is reduced.

3 Experiments and Results

In this section different experiments based in synthetic scenarios obtained from two public repositories are presented. The proposed GHSC is compared to different clustering solutions presented in the literature:

1. The K-means algorithm [20], which can be regarded as one of the most popular clustering algorithms. This approach requires the number of clusters K as an input parameter, and it obtains a partition of the data into K clusters. It is fairly well known that the K-means is simple and easy to use, which motivates its widespread use in a large variety of problems. However, it is important to remark that the K-means obtains poor results in problems where clusters have different sizes, densities and/or many outliers. Additionally, this deterministic algorithm has a strong dependency with the initialization of the centroids which can make the algorithm get local optimum solutions. As last claim, the required input of the number of clusters to be discovered by this clustering method limits significantly its practicality. In these experiments, the well-known elbow method is used to automatically set the number of clusters K.

2. Density Based Spatial Clustering of Applications with Noise (DBSCAN [21]), which finds the number of clusters based on the assumption that the spatial density of samples belonging to a certain cluster should be higher than a predefined density-reachability threshold parameter (ϵ). DBSCAN requires an additional input parameter to set the minimum number of points N_{min} required to form a cluster. As points are assigned to clusters based on ϵ, N_{min} and the distance between samples $d(\cdot,\cdot)$, this algorithm does not need to know a priori the number of clusters to be sought. Additionally, DBSCAN is able to recognize clusters with different shapes and sizes, and takes into account the notion of noise, in such a way that outlier samples do not influence the algorithm's performance. However, it performs poorly in clustering problems over spaces with areas characterized by significant density differences, and still requires two input parameters (ε and N_{min}) to be configured.

3. Grouping Genetic Algorithm (GGA) [12], which shares the same encoding with the CHS with the exception that GGA also utilizes the s_y part. In connection to the general scheme of the GHSC detailed in Sect. 2.2 the initialization step (A), local search (C), metric evaluation (D) and stop criterium (E) are kept the same in the GGA approach. However, the GGA operators follow the selection, crossover and mutation mechanisms featured by evolutionary algorithms: first, a rank-based roulette wheel mechanism is adopted for the *selection* of the individuals to be mated. It is important to note that this rank-based selection mechanism is static, in the sense that probabilities of survival do not depend on the generation, but on the position of the individual in the list. Likewise, the *crossover* operator implemented in GGA is a modified version of the one initially proposed by Falkenauer [10], adapted to the clustering problem to remove empty clusters and fulfill the redundancy-minimizing encoding strategy explained in Sect. 2.1. Additionally, the GGA implements two different types of mutation: (1) cluster splitting, by which a selected cluster is split in two, and (2) clusters merging, which gathers two randomly selected groups into one.

3.1 Results and Discussion

This section discusses results obtained by the above algorithms in datasets with diverse characteristics in terms of density, size and/or shape, which eases the understanding and assessment of the weaknesses and strong points of all the algorithms within the benchmark. Besides these synthetic scenarios, experiments with the widely utilized `Iris` and `Wine` databases are also presented. All reported scores have been computed over 20 Monte Carlo experiments and $\mathtt{I} = 100$ iterations. Forthcoming discussions are held on the *average* Rand Index (avg-RI) values, a measure of similarity between two clustering arrangements that in this case, is computed between the clustering produced by each algorithm and the *gold standard* known for every dataset. The proposed GHSC approach is configured with $\varphi = 50$, HMCR $= 0.3$, PAR $= 0.1$ and RSR $= 0.2$.

As anticipated above the discussion is focused on analyzing how GHSC performs in several feature spaces with different cluster properties, namely:

- `Spherical`, which is composed by $N = 300$ samples drawn uniformly at random from 8 independent Gaussian distributions with means $\mu_1 = (-1, 0), \mu_2 = (-1, -1), \mu_3 = (-1, -3), \mu_4 = (3, -1), \mu_5 = (-1, 1), \mu_6 = (2, -2), \mu_7 = (1, 2), \mu_8 = (3,\ 1)$, statistical independence between dimensions and standard deviation per dimension equal to 0.35.
- `Structured`, formed by $N = 400$ samples randomly generated using a Gaussian distribution from 3 classes with probability $p_1 = 0.5$, $p_2 = 0.33$ and $p_3 = 0.17$. Means for each class are $\mu_1 = (0, 2), \mu_2 = (-1, -1)$ and $\mu_3 = (2, -1)$, with the following covariance matrices:

$$\Sigma_1 = \begin{pmatrix} 1^2 & 0 \\ 0.8^2 & 0 \end{pmatrix}, \qquad \Sigma_2 = \begin{pmatrix} 0.6^2 & 0 \\ 0.4^2 & 0 \end{pmatrix}, \qquad \Sigma_3 = \begin{pmatrix} 0.3^2 & 0 \\ 0.5^2 & 0 \end{pmatrix}. \tag{4}$$

- `Unbalanced`, which comprises $N = 200$ samples randomly generated using a Gaussian distribution from 9 equiprobable classes, with means $\mu_1 = (1, -1), \mu_2 = (-1.5, 0), \mu_3 = (0, 1), \mu_4 = (-1, 1), \mu_5 = (2, -1), \mu_6 = (-2, -1), \mu_7 = (-0.5, 2), \mu_8 = (-1, -1)$ and $\mu_9 = (1.5,\ 0)$, statistical independence between dimensions and standard deviation per dimension equal to 0.2.
- `Iris`, which considers 3 classes formed by 50 samples each, totaling $N = 150$ instances. The challenge consists of differentiating among three different Iris plants (*Sentosa*, *Virginica* and *Versicolor*) based on 4 characteristics of the flowers: length and width of sepal and petal of the flower, all in centimeters.
- `Wine`, formed by 3 classes comprising 59, 41 and 78 samples each ($N = 178$ samples). These samples represent classes of wine from different regions of Italy that are defined by 13 analyzed chemical properties of the wines.

Table 1 summarizes the results obtained by the algorithms over the considered datasets. Results of the proposed GHSC and GGA are calculated by using the Davies-Bouldin (DB) index and the Silhouette (Υ) coefficient as their fitness function. It is important to recall that scores are given in terms of the Rand

Table 1. Comparison of the results obtained by the proposed GHSC algorithm and GGA with DB and Υ index, DBSCAN and K-means algorithms in the considered clustering problems. Best scores are highlighted in bold.

Algorithm	`Spherical`		`Structured`		`Unbalanced`		`Iris`		`Wine`	
	K	avg-RI	K	avg-RI	K	avg-RI	K	avg-RI	K	avg-RI
K-means	9	0.949	4	0.875	3	0.780	3	0.880	3	0.702
DBSCAN	7	0.955	1	0.376	8	0.395	2	0.777	2	0.349
GGA (DB)	8	0.981	3	0.917	9	**1.000**	3	0.873	3	0.731
GGA (Υ)	7	0.957	3	0.951	9	0.993	3	0.899	3	0.722
GHSC (DB)	8	0.994	3	0.942	9	**1.000**	3	0.901	3	**0.739**
GHSC (Υ)	8	**0.995**	3	**0.965**	9	0.976	3	**0.903**	3	0.733

Index, computed by assuming that the best data partition is the one corresponding to the original classes (distributions) of the dataset at hand. Focusing on the `Spherical` dataset, the solution provided by GHSC improves on average the best results obtained with the GGA, DBSCAN and K-means by 1.4%, 4% and 4.5%, respectively. Solutions obtained with the DB index with GGA and GHSC are similar to each other (just 1.3% in favour of GHSC), whereas this gap widens an additional 4% when utilizing the Silhouette index Υ. This slight margin improvement must be assessed jointly with the fact that just GGA (DB) and both GHSC proposals manage to infer the correct number of clusters within the data. Finally, when comparing the best approaches (GGA and GHSC), the latter attains the best results regardless the index utilized (DB or Υ).

In the `structured` case the best result is obtained by GHSC with Υ index as its fitness function. It is interesting to notice the poor result obtained by the DBSCAN approach. In addition, neither K-means nor DBSCAN determine the actual number of clusters given by the statistical distributions used for generating this dataset. When evaluating the DB and Υ indexes, the latter scores better average results for both GGA and GHSC, from which it is concluded that this index adapts better to non-compact clusters.

When it comes to the `Unbalanced` dataset both GHSC and GGA with the DB index are able to split the cluster space according to the distributions from which instances were produced (avg-RI equal to 1.000). The CHS driven by the Υ coefficient as its fitness function also attains a perfect cluster arrangement. In this case, K-means underestimates the number of clusters and ultimately produces bad performance scores. From these experiments two conclusions can be drawn: (1) meta-heuristic approaches are a good alternative to optimally detect the number of clusters; and (2) the DB index outperforms Υ, i.e. the best approach when dealing with non-overlapping cluster spaces is to balance inter-cluster and intra-cluster distances (as done by the DB index) instead of focusing on the cluster dispersion targeted by the Υ index.

For the `Iris` dataset the best result is given by GHSC with Υ index. Nevertheless, both GGA and K-means obtain good solutions, slightly worse (1%–3%) than the GHSC with the DB index, but better than the solution provided by GGA with the DB index. The DBSCAN approach, though, does not provide an accurate solution for this experiment. This dataset is composed by one compact, well-defined cluster (suitable for the DB index) and two more disperse additional groups (appropriate for the Υ index). There lies the rationale why both indexes attain similar scores for the proposed GHSC method. Finally, a same line of reasoning can be taken for the `Wine` setup: the GHSC approach with the DB index scores best for this dataset outperforming GGA, K-means and DBSCAN. From this discussion an essential conclusion can be obtained: in general all meta-heuristic approaches are able to determine optimally the number of clusters and outperform traditional clustering schemes, with the proposed GHSC technique as the prevailing option in the benchmark.

4 Concluding Remarks

In light of the obtained results the most straightforward conclusions point out that the K-means algorithm is likely to fall in local optimums and is mostly useful for problems where clusters have compact spherical shapes of uniform sizes and densities. On the other hand, DBSCAN is not able to differentiate accurately overlapped clusters. However, both GGA [12] and the proposed GHSC have proven to optimally infer the number of clusters and the partitions, being the latter slightly better than the former. One of the improvements provided by GHSC when compared to other algorithms from the literature is the inference of the number of clusters. Unlike other algorithms that require a dedicated step to this purpose or further information regarding the dataset at han, the proposed algorithm is able to determine both the optimal partitioning of the samples and the number of clusters for each dataset embedded within the search process. The algorithm begins with a number of randomly chosen clusters, and converges towards the optimum number of groups using a measure of clustering quality as the objective function to optimize.

In this context, the measures utilized for the experimental part of this work – namely, the DB index and the Silhouette index – focus on establishing compact clusters with delimited shapes and no overlap. As such, the DB index measures the average similarity between each cluster and the one that most resembles it based on the intra/inter cluster distance ratio. The Silhouette index is based on the concepts of average scattering for clustering and the total separation among clusters, which also reflect the compactness of clusters. Apart from these commonalities between both indexes, the first one works better with overlapped clusters and in general is more stable for all kind of data sets. Apart from failing with overlapped clusters, the related literature (see e.g. [22–24] and references therein) has often concluded that the DB index attains lower values in general cluster spaces than the Silhouette index. Although the study presented in this paper for testing the performance of GHSC is a small representation of all existing techniques and types of datasets, the conclusions drawn in this reduced experimental setup are congruent with those by previous references and span their applicability over the meta-heuristic field. Future developments will gravitate on the application of GHSC to real clustering problems, the use of alternative coefficients as an objective function for the heuristic search and the implementation of this algorithm in massively parallel Big Data computing architectures.

Acknowledgments. This work has been supported in part by the Basque Government through the ELKARTEK program (ref. KK-2015/0000080 and the BID3ABI project).

References

1. Xu, R., Wunsch, D.: Survey of clustering algorithms. IEEE Trans. Neural Netw. **16**(3), 645–678 (2005)
2. Johnson, S.C.: Hierarchical clustering schemes. Psychometrika **32**(3), 241–254 (1967)

3. Davidson, I., Wagstaff, K.L., Basu, S.: Measuring constraint-set utility for partitional clustering algorithms. In: Fürnkranz, J., Scheffer, T., Spiliopoulou, M. (eds.) PKDD 2006. LNCS, vol. 4213, pp. 115–126. Springer, Heidelberg (2006). doi:10.1007/11871637_15
4. Al-Shultan, K.S.: A tabu search approach to the clustering problem. Pattern Recogn. **28**(9), 1443–1451 (1995)
5. Du, K.L.: Clustering: a neural network approach. Neural Netw. **23**, 89–107 (2010)
6. Dhillon, I.S., Guan, Y., Kulis, B.: Kernel K-means: spectral clustering and normalized cuts. In: ACM SIGKDD, pp. 551–556 (2004)
7. Sarvari, H., Khairdoost, N., Fetanat, A.: Harmony search algorithm for simultaneous clustering and feature selection. In: International Conference of Soft Computing and Pattern Recognition, pp. 202–207 (2010)
8. Hoang, D.C., Kumar, R.: A Robust harmony search algorithm based clustering protocol for wireless sensor network. In: IEEE International Conference on Communications, pp. 1–5 (2010)
9. Senthilnath, J., Kulkarni, S., Raghuram, D.R., Sudhindra, M., Omkar, S.N.: A novel harmony-search approach for clustering problems. Int. J. Swarm Intell. 2(1) (2016)
10. Falkenauer, E.: A new representation and operators for genetic algorithms applied to grouping problems. Evol. Comput. **2**, 123–144 (1994)
11. Geem, Z.W., Kim, J.-H., Loganathan, G.V.: A new heuristic optimization algorithm: harmony search. Simulation **76**(2), 60–68 (2001)
12. Agustín-Blas, L.E., Salcedo-Sanz, S., Jiménez-Fernández, S., Carro-Calvo, L., Del Ser, J., Portilla-Figueras, J.A.: A new grouping genetic algorithm for clustering problems. Expert Syst. Appl. **39**(10), 9695–9703 (2012)
13. Falkenauer, E.: A hybrid grouping genetic algorithm for bin packing. J. Heuristics **2**(1), 5–30 (1996)
14. Falkenauer, E.: A genetic algorithm for bin packing and line balancing. IEEE Int. Conf. Rob. Autom. **2**, 1186–1192 (1992)
15. Landa-Torres, I., Manjarrés, D., Salcedo-Sanz, S., Del Ser, J., Gil-López, S.: A multiobjective grouping harmony search algorithm for the optimal distribution of 24-hour medical emergency units. Expert Syst. Appl. **40**(6), 2343–2349 (2013)
16. Manjarres, D., Landa-Torres, I., Gil-Lopez, S., Del Ser, J., Bilbao, M.N., Salcedo-Sanz, S., Geem, Z.W.: A survey on applications of the harmony search algorithm. Eng. Appl. Artif. Intell. **26**(8), 1818–1831 (2013)
17. Landa-Torres, I., Del Ser, J., Salcedo-Sanz, S., Gil-López, S., Portilla-Figueras, J.A., Alonso-Garrido, O.: A comparative study of two hybrid grouping evolutionary techniques for the capacitated p-median problem. Comput. Oper. Res. **39**(9), 2214–2222 (2012)
18. Davies, D., Bouldin, D.: A cluster separation measure. IEEE Trans. Pattern Anal. Mach. Intell. **1**(2), 224–227 (1997)
19. Rousseeuw, P.: Silhouettes: a graphical aid to the interpretation and validation of cluster analysis. J. Comput. Appl. Math. **20**, 53–65 (1987)
20. McQueen, J.: Some methods for classification and analysis of multivariate observations. In: Berkeley Symposium on Mathematics and Statistics, pp. 281–297 (1968)
21. Ester, M., Kriegel, H.P., Sander, J.: A density-based algorithm for discovering clusters in large spatial data bases with noise. In: International Conference on Knowledge Discovery and Data Mining, pp. 226–231 (1996)

22. Rendón, E., Abundez, I.M., Gutierrez, C., Zagal, S.D., Arizmendi, A., Quiroz, E.M., Arzate, H.E.: A comparison of internal and external cluster validation indexes. In: American Conference on Applied Mathematics, pp. 158–163 (2011)
23. Guerra, L., Robles, V., Bielza, C., Larrañaga, P.: A comparison of cluster quality indices using outliers and noise. Intell. Data Anal. **16**, 703–715 (2012)
24. Bandyopadhyay, S., Saha, S.: Unsupervised Classification: Similarity Measures, Classical and Metaheuristic Approaches, and Applications. Springer, Heidelberg (2013)

Joint Feature Selection and Parameter Tuning for Short-Term Traffic Flow Forecasting Based on Heuristically Optimized Multi-layer Neural Networks

Ibai Laña[1(✉)], Javier Del Ser[1,2,3], Manuel Vélez[2], and Izaskun Oregi[1]

[1] TECNALIA, 48160 Derio, Spain
{ibai.lana,javier.delser,izaskun.oregui}@tecnalia.com
[2] University of the Basque Country UPV/EHU, 48013 Bilbao, Spain
manuel.velez@ehu.eus
[3] Basque Center for Applied Mathematics (BCAM), 48009 Bilbao, Spain

Abstract. Short-term traffic flow forecasting is a vibrant research topic that has been growing in interest since the late 70's. In the last decade this vibrant field has shifted its focus towards machine learning methods. These techniques often require fine-grained parameter tuning to obtain satisfactory performance scores, a process that usually relies on manual trial-and-error adjustment. This paper explores the use of Harmony Search optimization for tuning the parameters of neural network jointly with the selection of the input features from the dataset at hand. Results are discussed and compared to other tuning methods, from which it is concluded that neural predictors optimized via the proposed heuristic wrapper outperform those tuned by means of naïve parametrized algorithms, thus allowing for longer-term predictions. These promising results unfold potential applications of this technique in multi-location neighbor-aware traffic prediction.

Keywords: Traffic forecasting · Neural networks · Bioinspired heuristics

1 Introduction

Forecasting traffic conditions is a key element in the development of Intelligent Transport Systems (ITS), providing the means to implement management (ATMS) and information (ATIS) systems for both road managers and users. Anticipating future traffic can aid the first to regulate signals, lanes and to cope with congestion, and the latter to plan travels and select the best routes to their destinations. For decades, researchers have built traffic models to predict volume, occupancy, speed, travel time or level of service, consisting of elements from time-series analysis in the beginning, and later evolving to non-parametric and machine learning models such as kNN, Artificial Neural Networks (ANN), Support Vector Machines (SVM), Bayesian Networks and Fuzzy Logic models,

J. Del Ser (ed.), *Harmony Search Algorithm*, Advances in Intelligent Systems and Computing 514, DOI 10.1007/978-981-10-3728-3_10

among others [1]. In the last decade an upsurge of traffic-related data has become available, which has lead, along with advances in computational technologies and machine learning techniques, to a noticeable research drift towards data-driven approaches, with more diverse and abundant data sources that conform large databases with hidden knowledge to be discovered by pattern recognition algorithms. Changes are also observable in predicted variables, which tend to become more user-friendly (travel time versus volume) and in the scope of predictions which are increasingly urban and network-wide [2].

In this context, ANNs and their combination with other methods have been extensively used with relative success over naïve (historic average and last measurement predictions) and time-series models [3–7], fueled by prior literature evincing that ANNs are more responsive to changes in data [8]. However, neural networks behave in a black-box manner that hinders their understanding. Furthermore, their internal structure and training process is known to be a slow and inefficient trial-and-error procedure. In this regard, seeking the optimal neural network structure started to be automated with bio-inspired heuristics [9]. In particular for the transport sector, Genetic Algorithms (GA) were first explored for refining the calibration parameters that enhanced neural network behavior without linking them to particular traffic characteristics, thus increasing generalization capacity of the model [10]. Researchers have used GA optimization [10–12] and other bio-inspired methods for configuring neural networks, such as Particle Swarm Optimization (PSO) in recent short-term traffic prediction literature [13,14], with significant results.

This research work will delve into the use of an specific meta-heuristic algorithms as a hyperparameter tuning wrapper for complex neural networks. To be concise the so-called Harmony Search (HS) algorithm [15] will be used to calibrate a multi-layer perceptron (MLP), which to the knowledge of the authors has not been so far used to this end. The proposed scheme will yield a set of neural network configurations for which the "best" one will balance the tradeoff between accuracy and MLP training time. The meta-optimized MLP model will be used to predict traffic flow in an urban center location in Madrid, where recent reports have revealed a high seasonal dependence and a stable behavior that complicates outperforming naïve approaches [16]. 15-min resolution traffic flow data of one entire year will be explored with different time windows and prediction horizons to show the promising performance of the models optimized by means of our proposed wrapper.

2 Materials and Methods

Flow is one of the most predicted traffic features [2] and embodies the data substrate on which most traffic models are built. Road traffic can be quantified by the amount of vehicles per hour crossing a certain link of the road network and it is measured, among others, with Automatic Traffic Recorders (ATR) sensors, magnetic loops embedded in roads that are able to count how many vehicles pass over them. Around 3700 of these ATRs are at the disposal of the Madrid

City Council, with readings taken every 5 min. Traffic flow and other metric measurements obtained from traffic counts are published in a live feed in the Madrid Open Data portal [17]. The portal also provides historically aggregated traffic flow data with 15 min granularity. From the latter data collection, a year worth of traffic flow data has been extracted and processed so as to compose the target dataset tackled in this work. These data correspond to the year 2015 for model building (training and validation), whereas the first three months of 2016 are left out to test the generalization performance of the optimized predictive model.

Table 1. Description of the loops under consideration

Loop	Location	Details
A	C. Alcalá and C. O'Donnell	City centre, intersection in a 4-lane road
B	Av. Monforte de Lemos and Av. Betanzos	Urban residential area

This research work focuses on two loops placed in a residential and a center area of Madrid (Table 1). Thus prediction models are built for locations with very dissimilar traffic and optimization results can be compared and assessed disregarding any particular set of traffic conditions. As noted in Fig. 1, there are substantial differences between both locations, introducing diversity to the study. Besides, a completeness criterion has been used to select the sensor locations: for all the loops deployed in this city, the available data are incomplete, and in some cases invalid, but some loops are considerable more complete than others. Both loops provide ca. 30000 valid flow readings corresponding to the whole year which, along with timestamps, are used to build the datasets based on three specifications: (1) step, i.e. the time between two readings (in this case 15 min); (2) depth or window size, namely, the time span of past observations that are used in each instance; and (3) prediction horizon, i.e. the number of steps into the future for which the prediction is made.

For a certain timestamp these readings configure a dataset with n features, which are the n observations prior to the timestamp, being n the window size. The target variable is defined by the observation y, corresponding to y slots after the timestamp, as explained in Fig. 3. Initially, a dataset for each loop has been created with default parameters. To begin with, the forecast horizon is set to 4 and 8 time steps (i.e. 1 and 2 h into the future), which is in accordance with the literature on short-term traffic prediction with horizons normally shorter than one hour [1,2]. The window size is fixed to 8 steps, so the predictions are based in the previous 2 h of observations. Indeed tailoring the window size is crucial for the predictive model, but unfortunately its value is strongly determined by the scenario at hand as the information provided by features in the window can be very divergent in different traffic areas [19]. Figure 2 shows the evolution of the average traffic over the slots of a day; in consonance with the stability of the observations, it is expected that for a residential loop this parameter becomes less relevant. Adjusting this window is a vibrant research topic [20] for which optimization strategies as the one presented in this paper take a key role in finding the best window sizes. In this research work window sizes of each location will be jointly optimized (along with regression model parameters) by the HS heuristic.

Once both datasets are created with above defined terms, they are splitted in 10 random parts, leaving 9 for training and 1 for test. This process is repeated 10 times, feeding each of the training subsets to the regression algorithm, and testing it against the test subset. A R^2 performance metric is obtained in each execution, and they are averaged to obtain the model performance, as depicted in Fig. 3.

2.1 Regression and Optimization

Our regression algorithm is a multilayer perceptron. Artificial Neural Networks (ANNs) have been extensively used for traffic forecasting applications, being the most extended non-parametric method used [1]. Multilayer perceptron falls under the feed-forward networks category [21], and is able to model highly non-linear patterns. An input vector is mapped to output through layers of weighted neurons with activation functions. In this research, logistic activation is used, whereas the rest of the MLP architecture is left undefined for its optimization through the proposed scheme. In the previous literature it is usual that the MLP configuration hinges on a trial-and-error process that does not always yield a configured MLP that outperforms other approaches such as ARIMA models.

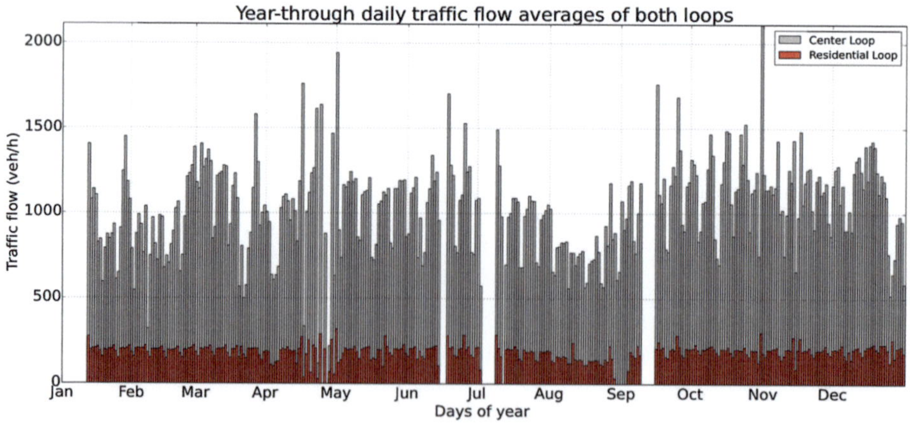

Fig. 1. Comparison of the day-average traffic flow registered by the two considered loops along the year. The x axis represents each one of the days in the sample. The loop in the city center has well-defined periods, while the loop in a residential area is more stable through the year. Empty days represent no available data.

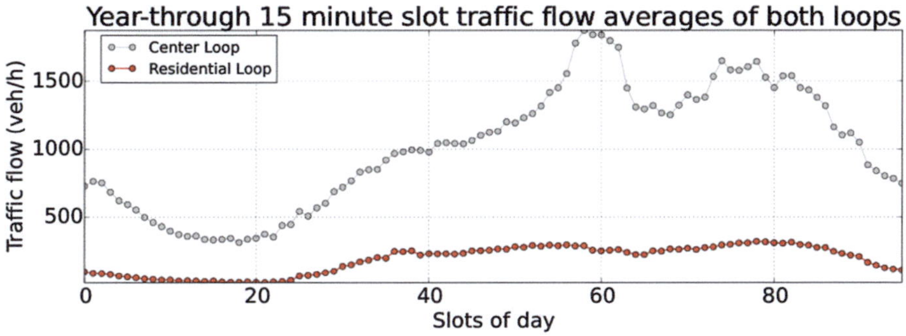

Fig. 2. Year-averaged traffic flow on each slot of the day for both loops.

For this reason a default parameter setting often used in other related contributions has been used as an initial baseline for comparison. Then, the regressor model is wrapped by means of an optimization algorithm that iteratively refines such a baseline configuration. In the last decade, optimization techniques have allowed a considerable improvement in ANNs performance [20]. However, to the best of the authors' knowledge there is no previous evidence of Harmony Search applied to the optimization of neural networks, nor has this heuristic been used to select features jointly with the model configuration.

Evolutionary algorithms are heuristic search techniques, and their operation grounds on the evolution of a group of candidate solutions towards progressively better individuals, with their *quality* defined by a fitness function. This evolutionary process finds its inspiration in the concept of natural evolution, encompassing selection, crossover and mutation of individuals [18]. Harmony Search

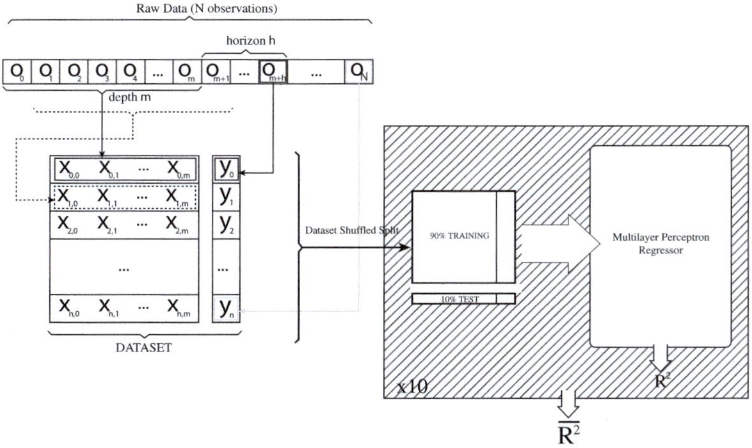

Fig. 3. Definition of the core predictive model.

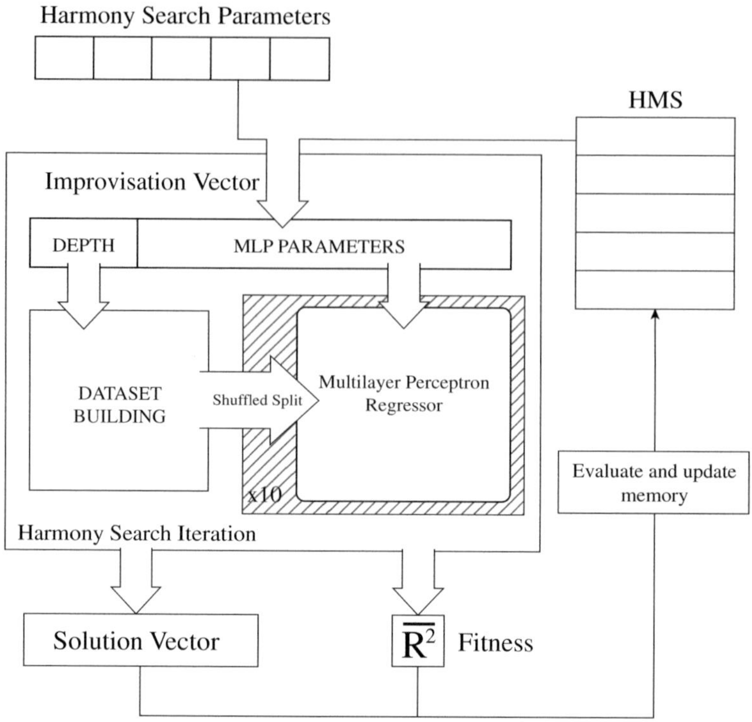

Fig. 4. Definition and operation of the HS wrapper.

(HS) can be thought of as being a specific yet differently motivated evolutionary solver, which imitates the seek of harmonies in a musical improvisation process to optimize a set of variables under a measure of quality. To do so, a vector of parameters and their boundaries are defined. Besides, HS wrapper is arranged to seek the optimal depth of each dataset. Figure 4 shows a harmony search implementation that jointly optimizes the feature selection and neural network parameter settings.

The HS algorithm requires input parameters that define its memory (hms), number of improvisations (I), and fitness function. At each iteration all raw data are used to build a new different dataset depending on the window size selected by the solution vector provided by HS, and evaluates the performance of the regression model by averaging over the scores produced by 10-fold cross-validation of the built dataset. The average R^2 metric obtained from the MLP over the 10 folds is the fitness function to maximize. In terms of computation a complete iteration takes around 2 min in an Intel i7 processor, which amounts up to an average of 720 iterations per day on a dedicated machine. We have set $hms = 25$ harmonies in the pool of candidate solutions, and a total of $I = 2000$ iterations. Besides these parameters we have set $hmcr = 0.75$, $par = 0.2$, $mpai = 3$ (maximum pitch adjustment for discrete variables) and $mpap = 0.25$ (maximum pitch adjustment for continuous variables).

Table 2. Optimized set of variables default values and boundaries.

Parameter	Type	Default	Min.	Max.
Depth	Discrete	8	4	48
Hidden layer sizes	Discrete	100	1	300
Alpha	Continuous	0.001	0.01	0.0001
Max. iterations	Discrete	200	10	1000
Tolerance	Continuous	0.0001	0.00001	0.0001
Learning Rate Init	Continuous	0.001	0.0001	0.01
Epsilon	Continuous	1e−08	1e−9	1e−07

Table 2 shows the MLP parameters chosen to optimize with their default values and maximum and minimum boundaries set for the experiment. Depth value has been set to take values between 4 and 48 steps, or 1 and 12 h of previous observations. For the neural network, default values are taken as a reference and for continuous variables, with maxima and minima one order of magnitude higher and lower, respectively. For discrete variables, boundaries are defined considering their purpose and operation.

After the whole process depicted in Fig. 4 is executed for both loops and horizons (4 and 8 steps), the obtained models are trained with the selected parameters and a dataset formed by the entire year 2015, and tested against the dataset corresponding to the first trimester of 2016.

3 Experimental Results

One of the main advantages of using HS to tune a predictive model is the possibility of adjusting its continuous parameters without discretizing them. In our model an automated exploratory search would require a discretization of four of the parameters that are to be estimated by the HS algorithm. With our set of parameters an exhaustive search would need a number of iterations given by

$$I = \left(\prod_{m=1}^{M_D} DPR_m\right) \cdot \left(\prod_{m'=1}^{M_C} CPR_{m'}\right), \tag{1}$$

where DPR_m is the number of possible values for the discrete parameter $m \in \{1, \ldots, M_D\}$, and $CPR_{m'}$ is the number of discretization steps for the continuous parameter $m' \in \{1, \ldots, M_C\}$, with M_D and M_C denoting the number of discrete and continuous parameters, respectively. In this case, only evaluating the combinations of discrete variables would require more than 26 million of iterations, each requiring an average of 2 min to be executed. HS, instead, has produced a result in less than three days.

Beyond the above gains in terms of computational effort, the tuning achieved by HS has provided improved performance results over the baseline configuration. Figure 5 shows scatter plots of both loops with both temporal horizons,

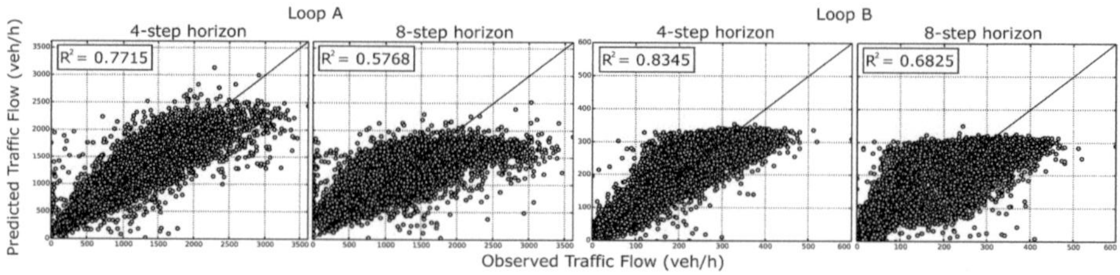

Fig. 5. Performance of the default MLP for both loops and temporal horizons.

Table 3. Optimized solution vectors for each loop and horizon

Loop	Horizon	Depth	H. L. Size	Alpha	Max. Iter.	Tolerance	L.R. Init.	Epsilon
A	4	35	285	2.96e−4	647	5.46e−5	4.94e−4	5.01e−9
A	8	42	160	9.25e−4	719	4.59e−5	5.26e−4	8.4e−8
B	4	40	262	3.67e−4	828	3.24e−5	8.54e−4	2.07e−8
B	8	45	81	6.84e−5	641	3.4e−5	3.4e−3	4.84e−8

and R^2 metric obtained for each dataset by using default parameters for the MLP and window size. In both loops, two hour horizon produces less accurate predictions (less fit to the line), and also, loop B, in a residential area, produces better overall performance, as a result of traffic stability there (Figs. 1 and 2).

After evolutionary search for optimal parameters, the algorithm has produced solution vectors presented in Table 3, which give rise to the predictive performance plots depicted in Fig. 6. All cases achieve an increment in R^2 metric, more noticeable in the 8-step prediction with almost 27% and 25% of R^2 performance relative gains. Due to their better performance with the default configuration, the improvement experienced by the MLP models is less significant – yet still notorious – in the case with 4-step horizon prediction. Window size is considerably larger than initially estimated, with up to 42 steps – 10.5 h – for the 8-step prediction in location A. This means that a forecast for a certain timestamp is based in values up to 12.5 h before, which according to Fig. 1 will always include values from both top and bottom ends of the curve. This result unveils that data are stable along days and that, taken the same range, the 42 steps that predict a value on e.g. a Saturday are similarly good to forecast the same timestamp value on a Monday.

Scatter plots are also helpful to visualize how predictions approach their exact values, specially for location B. Loop A, being placed in a very active zone of Madrid, produces more outliers which are difficult to predict, especially in the right part of the graphs where high traffic flow observed values are predicted lower. Indeed the regressor is unable to forecast a traffic peak flow of 3500 vehicles per hour, probably on account of an incident in the road at hand. All in all, these results buttress the need for an optimization wrapper when building a traffic forecasting model so as to configure it optimally.

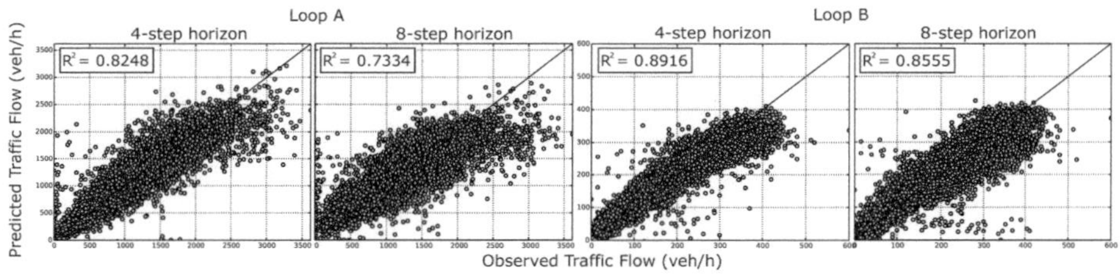

Fig. 6. Performance of optimized multilayer perceptron on both loops with both temporal horizons.

4 Conclusions and Future Work

Prediction horizons greater than 1 h are rarely explored in literature, being usually inefficient in urban areas as there are multiple immediate factors that influence traffic, ultimately making predictions useless. Nonetheless, this experiment has shown that in the absence of extraordinary circumstances, good prediction scores are attainable for longer-term predictions when the window sizes and algorithms are optimized. In this manuscript real traffic data in two different locations of Madrid (Spain) have been used to prove that the HS solver can efficiently optimize the parameters of a neural network and simultaneously select the depth of the input data to the model. This method allows for the automation of the tuning process of the algorithm, improving its overall performance in terms of computational complexity and predictive accuracy.

This work paves the way towards multi-location forecast. Network-wide traffic forecasting methods are gaining momentum, specially in urban environments. Part of the vehicles passing over one loop are likely to have passed through surrounding loops, thus the measurements taken in one loop might influence their neighboring loops. Without detailed origin-destination matrices, estimating this influence is an arduous task. The experiment presented in this paper can be extended to a multi-loop problem, where instances of the dataset are formed by observations of different loops, and the windows sizes of each loop are optimized. This extended implementation is expected to bring better performance levels, but also to unveil influence areas of each magnetic loop over the whole city.

Acknowledgments. This work has been supported by the Basque Government through the ELKARTEK program (ref. KK-2015/0000080 and the BID3ABI project).

References

1. Van Hinsbergen, C.P., Van Lint, J.W., Sanders, F.M.: Short term traffic prediction models. In: Proceedings of the 14th World Congress on Intelligent Transport Systems (ITS), Beijing, pp. 22–41 (2007)
2. Vlahogianni, E.I., Karlaftis, M.G., Golias, J.C.: Short-term traffic forecasting: where we are and where we're going. Transp. Res. Part C: Emerg. Technol. **43**, 3–19 (2014)

3. Dunne, S., Ghosh, B.: Regime-based short-term multivariate traffic condition forecasting algorithm. J. Transp. Eng. **138**(4), 455–466 (2011)
4. Yang, H., Dillon, T.S., Chen, Y.-P.P.: Evaluation of recent computational approaches in short-term traffic forecasting. In: Dillon, T. (ed.) IFIP AI 2015. IAICT, vol. 465, pp. 108–116. Springer, Heidelberg (2015). doi:10.1007/978-3-319-25261-2_10
5. Zhang, W., Xiao, R., Deng, J.: Research of traffic flow forecasting based on the information fusion of BP network sequence. In: He, X., Gao, X., Zhang, Y., Zhou, Z.-H., Liu, Z.-Y., Fu, B., Hu, F., Zhang, Z. (eds.) IScIDE 2015. LNCS, vol. 9243, pp. 548–558. Springer, Heidelberg (2015). doi:10.1007/978-3-319-23862-3_54
6. Meng, M., Shao, C.F., Wong, Y.D., Wang, B.B., Li, H.X.: A two-stage short-term traffic flow prediction method based on AVL and AKNN techniques. J. Cent. South Univ. **22**, 779–786 (2015)
7. Tang, J., Zou, Y., Ash, J., Zhang, S., Liu, F., Wang, Y.: Travel time estimation using freeway point detector data based on evolving fuzzy neural inference system. PloS one **11**(2), e0147263 (2016)
8. Smith, B.L., Demetsky, M.J.: Short-term traffic flow prediction: neural network approach. Transp. Res. Record **1453**, 98–101 (1994)
9. Abdulhai, B., Porwal, H., Recker, W.: Short term freeway traffic flow prediction using genetically-optimized time-delay-based neural networks. California Partners for Advanced Transit and Highways (PATH) (1999)
10. Vlahogianni, E.I., Karlaftis, M.G., Golias, J.C.: Optimized and meta-optimized neural networks for short-term traffic flow prediction: a genetic approach. Transp. Res. Part C: Emerg. Technol. **13**(3), 211–234 (2005)
11. Ishak, S., Kotha, P., Alecsandru, C.: Optimization of dynamic neural network performance for short-term traffic prediction. Transp. Res. Record J. Transp. Res. Board **1836**, 45–56 (2003)
12. Zhong, M., Sharma, S., Lingras, P.: Refining genetically designed models for improved traffic prediction on rural roads. Transp. Plan. Technol. **28**(3), 213–236 (2005)
13. Nagare, A., Bhatia, S.: Traffic flow control using neural network. Traffic **1**(2), 50–52 (2012)
14. Liu, S.Y., Li, D.W., Xi, Y.G., Tang, Q.F.: A short-term traffic flow forecasting method and its applications. J. Shanghai Jiaotong Univ. **20**, 156–163 (2015)
15. Geem, Z.W., Kim, J.H., Loganathan, G.V.: A new heuristic optimization algorithm: harmony search. Simulation **76**(2), 60–68 (2001)
16. Laña, I., Del Ser, J., Olabarrieta, I.: Understanding daily mobility patterns in urban road networks using traffic flow analytics. Network Operations and Management (NOMS), pp. 1157–1162. IEEE, Istanbul (2016)
17. Madrid Open Data portal. http://datos.madrid.es. Accessed 18 Nov 2016
18. Bäck, T., Schwefel, H.: An overview of evolutionary algorithms for parameter optimization. Evol. Comput. **1**(1), 1–23 (1993)
19. Hu, W., Liu, Y., Li, L., Xin, S.: The short-term traffic flow prediction based on neural network. International Conference on Future Computer and Communication, pp. 293–296. IEEE, Wuhan (2010)
20. Vlahogianni, E.I., Karlaftis, M.G., Golias, J.C.: Spatio-temporal short-term urban traffic volume forecasting using genetically optimized modular networks. Comput. Aided Civil Infrastruct. Eng. **22**, 317–325 (2007)
21. Jain, A.K., Mao, J., Mohiuddin, K.M.: Artificial neural networks - a tutorial. Computer **29**, 31–44 (1996)

A Heuristically Optimized Complex Event Processing Engine for Big Data Stream Analytics

Ignacio (Iñaki) Olabarrieta[1(✉)], Ana I. Torre-Bastida[1], Ibai Laña[1], Sergio Campos-Cordobes[1], and Javier Del Ser[1,2,3]

[1] TECNALIA, 48160 Derio, Spain
{ignacio.olabarrieta,isabel.torre,ibai.lana,sergio.campos, javier.delser}@tecnalia.com
[2] University of the Basque Country UPV/EHU, 48013 Bilbao, Spain
[3] Basque Center for Applied Mathematics (BCAM), 48009 Bilbao, Spain

Abstract. This paper describes a Big Data stream analytics platform developed within the DEWI project for processing upcoming events from wireless sensors installed in a truck. The platform consists of a Complex Event Processing (CEP) engine capable of triggering alarms from a predefined set of rules. In general these rules are characterized by multiple parameters, for which finding their optimal value usually yields a challenging task. In this paper we explain a methodology based on a meta-heuristic solver that is used as a wrapper to obtain optimal parametric rules for the CEP engine. In particular this approach optimizes CEP rules through the refinement of the parameters controlling their behavior based on an alarm detection improvement criterion. As a result the proposed scheme retrieves the rules parameterized in a detection-optimal fashion. Results for a certain use case – i.e. fuel level of the vehicle – are discussed towards assessing the performance gains provided by our method.

Keywords: Complex event processing · Big data · Optimization

1 Introduction

Wireless Sensor Networks (WSN) have become an ubiquitous way of getting all sorts of information in many application scenarios. Among the myriad of research initiatives around technological advances on WSN, the DEWI (Dependable Embedded Wireless Infrastructure) project aims to provide key solutions for wireless seamless connectivity and interoperability in smart cities and infrastructures by considering everyday physical environments of citizens in buildings, cars, trains and airplanes [1].

One of the specific industrial domains of the DEWI project is the automotive realm, with an emphasis on trucks. There are at least two main reasons for incorporating wireless sensors within this specific class of vehicles: the fact that

J. Del Ser (ed.), *Harmony Search Algorithm*, Advances in Intelligent Systems and Computing 514, DOI 10.1007/978-981-10-3728-3_11

there are no wires to connect imply savings in terms of installation time and cost of cabling. Moreover, the replacement of wireless devices are performed in a much simpler, less invasive manner that their wired counterparts. Indeed many studies highlight the key impact that the extensive use of WSN in the automotive realm can imprint on safety, maintenance, and energy efficiency of the vehicle [11].

This paper focuses on a platform developed within the above project for analyzing upcoming data streams from these sensors. Such a platform acquires and analyzes data from fuel, water level, electric suspension and brake lining wear captured by means of wireless sensors deployed at the corresponding parts of the vehicle. The information from the GPS device installed in the vehicle is also sent to the platform. From the perspective of computation and processing the treatment of large volumes of data streams (analyzing Big Data in motion) is nowadays an utmost necessity, and a large number of technologies have emerged to carry out this task (e.g. [9]). The analysis performed at the platform side consists of triggering different alarms when the level exceeds specific thresholds, when indicating a possible fuel theft from the truck or when the truck rides over a bumpy road. In this context the technology selected for alarm detection in the DEWI platform is one of the dominant technologies for stream processing, Complex Event Processing. Unfortunately, the challenge underlying this specific rule-based technology for stream processing is the appropriate configuration of its compounding rules, which should ensure the efficient and accurate detection of alarms in the vehicle.

There are multiple papers in the literature which aim to optimize CEP technologies in different aspects such as efficiency [14], performance [12], features [5,7,13] or special application fields [10]. In order to settle an argument for our study, we build on two of the above related works: [12] (complex event processing queries over real-time RFID streams events) and [14] (improvement of pattern queries performance). Both relate to the improvement of CEP engine efficiency and performance, but in our case the optimization is based on an improvement of the CEP rules and not on enhancing the detection or event pattern matching optimization techniques themselves. There are also scarce references dealing with the application of CEP techniques to vehicular scenarios [6]. However, interestingly within the scope of this manuscript to the knowledge of the authors there is no baseline literature where meta-heuristic algorithms are applied to improve the event detection rules of CEP technologies or to optimize the CEP engine itself. This paper covers this research gap by explaining a methodology for finding optimal rules based on already acquired data that have been properly labeled as alarm or not alarm by a technician using the DEWI platform. The method is based on the Harmony Search (HS) algorithm [8], a meta-heuristic optimization procedure that permits to efficiently seek optimal parameters for the rules. We will show how this nature-inspired solver can efficiently tackle the optimization of CEP rules by exploring its performance when utilized for detecting sudden drops in the fuel level of the monitored vehicle.

The rest of the paper is organized as follows: Sect. 2 provides an overview of the platform by explaining its compounding modules: the wireless network within the truck, the data acquiring platform and its Complex Event Processing (CEP) engine. Section 3 reviews the toy use cases and explain their implementation over CEP engines. The rule optimization methodology is explained in detail in Sect. 4. Finally, we present and discuss the results in Sect. 5, followed by Sect. 6 which concludes the work.

2 Platform Overview

The platform consists of three different elements: (1) the WSN, (2) the semantic middleware that receives the data from the sensors, and (3) the *Complex Event Processing* (CEP) functionality, which analyzes their captured information, and that is integrated with a sub module dedicated to the optimization of the definition of the rules. For testing purposes a simulator has been also developed to emulate output data from WSN. The output from the Complex Event Processing, namely, the computed alarms, is visualized in a tool specially deployed for this purpose. The connection from the WSN to the semantic middleware is currently done via TCP/IP connections, but other protocols can be potentially supported.

We assume that the WSN is deployed on the tractor unit of a Volvo FH-16 truck, whose trailer is not sensorized anyhow. Each of the deployed sensors measures a particular parameter for which it has been specially designed, along with their battery level and temperature. The specific signals that are measured by this mesh of sensors are brake lining wear (4 sensors), electronically controlled suspension level (another set of 4 sensors), the fuel tank level and the washer fluid level. All sensors label their measured samples with a timestamp, the last two bytes of the MAC address of the sensor for identification purposes, other sensor specific measurements, the voltage level of its battery and the value of its thermistor.

Aiming to assess the processing capabilities of the platform, a simulator has been developed to replicate the behavior of these sensors and the GPS signal of the truck in different scenarios. One of the scenarios emulates the normal use of the truck in the Göteborg area in which the fuel level, the washer fluid level and the brake lining wear decrease gradually over time, while the suspension sensor have small random variations. Other scenario that can be simulated by means of the developed simulator is the theft of fuel from the tank when the truck is stopped. This effect is replicated by a sudden reduction of the tank level. The truck driving through a rough road with uneven pavement is emulated by producing a high frequency and high amplitude random noise added independently to each of the sensors attached to the suspension.

The CEP engine used in this project has been developed using the EsperTech libraries [2]. Some of the advantages of using a complex event processing rely on its stream event management capabilities, the fact that it can analyze a massive amount of information really fast (e.g. we have been able to analyze the information arriving from over 10^5 simulated sensors per second) and without requiring to store any information in a database. Other appreciated characteristics of CEP are its scalability and a rich declarative language for behavioral configuration.

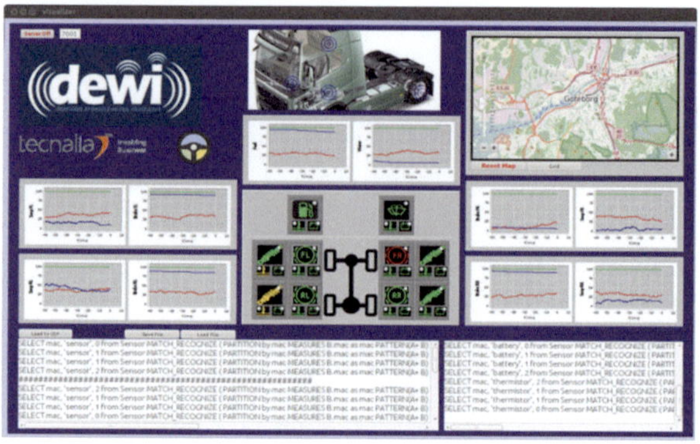

Fig. 1. Snapshot of the application tool to visualize the data from the sensors and the alarms. The visualization area has several areas: on the top right there is a map to visualize the current position of the truck, at the center there are icons denoting the level of alarm for each of the sensors, including the level of its battery and thermistor, around the center display the values acquired. Finally at the bottom of the visualization tool there is a text area where new EPL-coded rules can be added to the CEP.

The CEP is configured to detect different alarms and to forward them upon being triggered to a visualization tool for their output and inspection; a snapshot of the WSN monitoring system to visualize these alarms can be seen in Fig. 1. In order to interpret these alarms properly by the visualizer they are all defined to have a standardize data structure. The alarm is composed by two strings and an integer within the range $\mathbb{N}[0,6]$. The first string denotes the last two bytes of the MAC address associated with the sensor that triggers the alarm, the second string denotes the quantity associated to the alarm (three have been considered: "sensor", "thermistor" and "battery"), and finally an integer value denoting the level of alarm. The implementation of the rules which characterize the alarms defined in the previous paragraph is performed via the so-called Event Processing Language (EPL), which is a standard SQL language with extensions in order to provide aggregating functions, pattern matching, event windowing and joining.

Finally, the platform includes a labeling tool that allows tagging captured data. With the help of this tool a human technician can check, evaluate and properly review previously acquired data. These labels are crucial for the heuristic process described in Sect. 4.

3 Use Cases

We have considered two different use cases in order to test the ability of the platform to capture and analyze data and to assess the goodness of our rules optimization methodology. In the rest of this section we explain in detail the EPL rules for each of the scenarios that we want to detect. These involve detecting any fuel theft, when the truck drives along a rough road or when any of the sensor values changes above or below predefined thresholds. All the rules that we present here select the structure of the alarm that we mentioned above, two strings and an integer value, so they can be interpreted by the visualizer.

3.1 Fuel Theft Detection

The detection of fuel theft is performed by the following EPL rule:

```
SELECT mac, 'sensor', 4
FROM Sensor(mac="AAA8").win:length(10)
HAVING
  max(sensorValue)-min(sensorValue) > 10 & sensorValue < avg(sensorValue)
```

This rule filters out the information arriving to the stream of data Sensor by the MAC of the fuel sensor (i.e. MAC = "AAA8") and considers a data window of 10 measurements. The alarm is triggered in this case if the difference between the maximum and minimum values attained in the window is larger than ten units and the last measurement is less than the average within the window. Basically this rule tries to capture a sudden variation in the fuel level while the general trend is decreasing.

3.2 Bumpy Road Detection

In order to detect a road in bad shape the EPL rule reads:

```
SELECT mac, 'sensor', 5
FROM Sensor
(mac in("AAA4","AAA5","AAA6","AAA7")).
win:time(5 sec)
HAVING
(stddev(sensorValue,mac="AAA4") > 10 |
 stddev(sensorValue,mac="AAA5") > 10 |
 stddev(sensorValue,mac="AAA6") > 10 |
 stddev(sensorValue,mac="AAA7") > 10)
```

In this case the MAC addresses that are considered are the ones of the electronic controlled suspensions and what is measured is the standard deviation within a window of 5 s of duration. If the standard deviation for any of those sensors exceeds 10 units the alarm is deactivated.

4 Rule Optimization Procedure

The data flow entering the platform is defined by $F = \{f_j\}$, with $j \in (-\infty, ..., 0]$ and $f_j \in \mathbb{R}$ denoting the sample obtained at time index j. We consider j to be negative, the last obtained sample having index $j = 0$. In general the functions implemented in the CEP engine are applied not to all the history of data but to a limited time window. At a given time index j the time window w_j considered is defined by the measurements $\{f_i|_j \equiv f_{j-i}\}$ with $i \in [0, ..., W]$. We also assume that the human operator has labeled the inflow of data with its correspondent alarms $\{H_j\}$ such that:

$$H_j = \begin{cases} 1 & \text{if alarm should have been deactivated at time } j, \\ 0 & \text{else.} \end{cases} \quad (1)$$

Having a set of previously obtained data adequately labeled by $\{H_j\}$, and a parametric family of rules, the aim of our study is to design a heuristic procedure to obtain the parameters that fit best the alarms labeled by the operator. The method presented here relies on the Harmony Search (HS) algorithm. The underlying philosophy behind the general class of bio-inspired optimization algorithms (to which HS belongs) is to modify the group of candidates by operators that mimic processes observed in Nature. In these techniques candidate solutions compete, combine, mutate and evolve towards regions of the solution space of progressively increased optimality [4]. HS [8] is encompassed in this broad field of bio-inspired optimization, and is motivated by the idea of achieving harmony within the musical improvisation process undertaken by jazz bands. Harmonies represent sets of variables to optimize, whereas the aesthetic quality of the improvised harmony is given by the fitness function of the optimization problem at hand. In the rest of this section we provide two particular examples, two different families of rules, of the method applied to the fuel theft use case.

4.1 2-Parameter Rule

The first example considered in this work is an extension of the rule introduced in Sect. 3.1 with 2 parameters. Specifically we consider the rule $\mathcal{H}_j$ at time index j as:

$$\mathcal{H}_j(\{f_j\}) = \begin{cases} 1 & \text{if } (\max_{i \in W_j} \{f_i\} - \min_{i \in W_j} \{f_i\} > \mu) \text{ and } f_j < \beta \langle f_i \rangle_{i \in w_j}, \\ 0 & \text{otherwise,} \end{cases} \tag{2}$$

where $\mu \in \mathbb{R}[0, \max\{f_j\})$ and $\beta \in (0, ..., 1]$ are the parameters to be optimized. We assume the processing window length W_j to be fixed and equal to 10. Note that the rule introduced in Sect. 3.1 fits into this family with $\beta = 10$ and $\mu = 1$. The optimization process of this rule relies on the maximization of the F_1 score [3]:

$$F_1 \triangleq 2\,(\mathcal{P} \cdot \mathcal{R})/(\mathcal{P} + \mathcal{R}) \tag{3}$$

where $\mathcal{P}$ and $\mathcal{R}$ are the precision and the recall given by

$$\mathcal{P} = t_p/(t_p + f_p), \mathcal{R} = t_p/(t_p + f_n) \tag{4}$$

with t_p being the number of true positives (alarms that have been properly captured by the rule), f_p the false positives (number of times that the rule has predicted an alarm erroneously) and f_n is the number of false negatives (ratio of occurrences when the rule does not declares any alarm and there is one). These can be expressed as:

$$t_p = |\{\mathcal{H}_j = 1 \ \&\ H_j = 1\}|,\ f_p = |\{\mathcal{H}_j = 1 \ \&\ H_j = 0\}|,\ f_n = |\{\mathcal{H}_j = 0 \ \&\ H_j = 1\}|, \tag{5}$$

where $|\cdot|$ denotes set cardinality. The F_1 score can be thought of as a weighted average of $\mathcal{P}$ and $\mathcal{R}$. Alternatively, depending on the particular application of the rule, a different fitness function can be used such as the maximization of the precision $\mathcal{P}$, the minimization of the recall $\mathcal{R}$ or any other metric suited to the detection requirements of the specific scenario under scope.

4.2 Pattern Correlation Rule

We will consider a different family of rules in our second example. In this case the aim is to compare the incoming flux of data F with a pattern $\mathcal{A}(t)$ modeled as

$$\mathcal{A}(t) = \sum_{i=0}^{W} \alpha_i \cdot \delta(t - i \cdot \Delta t), \tag{6}$$

where $\delta(t)$ is the Dirac's delta function and we assume that the number W of parameters α_i agrees with the size of the CEP processing window. We will consider $\alpha_i \in [0, ..., 1]$, therefore $\mathcal{A}(t) \in [0, ..., 1]$. The rule family can be written in the following way:

$$\mathcal{H}_j = \begin{cases} 1, & \text{if } W^{-1} \sum_{i \in w_j} \left(\tilde{f}_i|_j - \alpha_i \right)^2 \leqslant \epsilon \\ 0, & \text{otherwise,} \end{cases} \tag{7}$$

where ϵ has an small value. In this case $\mathcal{H}_j$ fires and alarm whenever the pattern function $\mathcal{A}$ is close to the transformed incoming flux $\{\tilde{f}_i\}_{i \in w_j}$. We consider that $\{\tilde{f}_i\}_{i \in w_j}$ are related to $\{f_i\}_{i \in w_j}$ by the transformation:

$$\tilde{f}_i|_j = \begin{cases} a f_i|_j + b \text{ if } (1/a) \triangleq \max_{i \in w_j}(\{f_i\}) - \min_{i \in w_j}(\{f_i\}) > \mu, \\ 0 \qquad \text{otherwise,} \end{cases} \tag{8}$$

with $b = -a \cdot \min_{i \in w_j}(\{f_i\})$. This transformation is realized in order to normalize the incoming flux to the range $[0, ...1]$ and therefore to remove the dependence on the actual range of F. In order to accommodate small disturbances in the incoming flux we have included a parameter $\mu > 0$. The rule defined by Expression (7) compares the shape of the incoming flux with the shape of the pattern $\mathcal{A}$ as long as the variation of F is larger than μ in a particular processing window.

As it is explained in Sect. 4.1, we could fix ϵ to a particular value and use the F_1 score as the maximization criterion to find the pattern function parameters $\{\alpha_i\}$. Instead we choose to find $\{\alpha_i\}$ by minimizing the distance of $\mathcal{A}$ to the transformed incoming flux whenever the alarm should be deactivated. The resulting fitness function to be minimized can be written as

$$\mathcal{E}(\{\alpha_i\}) = \sum_j H_j \sum_{i \in w_j} \left(\tilde{f}_i|_j - \alpha_i \right)^2. \tag{9}$$

The benefit of using this particular equation for the fitness is that as a byproduct the befitted $\bar{\epsilon}$ to be used in Expression (7) can be estimated by computing

$$\bar{\epsilon} = \max_j \left\{ H_j \sum_{i \in w_j} \left(\tilde{f}_i|_j - \bar{\alpha}_i \right)^2 \right\}, \tag{10}$$

where $\bar{\alpha}_i$ are the values that minimize (9). Note that, in general, by minimizing (9) we minimize $\bar{\epsilon}$. On the other hand (9) has a remarkable caveat: it does not consider the behavior when $H_j = 0$, which implies that if the pattern $\mathcal{A}$ realized by $\{\alpha_i\}$ is found in the data when $H_j = 0$, $\mathcal{E}$ will not penalize those values. In other words f_p is not considered at all in finding $\{\bar{\alpha}_i\}$.

5 Experiments and Results

In order to test the methodology we have considered the fuel level data shown in Fig. 2. The data represents approximately 2600 captures of the level of the fuel tank. For $j \in [-2600, -1700]$ normal operation with a small decrement of the level is shown. At $j \approx -1700$ and ≈ -1650 there are two sudden reductions which are labeled by an alarm (in red), after the second reduction again normal operation occurs. Two more alarms are labeled happening at $j \approx -900$ and ≈ -400. Refueling can be noted at $j \approx -1300$ and ≈ -350.

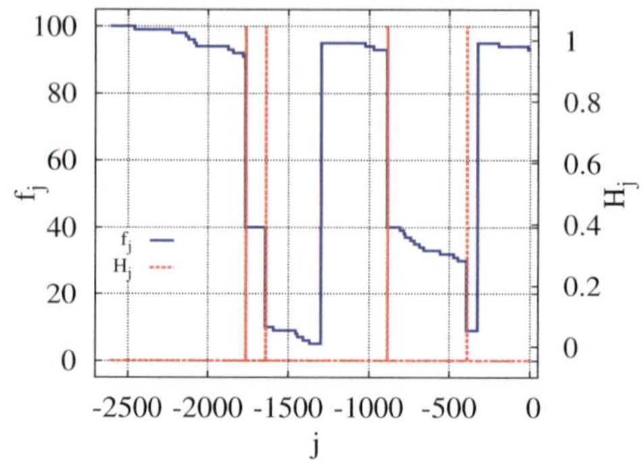

Fig. 2. Data obtained from the wireless fuel sensor f_j and the labels H_j denoting the fuel theft events. In these data 4 alarms can be distinguished. While a sudden decrease of the level of fuel occurs, these correspond to fuel theft events. In between the normal usage, small decrements of the level by the normal use of the truck can be noted plus two instances when the tank is refueled.

5.1 2-Parameter Rule

We have used the previous data in order to obtain the best parameters β and μ. An example of how the HS optimization process converges is shown in the left side of Fig. 3 where it can be observed how the fitness $\mathcal{E}$ goes to 1 as the iterative process evolve and the parameters reach some values in the solution space. One of things that is worth mentioning is the fact that there is no single solution in the solution space, actually we found that there exits a range of solution for this particular problem. The HS optimization process obtains a different solution depending on the initial random values choose for the parameters. In order to obtain a sense of the solution space we have performed thousands of HS optimization varying randomly the initial parameters. The region of the solution space can be hinted in the right subplot of Fig. 3 $\beta \times \mu \in \mathbb{R}\,(0.887, 0.99994) \times \mathbb{R}\,(1.003, 20.9993)$. The average number of iterations to attain a fitness of 1.0 is equal to 60 ± 1.

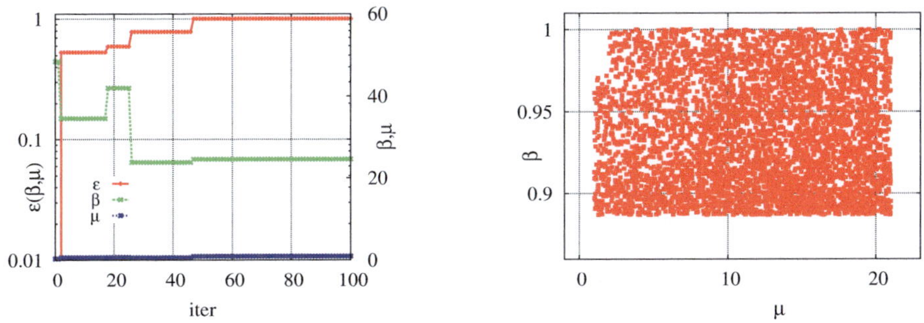

Fig. 3. On the left, an example of how the fitness $\mathcal{E}\{\beta, \mu\}$ converge to 1 (in red against the left axis) and the parameters β and μ (green and blue against the right axis) reach a point in the solution space. For a given set of training data the actual values obtained for the parameters depend on the initial seed used for the HS optimization process. On the right hand side there is a collection of solutions in the $\mu - \beta$ space giving hints of the range of the solution space.

5.2 Pattern Correlation Rule

In this case we use the same set of data to optimize the fitness derived for the pattern correlation rule. If we assume that in normal operation the level of the fuel does not decrease, the optimal solution corresponding to decreasing jumps in the fluid level corresponds to $\{\alpha_i\}_{i=0}^{3} = 1$ and $\{\alpha_i\}_{i=4}^{7} = 0$. Despite unrealistic it is logical to assume that the parameters for a real solution (meaning that in normal operation the fuel decreases smoothly) approach these values. The results for a single solution are presented in Fig. 4. Note that the HS optimization process takes a significant larger amount of iterations to converge, ∼8000

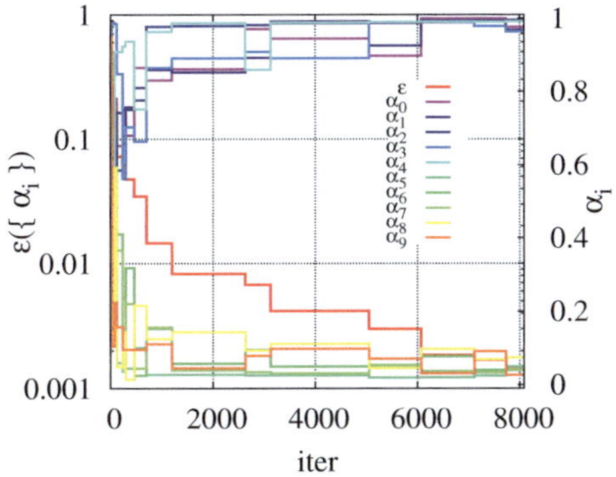

Fig. 4. Convergence of the fitness $\mathcal{E}(\{\alpha_i\})$ (in red and against the left axis) to 0 and the $\{\alpha_i\}$ (other colors and against the right axis) to their solution values.

iterations to reduce the fitness 3 orders of magnitude. The actual values that we found for $\{\alpha_i\}$ corresponds to:

$$\{\alpha_i\}_{i=0}^{7} = \{0.976, 0.995, 0.968, 0.963, 0.991, 0.050, 0.040, 0.041, 0.074, 0.026\}, \quad (11)$$

which qualitatively agree with the logic inferred from the ideal solution.

6 Conclusion

In this paper we have presented a platform that enables analyzing data originated in sensors within a truck tractor unit for the fuel level, the washer fluid level, the suspension and the brake lining wear. The platform is capable of detecting different complex alarms as proven using the data streamed by the simulator for several toy examples. A procedure to find optimal parametric rules based on HS optimization has been developed which has been tested for the fuel theft use case using two different families of parametric rules and different fitness functions, proving itself useful.

The next step within the project involves completing the connection of the actual sensors installed within the truck to the platform. Moreover a key ingredient in working with real signals will be to analyze the real signals and to develop machine learning algorithms capable of the finding the specific patterns which characterize the scenarios that want to be detected. Steps are already being taken to complete the overall system.

Acknowledgments. The research from DEWI project (www.dewi-project.eu) [1] leading to these results has received funding from ARTEMIS Joint Undertaking under grant agreement no. 621353. The authors want to specially thank Parthasarathy Dhasarathy, from Volvo Technology AB, who helped in the use cases and the platform infrastructure definition.

References

1. Dewi project. http://www.dewiproject.eu. Accessed 15 Oct 2016
2. Espertech. http://www.espertech.com/. Accessed 15 Oct 2016
3. F1 score. https://en.wikipedia.org/wiki/F1_score. Accessed 15 Nov 2016
4. Bäck, T., Schwefel, H.P.: An overview of evolutionary algorithms for parameter optimization. Evol. Comput. **1**(1), 1–23 (1993)
5. Björne, J., Heimonen, J., Ginter, F., Airola, A., Pahikkala, T., Salakoski, T.: Extracting complex biological events with rich graph-based feature sets. In: Proceedings of the Workshop on Current Trends in Biomedical Natural Language Processing: Shared Task, pp. 10–18. Association for Computational Linguistics (2009)
6. Bruns, R., Dunkel, J., Billhardt, H., Lujak, M., Ossowski, S.: Using complex event processing to support data fusion for ambulance coordination. In: 2014 17th International Conference on Information Fusion (FUSION), pp. 1–7. IEEE (2014)
7. Ding, L., Chen, S., Rundensteiner, E.A., Tatemura, J., Hsiung, W.P., Candan, K.S.: Runtime semantic query optimization for event stream processing. In: 2008 IEEE 24th International Conference on Data Engineering, pp. 676–685. IEEE (2008)

8. Geem, Z.W., Kim, J.H., Loganathan, G.: A new heuristic optimization algorithm: harmony search. Simulation **76**(2), 60–68 (2001)
9. Hirzel, M., Andrade, H., Gedik, B., Jacques-Silva, G., Khandekar, R., Kumar, V., Mendell, M., Nasgaard, H., Schneider, S., Soulé, R., et al.: IBM streams processing language: analyzing big data in motion. IBM J. Res. Dev. **57**(3/4), 7:1–7:11 (2013)
10. Liu, H.L., Chen, Q., Li, Z.H.: Optimization techniques for RFID complex event processing. J. Comput. Sci. Technol. **24**(4), 723–733 (2009)
11. Lu, N., Cheng, N., Zhang, N., Shen, X., Mark, J.W.: Connected vehicles: solutions and challenges. IEEE Internet Things J. **1**(4), 289–299 (2014)
12. Wu, E., Diao, Y., Rizvi, S.: High-performance complex event processing over streams. In: Proceedings of the 2006 ACM SIGMOD International Conference on Management of Data, pp. 407–418. ACM (2006)
13. Yan, Y., Yang, Y., Meng, D., Liu, G., Tong, W., Hauptmann, A.G., Sebe, N.: Event oriented dictionary learning for complex event detection. IEEE Trans. Image Process. **24**(6), 1867–1878 (2015)
14. Zhang, H., Diao, Y., Immerman, N.: On complexity and optimization of expensive queries in complex event processing. In: Proceedings of the 2014 ACM SIGMOD International Conference on Management of Data, pp. 217–228. ACM (2014)

Numerical Solution of Boundary Value Problems Using Artificial Neural Networks and Harmony Search

Neha Yadav[1], Thi Thuy Ngo[1], Anupam Yadav[2], and Joong Hoon Kim[1](✉)

[1] Department of Civil, Environmental and Architectural Engineering, Korea University, Seoul 02841, South Korea
nehayadav441@gmail.com, jaykim@korea.ac.kr

[2] National Institute of Technology Uttarakhand, Garhwal, Srinagar, India
anupam@nituk.ac.in

Abstract. In this paper, we present an algorithm based on artificial neural networks (ANNs) and harmony search (HS) for the numerical solution of boundary value problems (BVPs), which evolves in most of the science and engineering applications. An approximate trial solution of the BVPs is constructed in terms of ANN in a way that it satisfies the desired boundary conditions of the differential equation (DE) a utomatically. Approximate satisfaction of the trial solution results in an unsupervised error, which is minimized by training ANN using the harmony search algorithm (HSA). A BVP modeling the flow of a stretching surface is considered here as a test problem to validate the accuracy, convergence and effectiveness of the proposed algorithm. The obtained results are compared with the available exact solution also to test the correctness of the algorithm.

Keywords: Artificial neural network · Harmony search algorithm · Boundary value problems · Differential equations · Approximate solution

1 Introduction

Mathematical modeling of most of the science and engineering applications has led to the development of differential equations (DEs). A large variety of DEs exists in the literature for which analytical solutions are not available and to calculate their approximate solution is a complicated task. Thus, various numerical methods are developed in the literature, namely Finite difference method (FDM), Finite element method (FEM), Finite volume method (FVM) and Boundary element method for approximate solution of DEs. However, each of these numerical methods has its own advantages and limitations which strictly confine their functioning. Specifically, most of these methods require discretization of the domain and special treatment for nonlinear DEs.

Nowadays, artificial intelligence (AI) provides the powerful algorithms for solving the most complicated problem effectively. Thus, to avoid the difficulty

J. Del Ser (ed.), *Harmony Search Algorithm*, Advances in Intelligent Systems and Computing 514, DOI 10.1007/978-981-10-3728-3_12

arising in the approximate solution of DEs, artificial neural network (ANN) method is proposed in the literature. A sufficient amount of literature can be found which presents the numerical solution of DEs using ANN methods [1–6]. The Backpropagation (BP) learning has become the most popular method for training ANN in many problem domains. However, BP algorithm is a gradient descent technique, which is analogous to an error minimizing process. ANN generates complex error surface with multiple local minima and BP tend to become trapped in local solution instead of global.

Thus, many global search optimization algorithms have been adopted in the literature for training ANN for solving DEs [7,8] and references therein. Most of these methods draw their inspiration from biological processes like Genetic algorithms (GA) [9], Ant colony optimization algorithms (ACO) [10] and Particle swarm optimization [11]. These training methods overcome the aforementioned deficiencies of the BP algorithm by avoiding the local minima frequently by promoting the exploration of the search space.

A relatively young metaheuristic global search algorithm is a harmony search (HS), which draws its inspiration from the improvisation process of musicians. It can be seen in the literature that HSA performs better than the BP algorithm for training a feed forward ANN [12]. The HSA has also been shown to be comparatively simpler and more efficient than other global optimizers [12,13]. Thus, in this paper, we adopted HSA for updating ANN weight parameters while solving DEs.

The remainder of the paper is organized as follows. Section 2 describes the general introduction to the HSA. ANN method for solving DEs is presented in Sect. 3. Section 4 explains the chosen test problem along with results and discussion. Finally, Sect. 5 concludes the paper.

2 Harmony Search Algorithm

The HSA is a population based metaheuristic algorithm inspired by the improvisation process adopted by musicians and was first developed by Geem et al. [13]. In the HSA, an initial population of harmonies is randomly generated and stored in a harmony memory (HM). A new candidate harmony is then generated from all the solutions in the HM using harmony memory consideration rate (HMCR), pitch adjustment rate (PAR), and random re-initialization. The HM is then updated by replacing the worst harmony vector in HM with the new candidate harmony vector if it is better. The HSA consists of three basic phases, namely, initialization, improvisation of a harmony vector, and updating of the HM. The basic parameters required in the HSA are harmony memory size (HMS), HMCR, PAR, and the number of improvisations (NI). Both HMCR and PAR take a value between zero and one and control the balance between exploration and exploitation. The HMCR parameter presents the information sharing capacity of HM with harmony improvisation; PAR determines the step size of the improvisation process within the bandwidth (BW). The computational procedure for the HSA is given in Table 1:

Table 1. Pseudo code of the HSA

Step 1: Parameter settings: HMS, HMCR, PAR, BW, and NI.
Step 2: Initializing the HM and calculating the fitness function value $X_j^i = L_j + a \times (U_j - L_j)$ for $j = 1, 2, \ldots, N$; $i = 1, 2, \ldots, \mathrm{HMS}$, where $a \in (0, 1)$
Step 3: Improvisation of new harmony (X_{new}) from HM as:
For ($k = 1\ to\ n$) do
If ($a_1 < HMCR$)
$X_{new}(k) = X_o(k)$; $where o \in (1, 2, \ldots, HMS) and a_1 \in (0, 1)$
If ($a_2 < PAR$)
$X_{new}(k) = X_{new}(k) \pm a_3 \times BW$; $where\ a_2, a_3 \in (0, 1)$
else
$X_{new}(k) = L_j(k) + a \times (U_j(k) - L_j(k))$; $where\ a \in (0, 1)$
End for
Step 4: If $f(X_{new})$ is better than $f(X_{worst})$, update HM as $X_{worst} = X_{new}$
Step 5: Repeat steps 3 and 4 until NI is reached. Harmony vector X_{best} in the HM is the solution of the defined problem.

3 Artificial Neural Network Method

To describe the ANN method, let us examine a general DE of the form [5,6].

$$G(y(x)) = f(x) \tag{1}$$

where G is some differential operator over the domain Ω, $x \in R^n$ with y represents the exact solution subject to the following Dirichlet or Neumann BCs

$$y = g(x), \quad \forall x \in \partial\Omega \tag{2}$$

$$\hat{n}(x) . \nabla y(x) = g(x), \quad \forall x \in \partial\Omega \tag{3}$$

where $\hat{n}(x)$ represents a unit vector function that returns the boundary normal, while $g(x)$ returns the BC $\forall x$ on a Neumann boundary. The first step is to construct a trial solution that satisfies the desired BCs as

$$y_T(x, p) = A(x) + L(x)\, N(x, p) \tag{4}$$

where the first term$A(x)$represents a continuous function which is developed in a way that it satisfy the BCs and$N(x, p)$represents an ANN with input vector x and network parameters p. The length factor Lin the second term represents a measure of the distance from the boundary in a way that its value is zero $\forall x$ on the boundary and nonzero within the domain. The trial solution defined in Eq. (4) satisfies the desired BCs, regardless of the ANN output$N(x, p)$. In a similar way, the ANN output $N(x, p)$ optimizes the value of trial solution within the domain and is controlled by minimizing the DE error function

$$E(x, p) = f(x, G(y_T(x))) \tag{5}$$

by updating network weight parameter$p$ iteratively by using HSA. The most common ANN architecture is the multilayer perceptron network (MLP), which has the capability to approximate any Borel-measurable function defined on a hypercube. In this work, the training of ANN for optimizing network parameters is performed using the HSA.

4 Test Problem

A steady two-dimensional flow of an incompressible, electrically conducting viscoelastic fluid of the type, "Walter's liquid B past", an impermeable stretching sheet is considered as our test problem. The flow of a stretching surface is an important problem in many engineering applications such as extrusion, melt spinning, hot rolling, wire drawing, glass-fiber production, and manufacturing of plastics [14]. The governing equation for flow analysis can be written in terms of the following equation [15]:

$$y''' - y'^2 + y' . y'' = Q y' + k_1(2y' . y''' - y' . y^{iv} - y''^2) \tag{6}$$

together with the following BCs:

$$\begin{aligned} y = 0 \ , \ y' = 1 , \quad \text{at } \eta = 0, \\ y' \rightarrow 0 \ \ as \ \eta \rightarrow \infty \end{aligned} \tag{7}$$

where k_1 is the viscoelastic parameter and Q is the Chandrasekhar number. The exact solution of Eq. (6) together with BCs in Eq. (7) is given by Rajagopal [16] as follows:

$$y = \frac{1 - e^{-\alpha \eta}}{\alpha}, \tag{8}$$

where $\alpha = \sqrt{\frac{1+Q}{1-k_1}}$; $(1 \leq \alpha < \infty)$. Equation (6) is solved for the case $Q = k_1 = 0$. The trial solution is constructed for Eq. (6) in terms of neural network in the form mentioned in Eq. (4). ANNs are then trained by the HSA using the constructed trial solution with different values of hidden nodes $h = 5, 10, 15, 20, 25, 30$ and 100 random starting weights to minimize the approximation error. The statistical parameter mean absolute error (MAE) is chosen to calculate the best number of hidden nodes in ANN model for solving Eq. (6). The best performing hidden node in ANN model is selected on the basis of lowest STD in DE error. The MAE in the solution calculated while solving Eq. (6) is presented in Fig. 1 with different number of hidden nodes $h = 5, 10, 15, 20, 25, 30$. From the Fig. 1, $h = 15$is chosen as the best number of hidden nodes in ANN model for solving Eq. (6). The exact and ANN solutions are also presented to test the accuracy of the ANN solution inside the chosen domain [0,1], and are depicted in Fig. 2.

Statistical results obtained in optimizing ANN parameters using HSA are also provided to present the applicability of training ANN using HSA. The task of optimizing network parameters was carried out using 100 independent runs.

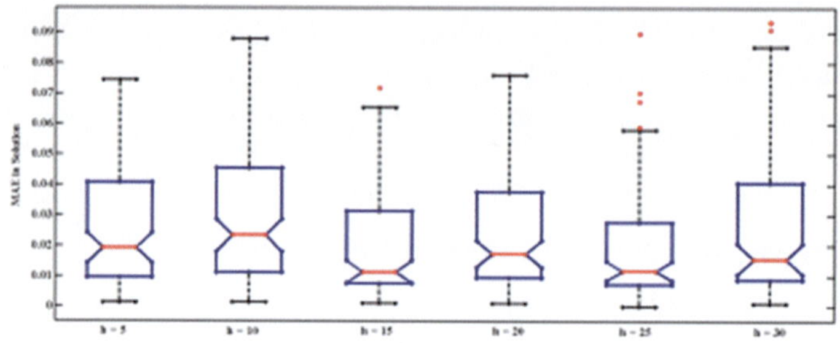

Fig. 1. MAE in the solution for 100 runs with $h = 5, 10, 15, 20, 25, 30$

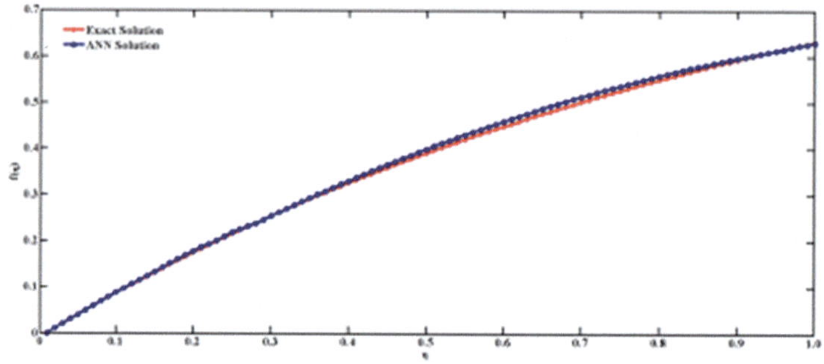

Fig. 2. Exact and ANN solutions for test problem

Table 2. Statistical optimization results for test problem using the HSA

Initial/BCs	Number of hidden nodes	Worst EF	Average EF	Best EF	SD EF
$y = 0$, $y' = 1$, at $\eta = 0$, $y' \to 0$ as $\eta \to \infty$	$h = 5$	9.13E−03	6.43E−03	4.28E−03	1.60E−03
	$h = 10$	9.93E−03	6.26E−03	4.18E−03	1.72E−03
	$h = 15$	9.80E−03	6.28E−03	4.21E−03	1.59E−03
	$h = 20$	9.84E−03	6.44E−03	4.26E−03	1.61E−03
	$h = 25$	9.93E−03	6.41E−03	4.26E−03	1.76E−03
	$h = 30$	9.89E−03	6.28E−03	4.18E−03	1.52E−03

The values for the user parameters maximum iteration, HMS, HMCR, and PAR of the HSA are chosen as 1000, 20, 0.9 and 0.05 respectively. Initial/boundary conditions of the DEs are chosen as the bounds of decision variable and bandwidth (BW) is represented by range of the problem domain. Obtained statistical results for the error functions (EFs) for the test problem using the HSA are presented in Table 2. A summary of minimization of EFs using different number of hidden nodes over number of iterations is shown in Fig. 3.

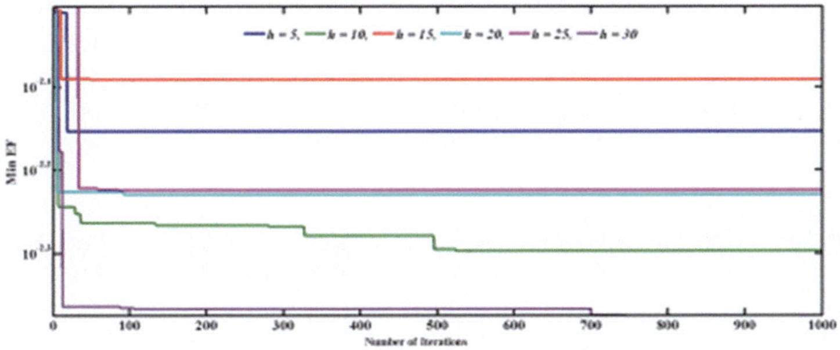

Fig. 3. Minimization of EFs over the number of iterations for $h = 5, 10, 15, 20, 25, 30$

5 Conclusion

An algorithm based on the ANN and HSA is presented in this paper for numerical solution of BVPs. The performance of the algorithm was checked by considering a test problem which models the flow of a stretching surface. On the basis of the simulation results, it can be concluded that the algorithm based on the ANN-HS algorithm is capable of solving BVPs with great effectiveness. The optimization results presented in this paper show that HSA can train the neural network parameters for solving DEs like any other optimizers and achieves faster convergence. The reliability and accuracy of the algorithm was validated by comparing the obtained ANN solution to the exact solution.

Acknowledgements. This work was supported by a grant from the National Research Foundation (NRF) of Korea, funded by the Korean government (MSIP) under grant number NRF-2016R1A2A1A05005306, and a Brain Korea 21 (BK-21) fellowship from the Ministry of Education of Korea.

References

1. Lagaris, I.E., Likas, A., Fotiadis, D.I.: Artificial neural networks for solving ordinary and partial differential equations. IEEE Trans. Neural Netw. **9**(5), 987–1000 (1998)
2. Malek, A., Beidokhti, R.S.: Numerical solution for high order differential equations using a hybrid neural network-optimization method. Appl. Math. Comput. **183**(1), 260–271 (2006)
3. McFall, K.S., Mahan, J.R.: Artificial neural network method for solution of boundary value problems with exact satisfaction of arbitrary boundary conditions. IEEE Trans. Neural Netw. **20**(8), 1221–1233 (2009)
4. Kumar, M., Yadav, N.: Multilayer perceptrons and radial basis function neural network methods for the solution of differential equations: a survey. Comput. Math. Appl. **62**(10), 3796–3811 (2011)
5. Yadav, N., Yadav, A., Kim, J.H.: Numerical solution of unsteady advection dispersion equation arising in contaminant transport through porous media using neural networks. Comput. Math. Appl. **72**(4), 1021–1030 (2016)

6. McFall, K.S.: Automated design parameter selection for neural networks solving coupled partial differential equations with discontinuities. J. Franklin Inst. **350**(2), 300–317 (2013)
7. Raja, M.A.Z., Khan, J.A., Qureshi, I.M.: A new stochastic approach for solution of Riccati differential equation of fractional order. Ann. Math. Artif. Intell. **60**(3–4), 229–250 (2010)
8. Raja, M.A.Z., Khan, J.A., Siddiqui, A.M., Behloul, D., Haroon, T., Samar, R.: Exactly satisfying initial conditions neural network models for numerical treatment of first Painlevé equation. Appl. Soft Comput. **26**, 244–256 (2015)
9. Kim, D., Kim, H., Chung, D.: A modified genetic algorithm for fast training neural networks. In: International Symposium on Neural Networks, pp. 660–665 (2005)
10. Wei, G.: Study on evolutionary neural network based on ant colony optimization. In: International Conference on Computational Intelligence and Security Workshops, pp. 3–6 (2007)
11. Yu, J., Wang, S., Xi, L.: Evolving artificial neural networks using an improved PSO and DPSO. Neurocomputing **71**(4), 1054–1060 (2008)
12. Kattan, A., Abdullah, R., Salam, R.A.: Harmony search based supervised training of artificial neural networks. In: International Conference on Intelligent Systems, Modelling and Simulation, pp. 105–110 (2010)
13. Geem, Z.W.: Music-Inspired Harmony Search Algorithm: Theory and Applications. Springer, Berlin (2009)
14. Khan, W.A., Pop, I.: Boundary-layer flow of a nanofluid past a stretching sheet. Int. J. Heat Mass Transf. **53**(11), 2477–2483 (2010)
15. Abel, M.S., Mahesha, N.: Heat transfer in MHD viscoelastic fluid flow over a stretching sheet with variable thermal conductivity, non-uniform heat source and radiation. Appl. Math. Model. **32**(10), 1965–1983 (2008)
16. Rajagopal, K.R.: On the boundary conditions for fluids of the differential type. In: Navier-Stokes Equation and Related Nonlinear Problems, pp. 273–278 (1995)

Harmony Search and Scheduling/Planning

Scheduling Optimisation of a Manufacturing Design Area: A Case Study

C.A. Garcia-Santiago(✉), Anna Rotondo, and Fergus Quilligan

Irish Manufacturing Research, Dublin, Ireland
carlos.garcia@imr.ie

Abstract. This paper presents the work implemented to improve the production scheduling of a real life manufacturing plant in Ireland. Since the scheduling algorithms are integrated in a wider cognitive system, where human and machine intelligence collaborate, an important constraint will be the maximum allowed computational time. This paper will also demonstrate the impact of using heuristic rules for the generation of initial solutions and provide a comparison between Hill Climbing, Harmony Search and Simulated Annealing.

Keywords: Optimisation · Scheduling · Metaheuristics · Harmony search

1 Introduction

This paper presents the work developed and implemented in a tooling design department of a high-tech manufacturing company in Ireland that designs and manufactures tooling solutions for the pharmaceutical sector. In this department, 28 engineers design the parts that will be later manufactured on site. Each tooling to manufacture is unique, and the design process is the first step of the production, followed by the manufacturing, assembly and testing of the final products. In this paper we will consider a job as the set of 3 operations with precedence constraints that will ultimately output the list of parts to manufacture. The first operation in the job checks that all necessary information for the correct completion of the design is available. If something is missing or badly understood the job will not continue until the necessary information is corrected, and in this case the job will start again. This pre-check operation is always assigned a fixed 15 min slot, and is always performed by one of the 4 most experienced designers in the department. There is an additional constraint in the number of pre-check operations that any given designer can perform daily. The second operation is the actual design process, which takes the greatest time to perform. Here one designer uses a 3D CAD software to produce the blueprints for the tooling. Finally, the third operation is the checking of the previous design step, trying to find possible defects or omissions. The main constraint for this check operation is that the designer performing it must be different from the one who made the design.

J. Del Ser (ed.), *Harmony Search Algorithm*, Advances in Intelligent Systems and Computing 514, DOI 10.1007/978-981-10-3728-3_13

The assignment of operations to resources is done manually by the department manager. Some designers are only permitted to perform a subset of the operations. This measure is usually taken to ensure that the more junior designers only perform the design operation, or for some very senior designers to perform only the checking operation. For the research in this paper we will also consider that each designer can only perform one operation at any given time, and that once started, the designer cannot perform any other operation until the current one is completed.

Usually the design department is overloaded and can cause a bottleneck for the rest of the production line. To ensure the correct balance of all the line, each of these design jobs have an internal due date that must be met to avoid later delays in the manufacturing line. The main constraint of the process is satisfying the customer demand on time, by (i) minimizing of the number of jobs missing the due date and (ii) for those jobs that are late minimise the number of days past the due date. A secondary goal is balancing the workload of the designers. Workload is quite unbalanced in the original schedule, primarily due to differences between designers regarding skill and experience. The skill of an engineer designing a particular piece of machinery is a very important factor in the quality of the final product. If the design is not correct, the manufacturing phase can be impossible to do, causing the job to go back to design phase, or the final customer will return the part due to imperfections or malfunctioning.

The research presented in this paper consists of finding new ways to help the planning and scheduling department of the company by rescheduling the production upon introduction of new constraints by the design department manager. The optimisation algorithms work as part of a cognitive system where human and machine intelligence cooperate to find optimal schedules in the event of unexpected disruption events like resource unavailability or new orders in the system. In this scenario, an important requirement is the speed of results, necessary to allow a "communication" between user and machine intelligence. In this research we follow the results of recent studies suggesting that the human attention timespan is decreasing in last years, from 12 to 8 s in just a 10 years time span, see [19].

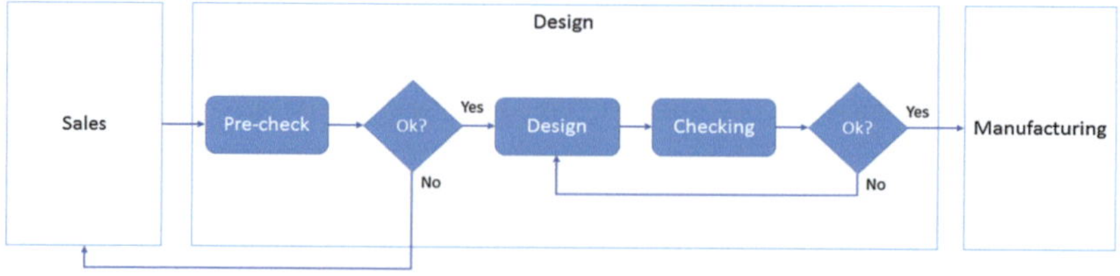

Fig. 1. Design process

1.1 Ensuring Quality

In many scheduling problems a very important factor that production mangers need to take into consideration is the quality of the performed work. The final product quality in the present case is heavily dependent on the skill of the designers. This issue has been studied in the past, for example in the framework of the Resource-Constrained Project Schedule Problem. In the literature we find two main methods to deal with quality issues; the first one modifies the activity duration depending on the skill of the assigned resource [1], while in other cases an additional re-work activity is scheduled if the skill of the resource assigned to the original activity is not high enough for the activity complexity [2]. In this use case quality was incorporated by the company by developing a job risk vs resource grade heuristic and by always having a check activity after every design activity, to ensure that the quality of the work performed by the first designer was good enough. On top of it, the manager sometimes introduces new constraints by selecting the designers that are available for the check operation among those with most experience.

The skill of a designer is an indicator of the experience that he has had in the past designing a particular tooling. This experience will not only have a positive impact on the time required to perform the operation, but will also decrease the number of non conformities or errors found by customers in the final product, as an unskilled designer is more prone to make errors in the design (faulty product), and as such is a very important key performance indicator (KPI) for the plant managers. Obviously it would be preferable if we could always assign each operation to the designer with highest skill in it, but the constraints on the availability of resources means this would often have negative consequences in the balance of workload.

The system implemented in the use case company to ensure the best possible quality in the jobs is based around two different values. The first one is a 'risk value' assigned to each job, depending on several factors including similarity with past jobs, type of tool, the machine where the tool will be installed, etc. Against this risk we map another value known as 'grade', which is manually assigned to each designer by the design department manager as an indication of his seniority and capacity to take on difficult jobs. There are currently four grades in use.

The system works as follows:

1. First, for each operation the available designers are selected based on the risk of the job and on their grade. In other words, we find the subset of designers with grade equal or greater than the minimum required for the job's risk value.
2. Second, from this first subset we filter out those designers without any skill in the operation.
3. In the case that there's no element in this final subset (i.e., no one has the necessary grade and experience for the operation), we will default to the subset of designers with the highest grade.

2 The Algorithms

In this section we will review the two metaheuristics used in this research. A metaheuristic is an algorithm that uses other algorithms iteratively, to search for the global optimum of a function across all the search space. Every metaheuristic starts exploring the search space in one particular point (in the case of single solution metaheuristics) or in several starting ones (population based). Though the standard approach is using a random initial solution(s), another approach is to use some heuristic to create the initial solution. Another important consideration in all metaheuristics is how to handle the constraints. In this section we will see how the research tackles these two issues, and also an introduction to the two metaheuristics used.

2.1 Representation

The encoding of the harmonies is always an important decision when designing the HS algorithm. In this paper we apply the Random Keys encoding, which was first described by James C. Bean in [9]. The main reason for using this representation is to use a representation that maintains the feasibility of any solution under any permutation, and also avoids the eventually increased computational load that specialized repair methods or solution discarding approaches would require. Under this representation the resource where an operation is to be processed and the sequence of the operation in that specific resource are jointly described by a single real number. The integer part is used to identify the resource, while the fractional part provides the sequencing of operations in every resource. Thus, taking a simple example with two resources and four operations, the following harmony

$$[1.05][2.86][1.87][2.19]$$

corresponds to the schedule

Resource 1: [Operation 1] - [Operation 3],
Resource 2: [Operation 4] - [Operation 2],

2.2 Initial Solution

Metaheuristic optimisation methods aim at effectively and efficiently exploring the solution domain of a combinatorial optimisation problem so as to refine the search directions and focus on domain areas characterised by better performing solutions, that is areas where the optimum solution is expected to lie. When dealing with the problem of identifying the initial search point across the entire solution space, two main approaches are generally considered. The ancestor, that is the initial solution, can either consist of a randomly generated solution or, alternatively, problem specific logics can be used to generate one. In the latter case, the ancestor is expected to perform better than the randomly generated ancestor as it corresponds to a heuristic solution to the problem. As a result, using the logical ancestor, the metaheuristic approach starts from a better solution, however, the likelihood of terminating the search at local optima increases.

The performance of the two approaches applied to the problem investigated here is compared in this section. The random initial solution is generated by assigning random feasible resources to each operation while taking all applicable constraints into account. On the other hand, the heuristic based solution is constructed using the Earliest Finish (EF) concept [20]. In the EF heuristics, jobs are first ordered by increasing due date. Then, the heuristic considers one job at a time and tentatively assigns the corresponding operations to eligible resources; the operations are considered one at a time following precedence constraints. The resource able to complete the operation at the earliest time is chosen and its availability is updated accordingly. In order to correctly calculate completion times of an operation, all events limiting the resource availability are considered; these events include previously assigned operations, time dedicated to other tasks, such as training or management meetings, and holidays. A non pre-emptive approach is used to calculate completion times in the sense that an operation cannot be interrupted by another operation; however, operations can be split across non consecutive available time slots to accommodate for non-production related tasks and holidays. This assignment logic is exemplified in Fig. 1 where the operation marked in dark blue (i.e. operation X) is tentatively assigned to resource A, B and C, respectively, whose schedule is partially built. Resource A can complete operation X at time 25; the slot between time 9 and 13 is too short to complete the operation and is bounded by another two operations; hence, operation X is moved to the next available time slot. Resource B can complete operation X at time 20; in this case, the operation is interrupted at time 13 to allow resource B to complete a non-productive task and is resumed soon after. As regards resource C, the operation cannot be interrupted at time 14 since the non-productive task starting at time 14 is followed by a previously assigned operation; hence, the pre-emption assumption would be violated. As a result, resource B is chosen for assignment (Fig. 2).

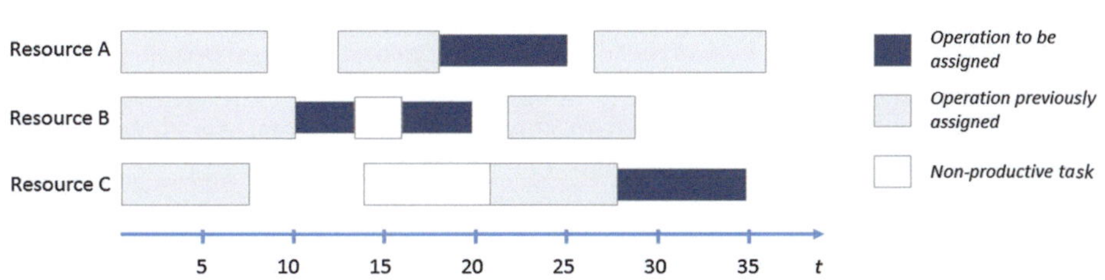

Fig. 2. Tentative assignment of an operation to eligible resources

The assignment algorithm described here can be implemented by using arrays of ordered available time slots for each resource and scanning through them until sufficient cumulative duration is found. When an operation is assigned to a resource, the time slot corresponding to the operation processing is deducted from the array [1].

2.3 Constraint Handling

In cognitive systems, where humans and automated agents share the responsibility of control, system observability and usability are two major concerns. One of the most important characteristics that a system like this must have to overcome those issues is responsiveness. To be able to cooperate, the automated system, must present a solution to a new input from the user in a short time. From an algorithmic point of view, this responsiveness represents a main constraint: the maximum allowed running time before a solution is presented to the user.

As mentioned in the introduction, the time to perform computations is limited by the user attention timespan. In this study we refrain from using constraint handling alternatives that might generate a great number of infeasible solutions, since this would require the evaluation of many invalid solutions. As a result, we won't use penalty functions or repair algorithms, but will rely on in specific operators that will generate only feasible solutions.

Using these operators ensures that every solution evaluated in the simulation represents a feasible schedule, see [3].

In the case of Simulated Annealing the only operator we need is the mutation. The standard mutation operator of genetic algorithms and other similar metaheuristics act by modifying a single value of the solution. In Boolean representations it changes one boolean value, while in real-valued encodings it modifies the value by a small random value.

1. For each chromosome, randomly select one gene to change.
2. For the selected gene, find the subset of designers that are valid.
3. Select a random designer from the subset.

In Harmony Search, to provide a more meaningful comparison, we will use the same mutation operator as Random operator.

2.4 Simulated Annealing

Simulated Annealing was developed by various researchers, in particular Kirkpatrick et al. [4] and V. Černý [6], for combinatorial optimization problems, though it has been later extended for its use in continuous optimization.

Mathematically SA is derived from the Metropolis algorithm, developed by Nicholas Metropolis et al. [7]. The main characteristic of Simulated Annealing is that sometimes, in the process of improving the candidate solution, it will accept a successor S with a worse fitness than its parent R, with probability

$$P\left(t,R,S\right)=e^{\frac{fitness(R)-fitness(s)}{t}}, \tag{1}$$

where t represents the temperature in the physical process, and acts as a tunable parameter. When t is high, the probability of accepting a worse solution than its parent is close to 1, while the probability is almost 0 when t approaches 0. When t is set at a high value at the start of the algorithm, and then allowed to decrease with every iteration, the algorithm will exhibit a explorative behaviour at the start and an exploitative one at the end. An important caveat with SA is that, as we decrease the value of t, the algorithm is more exploitative and more similar to a hill-climbing procedure. This is important in our case, as the limited amount of time to perform computations does not allow for a highly explorative search.

2.5 Harmony Search

Where SA only uses a single solution that evolves over time, Harmony Search [8] uses a set (called harmony memory) of solutions.

Harmony Search has been shown to outperform other meta-heuristic solvers in a wide range of application scenarios such as ground water engineering [10], localization [11], structural optimization [12], radar [13], telecommunications [14], music composition [15], power saving in manufacturing machines [16] and artificial vision [17], among many others. A thorough survey on applications of the HS algorithm can be found in [18].

The algorithm works iteratively by evolving an initial population in the following steps: (1) an initial population of size HMS (Harmony Memory Size) is created by adding random solutions; (2) following a procedure involving the application of three different operators, a new solution is created and evaluated; (3) if this new solution possess a better fitness than the worst solution in the HM, the new solution substitutes the worst one in the population (in this way the population size never changes). The pseudo-code is presented next:

Algorithm 1. Harmony Search pseudocode

```
Initialize the harmony memory with HMS randomly generated solutions
while termination criteria not met do
    create a new solution
    if rnd < HMCR then
        use a value of one of the solutions in the harmony memory (selected uniformly at random)
        if rnd < PAR then
            change this value slightly
        end if
    else
        Generate a random note
    end if
    if new solution is better than the worst solution in the harmony memory then
        replace the worst solution by the new one
    end if
end while
Return the best solution in the harmony memory.
```

The main parameters directing the creation of new solutions are usually only three: the harmony memory size, the Harmony Memory Considering Rate (HMCR) and the Pitch Adjustment Rate (PAR).

The Harmony Memory Considering Rate controls how much information from the harmony memory is used for the generation of a new solution. As such, it controls the rate of convergence of the algorithm. It does that by setting the probability $\varphi \in [0,1]$ that the new value for a certain note is taken uniformly at random from the values of this same note in all the other solutions in the HM. In case that (with a probability $1-\varphi$) the memory consideration step is not performed, the new value will be randomly chosen from their alphabet (the set of feasible values).

In the case that HMCR is performed, the probability $\vartheta \in [0,1]$ will control the application of the Pitch Adjusting Rate. This operator controls the frequency of adjustment of the note selected, by replacing the note by a neighbouring value. In the present paper, this operator acts by swapping the operation in the same resource with one of its neighbours.

2.6 Fitness Function

Three objectives are considered in the fitness function, and we use a weighted sum approach in order to prioritise them. The objectives are (i) the number of late jobs (ii), the summation of all the late days and (iii), the amount of idle time in all resources during the next 7 days, with respective weights of 5, 5 and 1 selected by the users.

Each objective is normalised by using a minimum and a maximum value for each one. In the case of late jobs, the minimum is the number of currently late jobs, while the maximum is taken as the total number of jobs to schedule. The minimum value of late days is taken as the current that each late job is already passed its due date, while the maximum was selected after running multiple simulations and taking the worst value. As for idle time, the minimum value was 0, while the maximum value is the total idle time after removing the length of the current in progress ops (given them time to be finished) and all the time that the designer will be doing other tasks during the next 7 days like holidays, training etc.

3 Computational Results

In this section we will present the results obtained in the computational experiments. These have been carried out in an Intel Core i7-5600 CPU @2.6 GHz, with 12 GB RAM, using only 1 core. All the experiments have been implemented in .NET framework. First we will look at the initialisation approaches, followed by the preliminary tests to find out the best possible parameters for each algorithm in the current scenario. Then we will compare the performance of SA and HS versus a hill climbing procedure.

The assumptions for all the tests are the following:

- Each test will run for 10 s, for the reasons explained in the introduction.
- Each test will be performed 30 times.
- Each run is independent. In particular, a new random seed is used for each run for the random number generator.
- First, for SA we will find the best temperature, by running the tests for different temperatures. Then we will run SA with both a random initial solution and an EF one.
- For HS we will experimentally find good values for HMS, HMCR and PAR.

The data we will use for the experiments is taken directly from the company's ERP[1] system, and represents a normal workload for the design department. The design department team is composed of 28 resources with varying levels of expertise. The work ready to schedule is composed by a total of 409 operations, belonging to 206 jobs. Of these jobs, 76 were already late as the due date was previous to the time the data snapshot was taken, for a total of 468 current days late. There were 23 constraints set by the manager on different operations, of which 6 set a starting time for operations, 1 modifies the run time of an operation and the rest are holidays and training times for the designers.

First we will look at the impact of using a good heuristic to find an initial solution for the algorithms. Table 1 shows the results obtained when using the two initialisation approaches applied to a problem instance; the random solution values are averaged across 100 repetitions. As expected, the EF approach provides a much better solution than a randomly generated ancestor with no significant increase in computational times. In particular, the effect on idle time is very noticeable.

Table 1. Comparison of initialization methods

	Late jobs	Late days	Total idle time (h)
Random	137	1124	264.7
EF	123	700	76.9

Next we present the results of the analysis on the impact of the most important parameters of the presented metaheuristics. In the case of SA, initial temperature in the solution fitness when constraint the computation time to 10 s. In Fig. 3 it is clear that the initial temperature must stay very small in order to perform well. Since the initial temperature is so small, we should expect that SA will perform quite similar to a plain hill climbing algorithm.

As for HS, we begin by performing a set of initial experiments to find a range of good parameters for the harmony search when applied to the current data set.

[1] Enterprise Resource Planning, the information system that manages production, sales, customers etc.

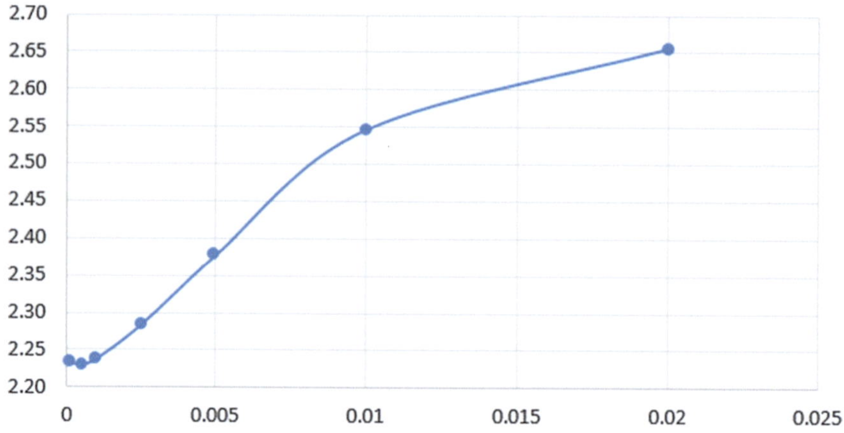

Fig. 3. Effect of SA initial temperature on fitness

In Table 3 the population is set to 10, with 7 random solutions and 3 EF solutions found by the heuristic described in Sect. 2.2. This table shows the relation between HMCR and PAR parameters.

The effect of the harmony memory size is studied in Fig. 4. This figure shows that smaller populations are preferable. The other consequence of varying the HMS is the number of iterations that the algorithm can go through in the fixed span of time, since the larger the population the more solutions need to be evaluated. This relation is shown in Table 2 (Fig. 5).

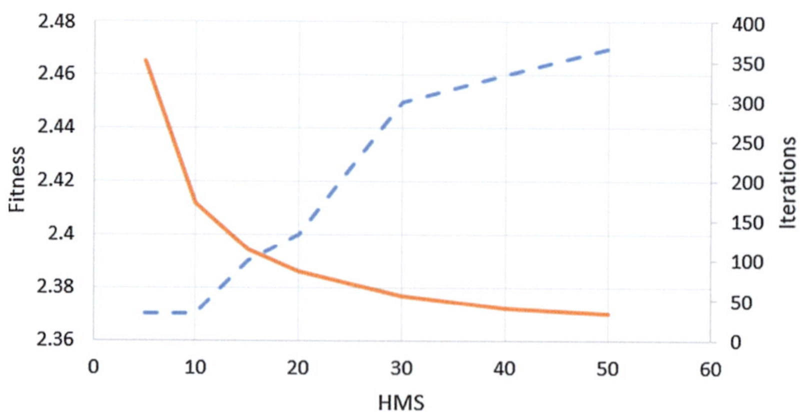

Fig. 4. Harmony memory size

Table 2. HMS vs. iterations

HMS	5	10	15	20	30	40	50
Fitness	2.37	2.37	2.39	2.4	2.45	2.46	2.47
Iterations	351	172	114	86	55	41	33

Table 3. HMCR vs. PAR

PAR	0.003	0.03	0.1	0.2	0.3
0.999	2.39	2.38	2.46	2.49	2.54
0.99	2.46	2.45	2.51	2.56	2.58
0.9	2.75	2.76	2.76	2.76	2.76

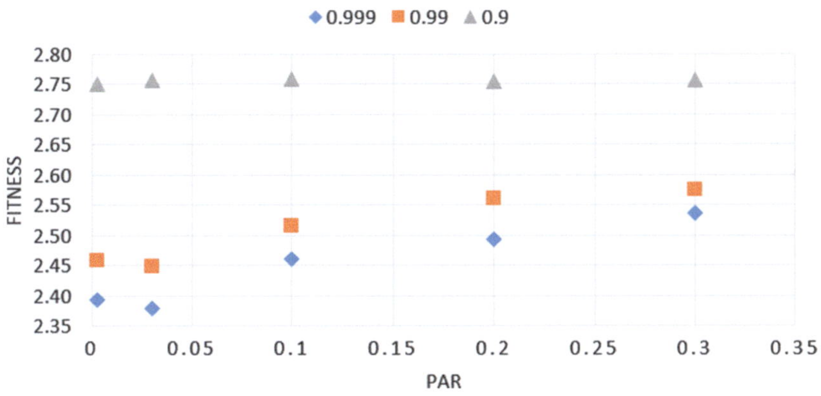

Fig. 5. HMCR vs. PAR

After these initial experiments, the parameters are thus set to:

- Simulated Annealing: $temperature = 0.0005$
- Harmony Search
  - $HMS = 10$
  - $HMCR = 0.999$
  - $PAR = 0.03$

With these parameters set, we run each algorithm 30 times, each time running for 10 s. The results in Table 4 are those obtained when running each algorithm with a Earliest Finish initial solution (3 EF solutions were used together with 7 random ones in case of the HS memory size), and Table 5 shows the results when the initial solution is a random one (100% random initial solutions in the HS memory).

As expected, the performance of simulated annealing when initial temperature is set to a very small value is very similar to that of the hill climbing procedure. Such initial temperature does not allow for a wide exploration of the solution space, and instead the algorithm mostly exploits the initial solution.

Table 4. EF initial solution

	Avg. fitness	Avg. late jobs	Avg. late days	Avg. idle time (h)
HC	2.22	117	692	107.1
SA	2.24	117	693	107.8
HS	2.38	118	695	142.1

Table 5. Random initial solution

	Avg. fitness	Avg. late jobs	Avg. late days	Avg. idle time (h)
HC	2.98	118	792	131.7
SA	3.05	118	799	142.3
HS	4.44	133	927	213.5

A follow-on experiment was performed, where 30 runs of simulated annealing were run with the stopping criterion of reaching a solution with the same fitness as that of the average hill climbing (2.22), in order to have an order of magnitude of the necessary time for SA. After 30 runs of SA, the average running time for reaching such fitness was 14.3 s (43% greater than the necessary time for the hill climbing).

4 Conclusions

The research presented in this paper aims at improving the quality of the production scheduling in real-life manufacturing scenarios. In those cases, where decision support systems need to react to unseen changes and in particular provide quick feedback to human users, the computational time becomes a limited resource. The experiments presented here show how an appropriate use of a good initialisation heuristic, like the earliest Finish approach presented in this paper, followed by an exploitation of the initial solution using simple hill climbing outperforms the use of more complex metaheuristics like Simulated Annealing or Harmony Search.

A following extension of this research will be the study of different local search strategies to work in conjunction with the initialisation heuristic to improve the quality of the solutions, always in scenarios where the time allowed for computations is limited by real-life user experience factors.

Acknowledgements. This work has been supported by Enterprise Ireland through the Innovation Partnership program.

References

1. Alcaraz, J., Maroto, C., Ruiz, R.: Solving the multi-mode resource-constrained project scheduling problem with genetic algorithms. J. Oper. Res. Soc. **54**(6), 614–626 (2003)

2. Tiwari, V., Patterson, J.H., Mabert, V.A.: Scheduling projects with heterogeneous resources to meet time and quality objectives. Eur. J. Oper. Res. **193**(3), 780–790 (2009)
3. Michalewicz, Z., Janikow, C.Z.: Handling constraints in genetic algorithms. In: ICGA, pp. 151–157 (1991)
4. Kirkpatrick, S.: Optimization by simulated annealing: quantitative studies. J. Stat. Phys. **34**(5–6), 975–986 (1984)
5. Foster, I., Kesselman, C.: The Grid: Blueprint for a New Computing Infrastructure. Morgan Kaufmann, San Francisco (1999)
6. Cerny, V.: Thermodynamical approach to the traveling salesman problem: an efficient simulation algorithm. J. Optim. Theor. Appl. **45**(1), 41–51 (1985)
7. Metropolis, N., et al.: Equation of state calculations by fast computing machines. J. Chem. Phys. **21**(6), 1087–1092 (1953)
8. Geem, Z.W., Kim, J.-H., Loganathan, G.V.: A new heuristic optimization algorithm: harmony search. Simulation **76**(2), 60–68 (2001)
9. Bean, J.C., Norman, B.A.: Random keys for job shop scheduling (1993)
10. Ayvaz, M.T.: Simultaneous determination of aquifer parameters and zone structures with fuzzy c-means clustering and meta-heuristic harmony search algorithm. Adv. Water Resour. **30**(11), 2326–2338 (2007)
11. Manjarres, D., Del Ser, J., Gil-Lopez, S., Vecchio, M., Landa-Torres, I., Lopez-Valcarce, R.: A novel heuristic approach for distance-and connectivity-based multihop node localization in wireless sensor networks. Soft Comput. **17**(1), 17–28 (2013)
12. Lee, K.S., Geem, Z.W., Lee, S.H., Bae, K.W.: The harmony search heuristic algorithm for discrete structural optimization. Eng. Optim. **37**(7), 663–684 (2005)
13. Gil-Lopez, S., Del Ser, J., Salcedo-Sanz, S., Perez-Bellido, A.M., Cabero, J.M., Portilla-Figueras, J.A.: A hybrid harmony search algorithm for the spread spectrum radar polyphase codes design problem. Exp. Syst. Appl. **39**(12), 11089–11093 (2012)
14. Del Ser, J., Matinmikko, M., Gil-Lopez, S., Mustonen, M.: Centralized and distributed spectrum channel assignment in cognitive wireless networks: a harmony search approach. Appl. Soft Comput. **12**(2), 921–930 (2012)
15. Geem, Z.W., Choi, J.Y.: Music composition using harmony search algorithm. In: Workshops on Applications of Evolutionary Computation, pp. 593–600 (2007)
16. Garcia-Santiago, C.A., Del Ser, J., Upton, C., Quilligan, F., Gil-Lopez, S., Salcedo-Sanz, S.: A random-key encoded harmony search approach for energy-efficient production scheduling with shared resources. Eng. Optim. **47**(11), 1481–1496 (2015)
17. Fourie, J., Mills, S., Green, R.: Harmony filter: a robust visual tracking system using the improved harmony search algorithm. Image Vision Comput. **28**(12), 1702–1716 (2010)
18. Manjarres, D., Landa-Torres, I., Gil-Lopez, S., Del Ser, J., Bilbao, M.N., Salcedo-Sanz, S., Geem, Z.W.: A survey on applications of the harmony search algorithm. Eng. Appl. Artif. Intell. **26**(8), 1818–1831 (2013)
19. Microsoft Canada: Attention Spans. https://advertising.microsoft.com/en/WWDocs/User/display/cl/researchreport/31966/en/microsoft-attention-spans-research-report.pdf
20. Rotondo, A., Young, P., Geraghty, J.: EF-based strategies for productivity improvements at wet-etch stations. Prod. Plann. Control **25**(11), 958–968 (2014)

Bat Algorithm for Coordinated Exploration in Swarm Robotics

Patricia Suárez[1] and Andrés Iglesias[2,3](✉)

[1] Faculty of Sciences, University of Cantabria, Avenida de los Castros s/n, 39005 Santander, Spain
[2] Department of Information Science, Faculty of Sciences, Toho University, Narashino Campus, 2-2-1 Miyama, Funabashi 274-8510, Japan
[3] Department of Applied Mathematics and Computational Sciences, University of Cantabria, Avenida de los Castros s/n, 39005 Santander, Spain
iglesias@unican.es
http://personales.unican.es/iglesias

Abstract. Bat algorithm is a powerful bio-inspired swarm intelligence method with remarkable applications in several industrial and scientific domains. However, to the best of authors' knowledge, this algorithm has not been applied so far to the exciting field of swarm robotics. This paper describes the first physical and computational implementation of the bat algorithm to a swarm of simple robotic units. The swarm consists of a set of identical wheeled robots equipped with simple yet powerful components that replicate the most important features of the bat algorithm by either hardware or software. The swarm has been applied to the problem of coordinated exploration, where the individual self-organizing robots generate an intelligent collective behavior emerging from the interactions between the robots and with the environment. A computational and real-world experiment has been carried out to check the feasibility and performance of this approach. Our experimental results show that the bat algorithm is extremely well suited for this task, actually leading to surprisingly intelligent behavioral patterns much better than expected.

Keywords: Swarm computation · Swarm robotics · Coordinated exploration · Bat algorithm · Collective behavior

1 Introduction

1.1 Swarm Intelligence

One of the most exciting advances in artificial intelligence during the last decades is the emergence of sophisticate behaviors arising from a collection of simple, unsophisticated agents collaborating together to solve a complex problem. This field, globally known as *swarm intelligence*, is overcoming the traditional mathematical approaches for solving optimization problems and laying the foundations

J. Del Ser (ed.), *Harmony Search Algorithm*, Advances in Intelligent Systems and Computing 514, DOI 10.1007/978-981-10-3728-3_14

for a new computational paradigm: the *swarm computation*. Under this new paradigm, there is no a centralized intelligence controlling the swarm, taking decisions, and sending orders to the swarm units about how to behave. Instead, the limited intelligence of swarm units is amplified by their (local or global) interactions. Members of the swarm have the ability to communicate with each other and with the environment, thus enhancing the global intelligence of the swarm. The interested reader is referred to [2,9] for a comprehensive overview about the field of swarm intelligence, its history, main techniques, and applications.

Nowadays, swarm intelligence is attracting increasing attention owing to its potential applications in several fields. For instance, military and civil applications related to the control of unmanned vehicles have been reported in the literature [10,11,13]. It has also been shown that self-organizing swarm robots can potentially accomplish complex tasks and thus replace sophisticated and expensive robots by simple inexpensive drones, a research subfield usually referred to as *swarm robotics* [1,3,12,14,20].

Swarm intelligence methods are based on the dynamics of natural groups. A typical example is the behavior of a flock of birds when moving all together following a common tendency in their displacements, an inspiration to the *particle swarm optimization* method. Other examples from nature include animal herding, ant colony, fish schooling, and many others. This leads to a valuable set of computational intelligence techniques known as *nature-inspired metaheuristic methods*, which are intensively used to solve optimization problems [4]. The reader is referred to [15] for a gentle introduction to several recent nature-inspired metaheuristic methods. Among them, the *bat algorithm* is an increasingly popular swarm intelligence algorithm originally proposed by Xin-She Yang in 2010 [16,18]. The algorithm is based on the peculiar behavior of microbats (see Sect. 2 for details). In our experience, the bat algorithm has shown to be a very effective method to solve complex multimodal nonlinear continuous optimization problems involving a large number of variables [5–8].

1.2 Aims and Structure of the Paper

Despite of its remarkable features, to the best of our knowledge, the bat algorithm has never been applied to the exciting field of swarm robotics. Aimed at filling this gap, in this paper we describe the first physical and computational implementation of the bat algorithm to swarm robotics. The swarm consists of a set of identical wheeled robots equipped with simple yet powerful components carefully designed to replicate the most important features of the bat algorithm by either hardware or software. The swarm is then applied to the problem of coordinated exploration in an unknown physical 3D environment.

The structure of this paper is as follows: in Sect. 2 we provide the reader with a gentle overview about the bat algorithm and its main rules and features. We also describe the algorithm through its pseudocode. In Sect. 3 we describe the main features of our implementation of the bat algorithm for swarm robotics in terms of hardware and software. The analogies between our computational and physical swarm robotics implementation, the real bats, and the bat algorithm are also discussed in that section. Finally, a computational and real-world experiment about coordinated exploration is reported in Sect. 4.

2 The Bat Algorithm

The *bat algorithm* is a bio-inspired swarm intelligence algorithm originally proposed by Xin-She Yang in 2010 to solve optimization problems [16–18]. The algorithm is based on the echolocation behavior of bats. The author focused particularly on microbats, as they use a type of sonar called *echolocation*, with varying pulse rates of emission and loudness, to detect prey, avoid obstacles, and locate their roosting crevices in the dark. The interested reader is referred to the general paper in [19] for a comprehensive, updated review of the bat algorithm, its variants and applications.

2.1 Basic Rules

The idealization of the echolocation of microbats can be summarized as follows (see [16] for details):

1. Bats use echolocation to sense distance and distinguish between food, prey and background barriers.
2. Each virtual bat flies randomly with a velocity $\mathbf{v}_i$ at position (solution) $\mathbf{x}_i$ with a fixed frequency f_{min}, varying wavelength λ and loudness A_0 to search for prey. As it searches and finds its prey, it changes wavelength (or frequency) of their emitted pulses and adjust the rate of pulse emission r, depending on the proximity of the target.
3. It is assumed that the loudness will vary from an (initially large and positive) value A_0 to a minimum constant value A_{min}.

In order to apply the bat algorithm for optimization problems more efficiently, some additional assumptions are strongly advisable. In general, we assume that the frequency f evolves on a bounded interval $[f_{min}, f_{max}]$. This means that the wavelength λ is also bounded, because f and λ are related to each other by the fact that the product $\lambda.f$ is constant. For practical reasons, it is also convenient that the largest wavelength is chosen such that it is comparable to the size of the domain of interest (the search space, for optimization problems). For simplicity, we can assume that $f_{min} = 0$, so $f \in [0, f_{max}]$. The rate of pulse can simply be in the range $r \in [0, 1]$, where 0 means no pulses at all, and 1 means the maximum rate of pulse emission. With these idealized rules indicated above, the basic pseudo-code of the bat algorithm is shown in Algorithm 1. It is described in next paragraphs.

Algorithm 1. Bat algorithm pseudocode

Require: (Initial Parameters)
Population size: $\mathcal{P}$
Maximum number of generations: $\mathcal{G}_{max}$
Loudness: $\mathcal{A}$
Pulse rate: r
Maximum frequency: f_{max}
Dimension of the problem: d
Objective function: $\phi(\mathbf{x})$, with $\mathbf{x} = (x_1, \ldots, x_d)^T$
Random number: $\theta \in U(0,1)$

1: $g \leftarrow 0$
2: Initialize the bat population $\mathbf{x}_i$ and $\mathbf{v}_i$, $(i = 1, \ldots, n)$
3: Define pulse frequency f_i at $\mathbf{x}_i$
4: Initialize pulse rates r_i and loudness $\mathcal{A}_i$
5: **while** $g < \mathcal{G}_{max}$ **do**
6:     **for** $i = 1$ **to** $\mathcal{P}$ **do**
7:         Generate new solutions by adjusting frequency,
8:         and updating velocities and locations //eqns. (1)–(3)
9:         **if** $\theta > r_i$ **then**
10:             $\mathbf{s}^{best} \leftarrow \mathbf{s}^g$ //select the best current solution
11:             $\mathbf{ls}^{best} \leftarrow \mathbf{ls}^g$ //generate a local solution around $\mathbf{s}^{best}$
12:         **end if**
13:         Generate a new solution by local random walk
14:         **if** $\theta < \mathcal{A}_i$ *and* $\phi(\mathbf{x_i}) < \phi(\mathbf{x}^*)$ **then**
15:             Accept new solutions
16:             Increase r_i and decrease $\mathcal{A}_i$
17:         **end if**
18:     **end for**
19:     $g \leftarrow g + 1$
20: **end while**
21: Rank the bats and find current best $\mathbf{x}^*$
22: **return** $\mathbf{x}^*$

2.2 The Algorithm

Basically, the algorithm considers an initial population of $\mathcal{P}$ individuals (bats). Each bat, representing a potential solution of the optimization problem, has a location $\mathbf{x}_i$ and velocity $\mathbf{v}_i$. The algorithm initializes these variables (lines 1–2) with random values within the search space. Then, the pulse frequency, pulse rate, and loudness are computed for each individual bat (lines 3–4). Then, the swarm evolves in a discrete way over generations (line 5), like time instances (line 19) until the maximum number of generations, $\mathcal{G}_{max}$, is reached (line 20). For each generation g and each bat (line 6), new frequency, location and velocity are computed (lines 7–8) according to the following evolution equations:

$$f_i^g = f_{min}^g + \beta(f_{max}^g - f_{min}^g) \tag{1}$$
$$\mathbf{v}_i^g = \mathbf{v}_i^{g-1} + [\mathbf{x}_i^{g-1} - \mathbf{x}^*]\, f_i^g \tag{2}$$
$$\mathbf{x}_i^g = \mathbf{x}_i^{g-1} + \mathbf{v}_i^g \tag{3}$$

where $\beta \in [0, 1]$ follows the random uniform distribution, and $\mathbf{x}^*$ represents the current global best location (solution), which is obtained through evaluation of the objective function at all bats and ranking of their fitness values. The superscript $(.)^g$ is used to denote the current generation g.

The best current solution and a local solution around it are probabilistically selected according to some given criteria (lines 8–11). Then, search is intensified by a local random walk (line 12). For this local search, once a solution is selected among the current best solutions, it is perturbed locally through a random walk of the form:

$$\mathbf{x}_{new} = \mathbf{x}_{old} + \epsilon \mathcal{A}^g \tag{4}$$

where ϵ is a random number with uniform distribution on the interval $[-1, 1]$ and $\mathcal{A}^g = <\mathcal{A}_i^g>$, is the average loudness of all the bats at generation g.

If the new solution achieved is better than the previous best one, it is probabilistically accepted depending on the value of the loudness. In that case, the algorithm increases the pulse rate and decreases the loudness (lines 13–16). This process is repeated for the given number of generations. In general, the loudness decreases once a bat finds its prey (in our analogy, once a new best solution is found), while the rate of pulse emission decreases. For simplicity, the following values are commonly used: $\mathcal{A}_0 = 1$ and $\mathcal{A}_{min} = 0$, assuming that this latter value means that a bat has found the prey and temporarily stop emitting any sound. The evolution rules for loudness and pulse rate are as follows:

$$\mathcal{A}_i^{g+1} = \alpha \mathcal{A}_i^g \tag{5}$$
$$r_i^{g+1} = r_i^0[1 - exp(-\gamma g)] \tag{6}$$

where α and γ are constants. Note that for any $0 < \alpha < 1$ an any $\gamma > 0$ we have:

$$\mathcal{A}_i^g \to 0, \quad r_i^g \to r_i^0, \quad \text{as } g \to \infty \tag{7}$$

In general, each bat should have different values for loudness and pulse emission rate, which can be computationally achieved by randomization. To this aim, we can take an initial loudness $\mathcal{A}_i^0 \in (0, 2)$ while the initial emission rate r_i^0 can be any value in the interval $[0, 1]$. Loudness and emission rates will be updated only if the new solutions are improved, an indication that the bats are moving towards the optimal solution. As a result, the bat algorithm applies a parameter tuning technique to control the dynamic behavior of a swarm of bats. Similarly, the balance between exploration and exploitation can be controlled by tuning algorithm-dependent parameters.

3 Bat Algorithm Implementation for Swarm Robotics

In this section we implement the bat algorithm described above to swarm robotics by using a swarm of simple robotic units. Our approach to this problem is a mixture of physical and computational components, as described in next paragraphs.

Fig. 1. Identical wheeled robots used for our implementation of the bat algorithm and main components: a single-board microcontroller *Arduino UNO* (for programming and connectivity), ultrasound sensors (for collision avoidance), magnetometer (for spatial orientation), and bluetooth card (for wireless data exchange and communication).

3.1 Hardware Implementation

Our robotic swarm consists of a set of identical wheeled robots. Figure 1 shows two robots of the swarm oriented in opposite directions for better visualization of their most distinctive features. Each robot is equipped with a kit of simple yet powerful components that replicate the most relevant features of the bat algorithm. The main hardware components are:

1. a *single-board micro-controller:* in our implementation we use the popular micro-controller *Arduino UNO*, since it provides a very good combination of small size, affordable price, and reasonable computing power. The micro-controller is responsible for all programming tasks and the connectivity among the different components;
2. two *ultrasound sensors:* they are used for collision avoidance with static and dynamic objects (including other robots in the swarm) as well as with the physical 3D environment;
3. a *magnetometer:* used for global spatial orientation purposes; and
4. a *Bluetooth card:* used for wireless communication and data exchange among the robots of the swarm and with a central server for tracking purposes.

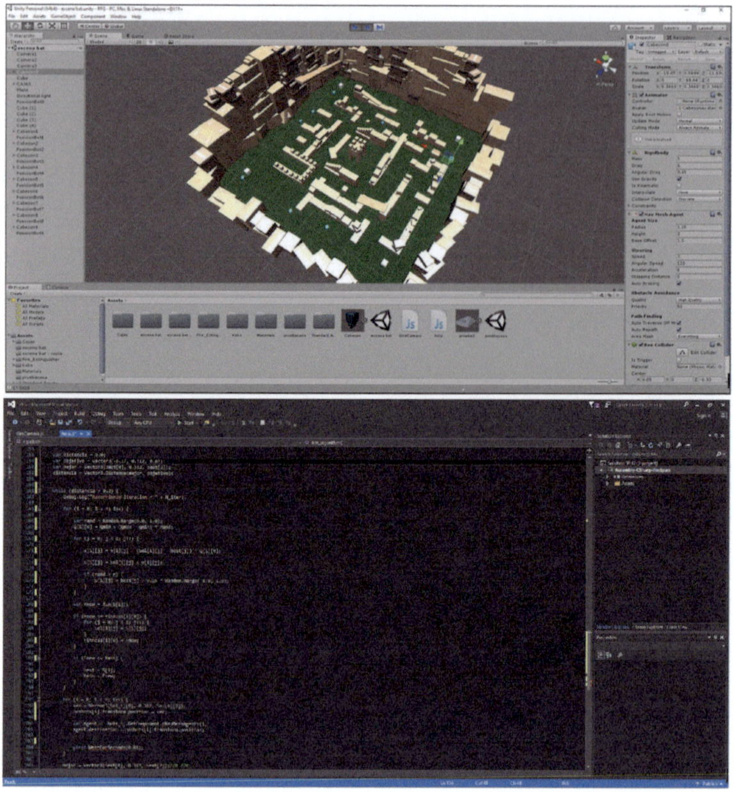

Fig. 2. Computational tools used in this paper: (top) graphical editor of game engine *Unity 5*; (bottom) programming framework for *JavaScript* in *Visual Studio.*

3.2 Software Implementation

The bat algorithm-based robotic swarm has been simulated graphically in *Unity 5*, a popular multi-platform game engine especially praised by its portability, supporting personal computers, video consoles, mobile devices and websites. It also provides a nice graphical editor shown in Fig. 2(top), with different windows and workspaces for the project, scene view, game view, hierarchy, toolbars, inspector, and many other features. It also supports the programming languages *C#* and *JavaScript.* All programming code in this paper has been created in *JavaScript* using the *Visual Studio* programming framework shown in Fig. 2(bottom).

3.3 Analogies with the Real Bats and the Bat Algorithm

As mentioned, all components of our implementation have been carefully chosen to replicate the most relevant features of the real bats and the bat algorithm by either hardware or software. The features replicated by hardware are:

1. As any other swarm intelligence approach, there is not a centralized system controlling the swarm behavior. This also happens in our implementation. There is a central server, but it does not control the system; it is used for tracking and monitoring purposes only. All robotic units move and behave on an individual (yet coordinated) basis.
2. Similar to some microbats, our robots are not provided with vision abilities at all; instead, they rely on echolocation based on ultrasounds for collision avoidance and interaction with other robots and the environment.
3. To this purpose, they have small ultrasound sensors sending an 8 cycle burst of ultrasound pulses, very similar to the typical 10–20 cycle burst of real bats.
4. Each ultrasound pulse of our robots operates at a constant frequency of 40 kHz, well within the typical range of 25–150 kHz of real bats.
5. The robot captures its echo with signals lasting in the order of milliseconds, as the real bats actually do.
6. The ranging accuracy of the sensor is about 3 mm, very similar to that of some real bats.
7. The traveling range of pulses in our robots is between 2–500 cm, similar to the range of a few meters of several bats.

Other bat algorithm features have been replicated by software to ensure the highest fidelity of our robotic swarm to the bats and the original bat algorithm:

8. The pulse rate of our ultrasound signals can be modified to adapt to different conditions, as it typically happens with bats in the real world.
9. The loudness can also be modulated in the source programming code.

4 Computational and Real-World Experiment

Our robotic swarm has been tested in an experiment about coordinated exploration on a physical 3D environment. The goal is to reach a target point without any knowledge about its location or the possible paths leading to it. The experiment has been conducted at both computational and real-world levels. We performed 100 computer executions for different random initial locations for the robots. Only 10 of those simulations have been replicated in the real-world experiment because of battery constraints. Figure 3 shows some screenshots of the computer experiment along with the pictures of their real-world counterpart. The environment is a closed storage room with several cardboard boxes stacked in a very messy way (even occasionally forming structures such as tunnels and passageways the robots can take advantage of). The different robotic units are highlighted with dashed circles of different colors for easier identification. We also marked the areas of the different robot locations with rectangles of different colors for better visualization of the regions of interest. We remark that the computational and the real-world simulations are performed in a completely independent way, but with a similar environment (the computer world is a replica of the real environment) and the same initial locations for the real and the simulated robots. In fact, the actual positions of the real and the simulated

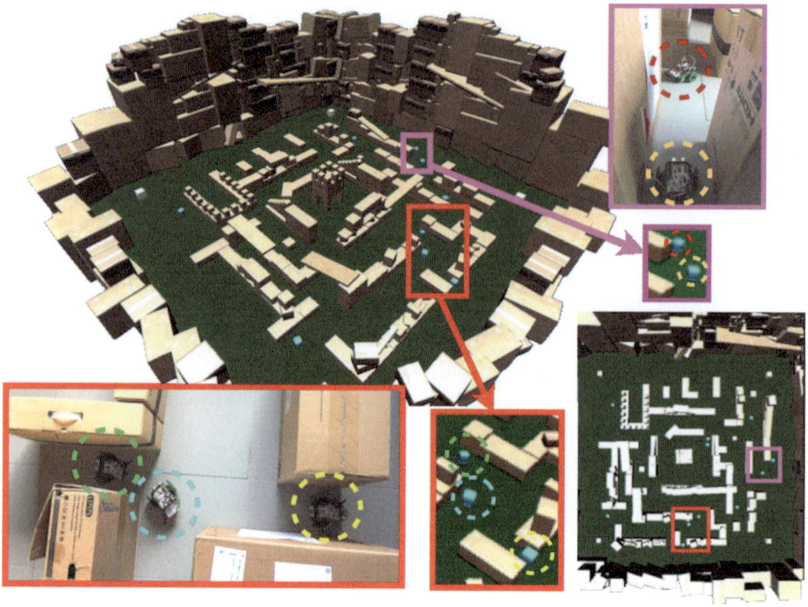

Fig. 3. Computational and real-world experiment about coordinated exploration for swarm robotics. General view (top-left), regions of interest and top view (bottom-right).

robots do not match exactly frame-by-frame because of the stochasticity of the algorithm. Yet, the general dynamics of the swarm is surprisingly similar.

Our experimental results have been amazingly good, actually much better than expected. In our prior opinion, the robots should get trapped in challenging configurations such as U and V shapes, where the robotic units would be forced to come back, in cases of traffic jam owing to overcrowding, or in cases of narrow passages and corridors where only a single robot could advance at once. However, we found surprisingly intelligent behaviors where some units in a formation leave the group to explore other alternative ways. Sometimes, the robots turn around without any apparent advantage for doing so; however, the reason becomes clear after some iterations, when we realized that the unit found a way to reach a better location than its robotic mates in the formation, even although it initially implies to move further away from better locations. Finally, to our delight, we did not find any configuration so far unsolvable for the robots, provided that at least one 2D solution exists (unfortunately, our robots do not fly!). This experiment shows that the bat algorithm is extremely well suited for this particular task, actually much better than we anticipated. We are planning now to improve our design for smaller robots and extend this analysis to other challenging tasks.

Acknowledgements. This research has been kindly supported by the Computer Science National Program of the Spanish Ministry of Economy and Competitiveness, Project Ref. #TIN2012-30768, Toho University, and the University of Cantabria.

References

1. Arvin, F., Murray, J.C., Licheng, S., Zhang, C., Yue, S.: Development of an autonomous micro robot for swarm robotics. In: Proceedings of the IEEE International Conference on Mechatronics and Automation - ICMA 2014, pp. 635–640 (2014)
2. Engelbretch, A.P.: Fundamentals of Computational Swarm Intelligence. Wiley, Chichester (2005)
3. Faigl, J., Krajnik, T., Chudoba, J., Preucil, L., Saska, M.: Low-cost embedded system for relative localization in robotic swarms. In: Proceedings of the IEEE International Conference on Robotics and Automation - ICRA 2013, pp. 993–998 (2013)
4. Iglesias, A., Gálvez, A.: Nature-inspired swarm intelligence for data fitting in reverse engineering: recent advances and future trends. Stud. Computat. Intell. **637**, 151–175 (2016)
5. Iglesias, A., Gálvez, A., Collantes, M.: Bat algorithm for curve parameterization in data fitting with polynomial Bézier curves. In: Proceedings of the Cyberworlds 2015, Visby, Sweden, pp. 107–114. IEEE Computer Society Press, Los Alamitos (2015)
6. Iglesias, A., Gálvez, A., Collantes, M.: Global-support rational curve method for data approximation with bat algorithm. In: Chbeir, R., Manolopoulos, Y., Maglogiannis, I., Alhajj, R. (eds.) AIAI 2015. IFIPAICT, vol. 458, pp. 191–205. Springer, Heidelberg (2015). doi:10.1007/978-3-319-23868-5_14
7. Iglesias, A., Gálvez, A., Collantes, M.: A bat algorithm for polynomial Bézier surface parameterization from clouds of irregularly sampled data points. In: Proceedings of the International Conference on Natural Computation 2015 - ICNC 2015, Zhangjiajie, China, pp. 1034–1039. IEEE Computer Society Press, Los Alamitos (2015)
8. Iglesias, A., Gálvez, A., Collantes, M.: Four adaptive memetic bat algorithm schemes for Bézier curve parameterization. In: Gavrilova, M.L., Tan, C.J.K., Sourin, A. (eds.) Transactions on Computational Science XXVIII. LNCS, vol. 9590, pp. 127–145. Springer, Heidelberg (2016). doi:10.1007/978-3-662-53090-0_7
9. Kennedy, J., Eberhart, R.C., Shi, Y.: Swarm Intelligence. Morgan Kaufmann Publishers, San Francisco (2001)
10. Sauter, J.A., Matthews, R., Parunak, H.V.D., Brueckner, S.A.: Effectiveness of digital pheromones controlling swarming vehicles in military scenarios. J. Aerosp. Comput. Inf. Commun. **4**(5), 753–769 (2007)
11. Saska, M., Vonasek, V., Krajnik, T., Preucil, L.: Coordination and navigation of heterogeneous formations localized by a hawk-eye-like approach under a model predictive control scheme. Int. J. Robot. Res. **33**(10), 1393–1412 (2014)
12. Tan, Y., Zheng, Z.Y.: Research advance in swarm robotics. Defence Technol. J. **9**(1), 18–39 (2013)
13. Vasarhelyi, G., Virágh, C., Tarcai, N., Somorjai, G., Vicsek, T.: Outdoor flocking and formation flight with autonomous aerial robots. In: Proceedings of the IEEE/RSJ International Conference on Intelligent Robots and Systems - IROS 2014, pp. 3866–3873 (2014)
14. Wagner, I., Bruckstein, A.: Special issue on ant robotics. Ann. Math. Artif. Intell. **31**, 1–4 (2001)
15. Yang, X.-S.: Nature-Inspired Metaheuristic Algorithms, 2nd edn. Luniver Press, Frome (2010)

16. Yang, X.-S.: A new metaheuristic bat-inspired algorithm. In: González, J.R., Pelta, D.A., Cruz, C., Terrazas, G., Krasnogor, N. (eds.) Nature Inspired Cooperative Strategies for Optimization. Studies in Computational Intelligence, vol. 284, pp. 65–74. Springer, Berlin (2010)
17. Yang, X.S.: Bat algorithm for multiobjective optimization. Int. J. Bio-Inspired Comput. **3**(5), 267–274 (2011)
18. Yang, X.S., Gandomi, A.H.: Bat algorithm: a novel approach for global engineering optimization. Eng. Comput. **29**(5), 464–483 (2012)
19. Yang, X.S.: Bat algorithm: literature review and applications. Int. J. Bio-Inspired Comput. **5**(3), 141–149 (2013)
20. Zahugi, E.M.H., Shabani, A.M., Prasad, T.V.: Libot: design of a low cost mobile robot for outdoor swarm robotics. In: Proceedings of the IEEE International Conference on Cyber Technology in Automation, Control, and Intelligent Systems - CYBER 2012, pp. 342–347 (2012)

Quantitative Analysis and Performance Study of Ant Colony Optimization Models Applied to Multi-mode Resource Constraint Project Scheduling Problems

Antonio Gonzalez-Pardo[1(✉)], Javier Del Ser[2,3,4], and David Camacho[1]

[1] Escuela Politécnica Superior, Universidad Autónoma de Madrid, 28049 Madrid, Spain
{antonio.gonzalez,david.camacho}@uam.es

[2] TECNALIA, 48160 Derio, Spain
javier.delser@tecnalia.com

[3] University of Basque Country EHU/UPV, 48009 Bilbao, Bizkaia, Spain
javier.delser@ehu.es

[4] Basque Center for Applied Mathematics, 48009 Bilbao, Bizkaia, Spain

Abstract. Constraint Satisfaction Problems (CSP) belongs to this kind of traditional NP-hard problems with a high impact in both, research and industrial domains. However, due to the complexity that CSP problems exhibit, researchers are forced to use heuristic algorithms for solving the problems in a reasonable time. One of the most famous heuristic algorithms is Ant Colony Optimization (ACO) algorithm. The possible utilization of ACO algorithms to solve CSP problems requires the design of a decision graph where the ACO is executed. Nevertheless, the classical approaches build a graph where the nodes represent the variable/value pairs and the edges connect those nodes whose variables are different. In order to solve this problem, a novel ACO model have been recently designed. The goal of this paper is to analyze the performance of this novelty algorithm when solving Multi-Mode Resource-Constraint Satisfaction Problems. Experimental results reveals that the new ACO model provides competitive results whereas the number of pheromones created in the system is drastically reduced.

Keywords: Ant Colony Optimization · Oblivion Rate · Resource-Constraint Project Scheduling Problems · Pheromone control

1 Introduction

One of the compounding paradigms within the set of NP-hard problems is related to as Resource-Constraint Project Scheduling Problem (RCPSP) [14]. This family of problems is defined by a set of variables that need to be assigned with a particular value taking into account a set of restrictions that establish constraints among the different values assigned to the variables. Therefore, any

J. Del Ser (ed.), *Harmony Search Algorithm*, Advances in Intelligent Systems and Computing 514, DOI 10.1007/978-981-10-3728-3_15

Constraint Satisfaction Problem (CSP) is represented with a triple (X, D, C) where $X = \{x_1, x_2, \ldots x_n\}$ represents the set of objects that composes the problem, $D = \{d_1, d_2, \ldots d_n\}$ is used to describe the domains that contain the different values for the objects described in X, and C represents the set of constraints that relates the objects with their values [3,14].

There is a wide number of complex research and industrial problems that can be modelled as a CSP, the main techniques, algorithms and methods obtained from this area have been applied in the last decades to real domains with an increasing level of complexity (i.e. scheduling and planning problems, energy optimization, man-power scheduling, travel and car routing optimization, etc...) [1,8,11,12,14].

Due to the inherent complexity of CSP problems, it is common to use Computational Intelligence algorithms (such as Ant Colony Optimization (ACO)) to solve these problems. In order to use ACO for solving CSP problems, the solution space must be represented as a graph (called *decision graph*) over which the ACO algorithm is executed. The standard approaches for building this *decision graph* presents several drawbacks that make it difficult to apply this approach to CSP instances of moderate to high dimensionality. To solve this problem, a new CSP-graph based model was proposed in [6], where the reduction in the size of the *decision graph* results in a fast growth in the number of pheromones. In order to control this increase rate of the number of pheromones created in the model, a new heuristic called *Oblivion Rate* was included in the model. This model has been applied to the *N-Queens problem* [6], the *Resource-Constraint Project Scheduling Problem* [7] and the *Lemmings Game* [9]. The contribution of this work is the analysis of the performance of the *Oblivion Rate* heuristic in a new family of RCPSP problems called, *Multi-Mode Resource Constraint Project Scheduling Problems*.

This paper is structured as follows: Sect. 2 contains the description for the RCPSPs. Section 3 details the implementation of the ACO model used to RCPSP, including the definition of the new decision graph, the behaviour of the ants, and the *Oblivion Rate* heuristic used in this work. The performance of the selected model is analyzed in Sect. 4 and the conclusions extracted from this work are outlined in Sect. 5.

2 Resource-Constraint Project Scheduling Problem

This work gravitates on the use of ACO algorithms to the Resource-Constraint Project Scheduling Problem (RCPSP) [2,4]. The goal of this class of problems is to find an optimal schedule of the activities that compose a project subject to the availability and demand of different resources required to undertake these tasks. In mathematical terms, a project is composed by a set of activities $\mathcal{J} = \{0, \ldots, n+1\}$, a set of resource types $\mathcal{Q} = \{1, \ldots, q\}$ and a specific number of resources for each resource type $r_q \forall q \in \mathcal{Q}$. A project composed by n activities has always $n+2$ activities in the set $\mathcal{J}$ because activity 0 and $n+1$ are dummy activities explicitly included to represent the start and end of the project and do not imply any duration nor need for resources.

Each activity can be executed in one or more different modes. If activities can be executed only in one mode, the problem is labeled as *Single-Mode*. Likewise, if activities can be executed in more than one mode, the problem is called *Multi-Mode*. The modes of a given activity represent different ways to execute this activity. For the same activity, modes differ in both the duration needed to complete the activity and the set of resources required for its accomplishment. Formally, the set of different modes of the activity j is denoted as $Mode_j$, the duration of activity j executed with mode m is denoted as d_{jm} and it requires r_{jmq} units of the resource $q \in \mathcal{Q}$. Moreover, s_j denotes the time when activity j started the execution, and f_j denotes the time when such an activity has finished. Note that $f_j = s_j + d_{jm}$ because the execution of any activity cannot be interrupted.

Each project may also contain *precedence constraints* that establish relations of time interdependence between the different activities that compose the project. If a given activity j has a precedence constraint with activity i, activity i cannot be executed until activity j has finished (i.e. $s_i \geq f_j$). By considering these constraints each activity can be assigned two lists, namely, $\mathcal{P}_j$ and $\mathcal{S}_j$, which contain its direct predecessors and successors. It is relevant to note that activity 0 is the only start activity and hence has no predecessors. Likewise, activity $n + 1$ is the only end activity and consequently, has no successors.

A solution for a RCPSP is schedule for the different activities that compose the project. This schedule is composed by the start time for all the activities that compose the project, $\mathbb{S} = \{s_x \mid \forall x \in \mathcal{J}\}$ and the different execution modes for the activities. For a given schedule, the start time is the initial time for activity 0 (s_0) and the finish time is the time for activity $n+1$ (f_{n+1}). The best solution is those with a minimum makespan [13], i.e. the difference between its finishing and starting times ($f_{n+i} - s_0$). A schedule will be declared *feasible* if it satisfies the following constraints:

- All the activities are scheduled, and each of them is executed once.
- Any activity must not be started before all its predecessors have finished. $s_i \geq f_j \mid \forall j \in \mathcal{P}_i, i \in \mathcal{J}$.
- At any time t, the sum of resources required for the activities in execution must not exceed the resource capacities of the project.

3 The Selected ACO Model for Constraint Satisfaction Problems

This section describes the model (first proposed in [6]) that allows a significant reduction in the size of the decision graph. The goal of this model is to create smaller graphs than the ones created in the literature. This section provides a detailed description of the different components that defines the model which are:

- A new decision graph, which is smaller than the ones created in the classical approaches.

- The new ants' behaviour: the reduction in the size of the graph yields a slightly more complex behaviour of the ants.
- The new heuristic called *Oblivion Rate* needed to control the number of pheromones created in the system.

The classical procedure to model any CSP as a graph is by creating as many nodes as pairs <*variable*, *value*> available in the problem, and connecting those nodes whose variables are different. More formally, the resulting graph is defined as $G = (V, E)$ where:

$$\begin{aligned} V &= \{\langle X_i, v\rangle | X_i \in X \text{ and } v \in D(X_i)\} \\ E &= \left\{(\langle X_i, v\rangle, \langle X_j, w\rangle) \in V^2 | X_i \neq X_j\right\} \end{aligned} \tag{1}$$

where nodes V represent the pairs <*variable*, *value*>, and E represents edges connecting those nodes whose associated variables X are different from each other.

There are several pitfalls regarding this representation but the most important are related to the size of the resulting graph and the type of CSPs that can be represented. In this sense, if the problem has N variables and each of them can take M different values, the resulting graph will contain $N \cdot M$ nodes. As the graph is almost fully connected, the number of edges is $(N \cdot M) \cdot (M \cdot (N-1)) = N^2 \cdot M^2 - N \cdot M^2 \cong N^2 \cdot M^2$. This observation implies that problems composed by many variables or by variables that could take on a high number of different values would become really difficult to model and almost computationally prohibitive to handle due to the size of their underlying graph.

The CSP graph representation selected in this paper was initially proposed in [6]. This representation focuses on the reduction of the graph size resulting from the modeling of the CSP as a graph. In this approach, the size of the resulting graph is drastically reduced because each variable in the problem is represented only by one node, independently of the number of values that can be assigned to this variable (as it is traditionally represented in CSP solvers). Therefore, given any problem composed by N variables whose value can be drawn from a set of M different values, the resulting graph will have only N nodes, instead of $N \cdot M$ nodes created in classical graph models. This representation was applied to the N-Queens Problem, though it can be used in other CSP-like problems such as video games [5].

The restrictions of the problem are represented in the edges of the graph. Two nodes will be connected if there is at least one restriction that involves the variables represented by the nodes. For example, given the nodes $\mathcal{N}_1$ (that represent the variable x_1) and $\mathcal{N}_2$ (correspondingly, variable x_2) there will be an edge connecting both nodes if there is at least one constraint involving the values of x_1 with the values of x_2. Using this representation, the number of edges is drastically reduced due to the decrease of the number of nodes.

This simplification in the graph size entails a change in the behavior of the ants. In classical ACO approaches ants have to select the next node to visit, because the node itself contains the value assignment. Ants only deposit a small

Algorithm 1. Ants' behavior needed in the selected graph representation.

1: $\mathcal{E}val\mathcal{V}al\mathcal{L}ist \leftarrow$ getEvaluatedValues($current\mathcal{N}ode$)
2: $\mathcal{P}her\mathcal{L}ist \leftarrow$ getPheromoneInformation()
3: $selected\mathcal{V}al \leftarrow$ selectValue($\mathcal{E}val\mathcal{V}al\mathcal{L}ist$,$\mathcal{P}her\mathcal{L}ist$)
4: updatePersonalAssignment($selected\mathcal{V}al$)
5: $\mathcal{D} \leftarrow$ getPossibleNodes($current\mathcal{N}ode$)
6: **if** ($\mathcal{D} \neq null$) **then**
7:     $node \leftarrow$ selectNextNode(D)
8:     $current\mathcal{N}ode \leftarrow node$
9: **else**
10:     resetAnt()
11: **end if**

quantity of pheromone on the graph and repeat the process until they finish their execution. When adopting the new representation the ant behavior becomes more complex because ants are in charge of selecting a specific value for the variable encoded in the node (see Algorithm 1).

In Algorithm 1 ants evaluate the different values that can be assigned to the variable encoded in the corresponding node (Line 1). This evaluation is performed by using the heuristic function defined for the specific problem. Then the pheromone information deposited in the graph is used in Line 2. Once the pheromone and the heuristic values are obtained, ants select one value for the variable encoded in the node (Line 3). Every ant updates its personal assignments, i.e. its local solution, and compute the possible nodes to visit taking into account their local solution built so far. If there is at least one possible destination, the ant selects one of them to visit in the next time step. Otherwise, the ant finishes its execution and goes back to the nest updating, at the same time, the pheromone information that has deposited through the graph (Line 10).

Another consequence of the graph reduction is the increase of the number of pheromones deposited in the graph. Pheromones are placed in the edges of the graph because the validity of a specific value in a node depends on the given values to the rest of variables in the other nodes. Thereby, the edge connecting nodes i and j stores all pheromones related to the variables and values for these nodes. Depending on the complexity of the problem being solved, the number of pheromones stored in the graph might saturate the system. The total number of different pheromones in an edge is proportional to the size of the domains of the variables involved in the constraint represented by the edge. That is, if $|D(var_s)|$ denotes the different values that the source variable can take, and $|D(var_d)|$ represents different values for the destination variable, the edge connecting source and destination node could store, a maximum of $|D(var_s)| \cdot |D(var_d)|$ different pheromones.

In order to reduce the number of pheromones stored in the graph an *Oblivion Rate* heuristic is incorporated to the system. This heuristic removes a subset of pheromones from the network. It is important to note that this heuristic must be carefully designed, because it affects directly on the system performance.

Consequently, the design of this heuristic depends on the problem being addressed. In this work, the selected *Oblivion Rate* is a dynamic function that depends on the number of pheromones created in the system to compute the number of pheromones that will be removed. This heuristic applied at step t is defined as:

$$R(t) = 1 - \frac{1}{t^{S(t)}}, \tag{2}$$

where $\dot{S}(t)$ represents the number of pheromones created in the graph at step t. Equation 3 defines this function, that depends on the number of pheromones created ($P(t)$) and the maximum number of pheromones that can be created ($MaxPher$), yielding

$$S(t) = \frac{P(t)}{MaxPher}. \tag{3}$$

In order to compute the maximum number of pheromones, Expression 4 provides an upper bound value using the classical graph-based representation previously described. This upper bound is computed by estimating the number of nodes and edges that the graph would contain by using the classical representation, i.e.

$$MaxPher(j, m) = j \cdot m \cdot (j - 1) \cdot m = j \cdot m^2 \cdot (j - 1). \tag{4}$$

4 Experimental Results

The main goal of the experimental results discussed in this section is to analyze the performance of the described ACO model when tackling RCPSP problems. Performance will be measured as the quality of the solutions found by the ACO model, as well as the number of pheromones stored in the system. The dataset used in this work has been extracted from the PSPLIB library [10] by selecting those problems where the number of execution modes are greater than 1, i.e. selecting the *Multi-Mode* problems (a description about the characteristics for the selected problems can be found in Table 1).

The configuration for the ACO algorithms carried out in this work is the same for all the experiments. The colony is composed by 100 ants that are executed during 100 steps. The evaporation rate is fixed to $\rho = 0.05$ whereas the values for α and β are 1 and 2 respectively.

The first experiment carried in this work analyzes the reduction in the number of pheromones created in the system. In order to do that, the different problems have been solved using the selected ACO model without Oblivion (Normal ACO) and using the *Dynamic Oblivion.* The number of pheromones created by both ACO models and the corresponding reduction percentage are shown in Table 2.

As it can be observed in this table, there is an important reduction in the number of pheromones of, at least, 94%. This is an important reduction because each pheromone is a structure stored in the memory of the system and it could be saturated. Nevertheless, pheromones are used to guide the colony to the optimal solutions. Thus, this reduction could affect to the quality of the solutions

Table 1. Description of the different problems available in the RCPSP dataset.

Problem	#Instances	#Activities	#Modes
j10.mm	536	10	3
j12.mm	547	12	
j14.mm	551	14	
j16.mm	550	16	
j18.mm	552	18	
j20.mm	554	20	
j30.mm	640	30	
m2.mm	481	16	2
m4.mm	555		4
m5.mm	558		5

Table 2. This table shows the maximum number of pheromones created in the system using the Dynamic Oblivion Rate, and without the Oblivion Rate (Normal ACO).

Problem	Normal ACO	Dynamic Oblivion	Reduction pct.
j10.mm	2023	68	96.63%
j12.mm	2695	83	96.92%
j14.mm	3246	101	96.88%
j16.mm	3841	120	96.87%
j18.mm	4315	142	96.71%
j20.mm	4865	168	96.54%
j30.mm	5922	276	95.33%
m2.mm	2092	112	94.64%
m4.mm	4891	268	**94.52%**
m5.mm	5645	121	**97.85%**

found by the ACO model. In order to measure whether this reduction affects to the quality of the solutions found, we have computed the average minimum makespan obtained by the different ACO algorithms and compared it against the average best makespan published by the research community.

Table 3 shows the average minimum makespan published by the research community and the average minimum makespan obtained by the selected model: without using the Oblivion Rate (Normal) and using the Dynamic Oblivion Rate for the Multi-Mode problems of the PSPLib dataset. As it can be seen in this table, the average minimum makespan obtained by our approach is really similar when the model is not using the Oblivion Rate, and when the Dynamic Oblivion Rate is used. These results are really promising if we take into account the strong reduction in the number of pheromones created in the system when

Table 3. This table shows the average minimum makespan published by the research community, the average minimum makespan obtained by the selected model without using the Oblivion Rate (Normal) and using the Dynamic Oblivion Rate for the Multi-Mode problems belonging to PSPLib.

Dataset	MinPubl.	Normal ACO	Dynamic Oblivion
`m2.mm`	30.16 ± 6.87	31.04 ± 7.42	31.04 ± 7.4
`m4.mm`	22.71 ± 7.3	26.69 ± 8.61	26.68 ± 8.51
`m5.mm`	21.16 ± 8.14	25.73 ± 9.4	25.76 ± 9.31
`j10.mm`	19.04 ± 6.21	19.69 ± 6.46	19.68 ± 6.46
`j12.mm`	21.34 ± 6.48	22.36 ± 6.63	22.25 ± 6.54
`j14.mm`	23.18 ± 6.14	25.23 ± 6.73	25.31 ± 6.79
`j16.mm`	24.93 ± 6.02	27.67 ± 7.03	27.76 ± 7.07
`j18.mm`	26.57 ± 6.47	29.99 ± 7.66	30.09 ± 7.54
`j20.mm`	27.71 ± 6.99	32.08 ± 8.62	32.08 ± 8.76
`j30.mm`	28.79 ± 7.44	36.16 ± 17.39	36.16 ± 17.44

the Oblivion Rate is used. The utilization of the Dynamic Oblivion Rate reduces at least the 94% of the pheromones for Multi-Mode problems, building solutions really close to the ones obtained by the system without controlling the number of pheromones, and thus having more information about the past of the algorithm. Finally, all these makespan values obtained by our approach are really close to the optimal makespan obtained by the research community. For previous reasons, we can conclude that the Oblivion Rate heuristic and the selected ACO model are a good approach for solving RCPSP problems because the system obtains solutions close to the optimal, and the number of pheromones created in the system has been extremely reduced.

5 Concluding Remarks

Constraint Satisfaction Problems (CSP) belongs to this kind of traditional NP-hard problems with a high impact in both, research and industrial domains. There are several problems that can be modelled as a CSP such as planning, scheduling, travel and car routing problems, videogames or energy, among others.

However, due to the complexity that CSP problems exhibits, researchers are forced to use heuristic algorithms for solving the problems in a reasonable time. One of the most famous heuristic algorithms is Ant Colony Optimization (ACO) algorithm, but the classical utilization of ACO algorithms build a decision graph composed by the same number of nodes as pairs <*variable*, *value*> available in the problem. Therefore the size of the resulting graph could be unmanageable depending on the number of variables and values of the selected problem.

In order to solve this problem, a new ACO model was proposed in [6]. This model is characterized by the utilization of a reduced decision graph and by the usage of a *Oblivion Rate* heuristic for controlling the number of pheromones created in the system. This paper studies the applicability of the novel approach to solve Multi-Mode Resource Constraint Satisfaction Problem. For evaluating the ACO model, we have selected the Multi-Mode instances belonging to the PSPLib dataset. The experimental results reveals that the ACO model is able to remove, at least, the 94% of the pheromones of the system without affecting to the quality of the solutions built by the ACO algorithm. This result reveals that the ACO algorithm is a good approach for solving RCPSP problems because it is able to guide ants to optimal, or sub-optimal solutions, maintaining in the system those pheromones created by the best solutions.

Acknowledgements. This work has been supported by the research projects: EphemeCH (TIN2014-56494-C4-4-P) Spanish Ministry of Economy and Competitivity, CIBERDINE S2013/ICE-3095, both under the European Regional De-velopment Fund FEDER, Airbus Defence &Space (FUAM-076914 and FUAM-076915), BID3ABI (Basque Government), and RiskTrack (JUST-2015-JCOO-AG-723180). Javier Del Ser is also grateful to the Basque Government for its support through the BID3ABI project.

References

1. Bell, J.E., McMullen, P.R.: Ant colony optimization techniques for the vehicle routing problem. Adv. Eng. Inform. **18**(1), 41–48 (2004)
2. Blazewicz, J., Lenstra, J.K., Kan, A.R.: Scheduling subject to resource constraints: classification and complexity. Discrete Appl. Math. **5**(1), 11–24 (1983)
3. Eiben, A.E., Ruttkay, Z.S.: Constraint Satisfaction Problems. IOP Publishing Ltd. and Oxford University Press, New York (1997)
4. Gao, K.Z., Suganthan, P.N., Pan, Q.K., Chua, T.J., Cai, T.X., Chong, C.S.: Pareto-based grouping discrete harmony search algorithm for multi-objective flexible job shop scheduling. Inf. Sci. **289**, 76–90 (2014)
5. Gonzalez-Pardo, A., Camacho, D.: Environmental influence in bio-inspired game level solver algorithms. In: Zavoral, F., Jung, J.J., Badica, C. (eds.) Intelligent Distributed Computing VII, pp. 157–162. Springer, Cham (2014)
6. Gonzalez-Pardo, A., Camacho, D.: A new CSP graph-based representation for ant colony optimization. In: IEEE Congress on Evolutionary Computation, pp. 689–696 (2013)
7. Gonzalez-Pardo, A., Camacho, D.: A new CSP graph-based representation to resource-constrained project scheduling problem. In: IEEE Congress on Evolutionary Computation, pp. 344–351 (2014)
8. Gonzalez-Pardo, A., Ser, J., Camacho, D.: On the applicability of ant colony optimization to non-intrusive load monitoring in smart grids. In: Puerta, J.M., Gámez, J.A., Dorronsoro, B., Barrenechea, E., Troncoso, A., Baruque, B., Galar, M. (eds.) CAEPIA 2015. LNCS (LNAI), vol. 9422, pp. 312–321. Springer, Heidelberg (2015). doi:10.1007/978-3-319-24598-0_28
9. Gonzalez-Pardo, A., Palero, F., Camacho, D.: An empirical study on collective intelligence algorithms for video games problem-solving. Comput. Inform. **34**(1), 233–253 (2015)
10. Kolisch, R., Sprecher, A.: PSPLIB-a project scheduling problem library: OR software-ORSEP operations research software exchange program. Eur. J. Oper. Res. **96**(1), 205–216 (1997)

11. Kumar, V.: Algorithms for constraint-satisfaction problems: a survey. AI Mag. **13**(1), 32 (1992)
12. Morin, S., Gagné, C., Gravel, M.: Ant colony optimization with a specialized pheromone trail for the car-sequencing problem. Eur. J. Oper. Res. **197**(3), 1185–1191 (2009)
13. Schirmer, A.: Case-based reasoning and improved adaptive search for project scheduling. Naval Res. Logistics (NRL) **47**(3), 201–222 (2000)
14. Tsang, E.P.K.: Foundations of Constraint Satisfaction. Computation in Cognitive Science. Academic Press, New York (1993)

Harmony Search and Telecommunications

An Analysis of Coalition-Competition Pricing Strategies for Multi-operator Mobile Traffic Offloading Using Bi-objective Heuristics

Jone Consul[1], Cristina Perfecto[1(✉)], Miren Nekane Bilbao[1], and Javier Del Ser[1,2,3]

[1] University of the Basque Country UPV/EHU, 48013 Bilbao, Spain
{cristina.perfecto,nekane.bilbao,javier.delser}@ehu.eus
[2] TECNALIA, 48160 Derio, Spain
[3] Basque Center for Applied Mathematics (BCAM), 48009 Bilbao, Spain

Abstract. In a competitive market relationships between telecommunications operators serving simultaneously over a certain geographical area are diverse and motivated by very different business strategies and goals. Such relationships ultimately yield distinct pricing portfolios depending on the contractual affiliation of the user being served. Furthermore a key role in the last decade is the concept of tethering (connection sharing) which, when controlled by the operator, may help alleviating the consumption of network resources in densely populated scenarios. In this work we investigate the application of bi-objective heuristics for the design of Pareto-optimal network topologies leading to an optimal Pareto between the revenue of the incumbent operators in the scenario and the quality of service degradation experienced by the end users as a result of tethering. Based on computer simulation this work unveils that such a Pareto-optimal set of topologies is strongly determined by the market relationships between such operators.

Keywords: Traffic offloading · Pricing · Multiobjective · Competition · Coalition

1 Introduction

In the last years the telecommunication sector has witnessed an upsurge of the number of operators concurrently offering services over the same geographical area [1]. This sharp increase has been specially notable in the provision of mobile services, mainly due to the liberalization of this market in several economies and the decoupling between service deployment and infrastructure management, the latter introducing the role of the so-called virtual operator [2]. As such, virtual operators do not follow the conventional approach of acquiring a spectrum license to deploy a mobile network and offer their services, but rather enter the market by exploiting the infrastructure of incumbent operators to offer Over-The-Top (OTT) services [3]. The coexistence of these market stakeholders with

J. Del Ser (ed.), *Harmony Search Algorithm*, Advances in Intelligent Systems and Computing 514, DOI 10.1007/978-981-10-3728-3_16

traditional network operators lay the basis for a strongly competitive mobile market analyzed from different techno-economic perspectives [4–9].

In this context the appearance of multiple virtual players in the telecommunication market has grown at a significantly higher pace than the deployment of new access infrastructure. Content providers and virtual operators have come along with new mobile usage patterns by end users, who have increased their demanded resources from the network as a result of lower prices of terminal devices and the ubiquitous access to high-quality contents. According to Cisco [10] an exemplifying growth of 74% in the global mobile data traffic has been registered in 2015. The need for allocating resources to accommodate the challenging data explosion in mobile networks is what has pushed operators not only to share not only infrastructure in terms of core and radio access networks (as done by the aforementioned virtual operators), but also to explore opportunistic methods that involve their users themselves.

To this end, traffic offloading refers to the family of mechanisms aimed at minimizing the amount of information delivered over licensed communication resources which among other strategies can be accomplished by opportunistically resorting to local wireless communication technologies and dynamic content caching [11]. The criteria adopted in mobile traffic offloading can be very diverse, from the most straightforward policies (i.e. the maximization of the end-user satisfaction or the minimization of network operating expenses) to more elaborated schemes dealing with energy consumption, offload/upload persistence or the social centrality of the user along its predicted mobility trace [12]. Disregarding the criterion adopted to this end, the result is that traffic is shifted to opportunistically set inter-device networks.

In this paper we postulate that opportunistic traffic offloading can also provide interesting benefits when implemented between users of different mobile operators. In particular our work can be framed in the context of user incentives, by which users allow sharing their bandwidth for either offloading traffic of the operator to which they are subscribed, or providing service to subscribers of other operators. In this latter case incentives must be provided at two different levels of the scenario: between different operators and from operators to those subscribers from their client portfolio that should act as opportunistic relays and share their resources. This manuscript will explore how such incentive agreements impact on the Pareto trade-off between the quality of service delivered to end-users and the costs incurred to implement such policies. In particular we will resort to multi-objective heuristics to seek the set Pareto-optimal multi-hop network configurations – i.e. which nodes should be promoted to relays– under different incentive ratios among the operators. We will show that when operators ally together by agreeing low resource sharing fees, their overall benefit increases with respect to a competitive scenario with higher sharing fees. This increased benefit, however, yields a degraded quality of service to the end-user which, in a practical scenario, should be lower bounded in a per application basis. The simulations results obtained from different scenarios and incentive policies will be discussed so as to support these conclusions.

The reminder of this paper is structured as follows: Sect. 2 will formulate the optimization problem that models the offloading of traffic among operators, which will be efficiently solved by means of the solution encoding approach and heuristic solver described in Sect. 3 and subsections therein. The performance of the overall scheme is analyzed and discussed in Sect. 4 based on computer simulations. Section 5 ends the manuscript by drawing several conclusions.

2 Problem Formulation

This section presents key concepts and introduces the notation used throughout the rest of the paper. Let us suppose an area of dimensions $X_{\max} \times Y_{\max}$, where N different operators offer their services so that the n-th operator – where $n \in \{1, \ldots, N\}$ – has $M(n)$ clients, i.e. each operator serves a total of $M(n)$ users with $\sum_{n=1}^{N} M(n) \triangleq M$ denoting the total number of users existing in the area. Each operator will have a constant and circular coverage area with radius $R(n)$, whose limits fall within the previously defined area $X_{\max} \times Y_{\max}$. Such operators are connected to a backhaul station as shown in Fig. 1. Nodes correspond to users' devices (e.g. smartphones, tablets, etc.) located at coordinates $\{(x_m, y_m)\}_{m=0}^{M}$, each with its contracted services signed with operator $O(m) \in \{1, \ldots, N\}$.

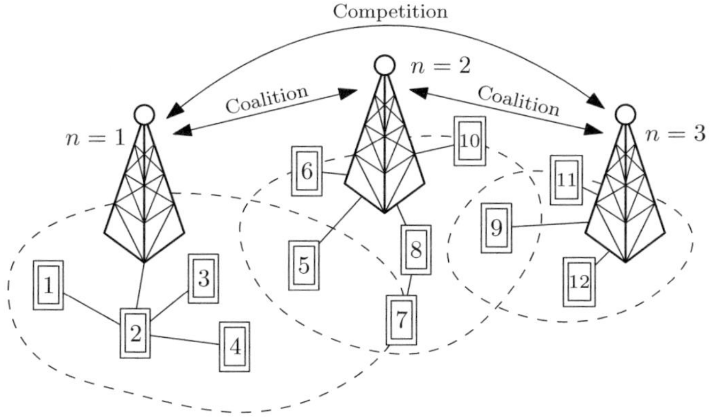

Fig. 1. Diagram of the scenario with $N = 3$ operators and $M = 12$ users. In this hypothesized situation user $n = 2$ is sharing his/her connection with users $n \in \{1, 3, 4\}$, which may (or may not) be clients of the same operator.

User devices are assumed to be equipped with at least two wireless network interfaces such that one of them connects to Internet contents through the operator's cellular base station (BS), whereas the other resorts to short-range wireless protocols (e.g. WiFi or Bluetooth) to share its bandwidth with other users as a tethering interface. All devices operate in full duplex mode so as to be able to send data to the BS and receive data from other nearby nodes if tethering is enabled. Users under service can be connected to the backhaul through three different ways, each characterized by diverse requirements regarding the perceived quality of service and the cost for both the operator and the user/client:

1. Via the operator contracted by the user: in this case the node will be connected directly to the operator with whom he/she has signed the contract. In this case the quality of service delivered to the user is expected to be contracted and ensured by the operator. For this rationale this direct connection policy yields the highest cost C_* for the operator among all the cases.
2. Via other third-party operator: in this second connection mode the node at hand will access the backhaul via an external operator with whom he/she has no signed contract. In this case a cost penalty must be taken into account to model the market relation between the operators in the management of each other's clients. A parameter $\alpha \triangleq C_*/C_{\circledast} \in [0,1]$ with $n \neq n'$ is defined such that
   - If $\alpha \approxeq 1$ the relationship between operators is a market coalition, i.e. the operational cost associated to the provision of services to any given user is approximately the same disregarding the operator with whom the user signed his/her contract (i.e. $C_* \approxeq C_{\circledast}$).
   - If $\alpha \ll 1$ operators will be assumed to compete under a hostile market relationship, namely, it will be significantly more costly for an operator to serve its contracted clients through the premises of the other operator than through its own network equipment.
3. Via tethering: in this last option a node will connect to its BS through the tethered connection open by another user by virtue of the short-range wireless interface of his/her device. This case is expected to maximize the benefit of the operator and the users: operators can serve the user via tethering at a reduced cost C_T modeling an incentive paid to the user sharing his/her connection, which is assumed to be lower than the cost C_* incurred when serving the user directly. However, the relayed connection is assumed to degrade the quality of service experienced by the tethered user in the form of processing delays and/or reduced connection bandwidth. The relay node may represent a client from the same operator or instead, belong to any other provider.

The network operator $n \in \{1, \dots, N\}$ obtains its benefit as the difference between the operational costs of providing service to its users and the monetary income yielded by their contracts $\{\Omega_m\}_{m=1}^{M}$. We simplify this cost as the sum of the costs associated to the price of the resources needed to establish the connection directly to the BS (collected in C_*), the price of serving a user through a connection of any other operator (resp. $C_{\circledast}$) and the cost of tethering a connection, in the form of an incentive C_T to the client owning the tethering device. If we denote such a profit as $B(n)$, we have that

$$B(n, \zeta) \triangleq \sum_{m \in M(n)} \Omega_m - C_m^{tot}(\zeta), \tag{1}$$

where the total cost per user $C_m^{tot}(\boldsymbol{\zeta})$ is given by

$$C_m^{tot}(\boldsymbol{\zeta}) \triangleq C_* I(\zeta(m) = \texttt{D}) + C_{\circledast} I(\zeta(m) = \texttt{O}) + C_T I(\zeta(m) = \texttt{T}), \tag{2}$$

with $I(\cdot)$ as an auxiliary indicator function taking value 1 if its argument is true (and 0 otherwise). In the above formulae $\zeta(m) \in \{\texttt{D}, \texttt{O}, \texttt{T}\}$ denotes the connection mode (`DIRECT`, `OTHER`, `TETHERED`) of user m as per the description above. This defined budget will encompass the first optimization goal tackled in this study via the heuristic refinement of the vector of connection modes $\boldsymbol{\zeta} \triangleq \{\zeta(m)\}_{m=1}^M$.

When the user at hand is tethered through the shared connection of any other user in the network, we will assume that the quality of service degrades in the form of a processing delay. The delay severity will be modeled as a numerical score whose value depends on (1) the number of hops from the user to the BS; (2) the number of simultaneously tethered users at the intermediate nodes that compose the path from the user to the BS. This numerical score $R(\boldsymbol{\zeta})$ will be set by the topology of the tree network modeling the connections between the underlying set of users and the BS deployed in the scenario, which in turn is given by the choice of connection modes $\boldsymbol{\zeta}$ for the compounding nodes of the network. This tree is rooted on the backhaul, with N first-level nodes representing the BS in the scenario, and intermediate/leaf nodes standing for the M users.

While other progression models for the delay can be assumed instinctively, in this work we will compute $R(\boldsymbol{\zeta})$ as the average of the transmission slots to be waited by every node in the network in the *best* and *worst* cases. Assuming a round robin scheduling policy among users tethered at the same device, the *best* case stands for the case where node m transmits directly through a multihop path to the BS without awaiting for any other's transmission. The latter (*worst* case) corresponds to the case when the node is scheduled for transmitting during the last transmission slot of every intermediate tethered set of users until the BS at hand. If $M_l^m(n)$ denotes the number of users at level l in the subtree rooted on BS $n \in \{1, \ldots, N\}$ such that $\sum_l \sum_m M_l^m(n) = M(n)$ and all nodes in $M_l^m(n)$ share the same parent node as m; and $l(m) \in \{1, \ldots, L_{\max}^n\}$ (with $n = O(m)$) is the level at which user m is located in the aforementioned subtree, this score particularized for BS n will be given as

$$R(n, \boldsymbol{\zeta}) = \frac{1}{M(n)} \sum_{m \in M(n)} \frac{(R_{wc}^m(\boldsymbol{\zeta}) + R_{bc}^m(\boldsymbol{\zeta}))}{2} = \frac{1}{M} \sum_{m=1}^{M} \frac{l(m) + M_1(n) \prod_{l'=2}^{l(m)} M_{l'}^m(n)}{2},$$

namely, as the average between the number of hops between m and its BS (best case) and the maximum number of scheduling slots that m should await for transmission (worst case). Based on this definition, the problem tackled in this paper can be formulated as the search for a K-sized set of optimal connection modes $\{\boldsymbol{\zeta}_k^\star\}_{k=1}^K$ such that the Pareto trade-off between the operators' revenue and the quality of service experienced by the users is differently balanced, i.e.

$$\{\boldsymbol{\zeta}^\star\}_{k=1}^K = \underset{\boldsymbol{\zeta} \in \{\texttt{D},\texttt{O},\texttt{T}\}^M}{\arg} [\max B(n, \boldsymbol{\zeta}), \min R(n, \boldsymbol{\zeta})], \tag{3}$$

which can be read as the maximization of the operators' benefit and the minimization of the delay experimented by the users. From an intuitive perspective such optimization objectives are conflicting: in order to reduce operational expenditures, an operator will prefer that their users share as much bandwidth as possible via tethering and a favorable mechanism for incentives. Likewise, direct connections will ensure a high quality of service for the users, but will go against the business goal of the operator at hand. By solving the above optimization problem the decision maker commanding the operator of the network can easily trade one objective for the other as a function of the business priorities and the contractual requirements established at the time. In the next section we will describe the heuristic solver designed to efficiently deal with the above problem.

3 Proposed Solver

To solve the problem formulated in the previous section we will resort to a multi-objective version of the Harmony Search (HS) algorithm, a music-based meta-heuristic optimization method first presented in [13] as a result of the observation of the music improvisation procedure undertaken by musicians. When seeking a perfect harmony musicians rely on both their memory of notes played in the past and random pitch variations, which are emulated in a computer program much alike crossover and mutation processes in Evolution are mimicked in Evolutionary Algorithms. Notes played by the musicians represent the values of the optimization variables, which are iteratively refined by applying the aforementioned operators until a stop criterion is met (e.g. a maximum number of iterations or a steady convergence of the fitness values along successive iterations). The good performance scores obtained by this heuristic has been evinced in many application scenarios [14], with several prior contributions in the telecommunication domain [15–18].

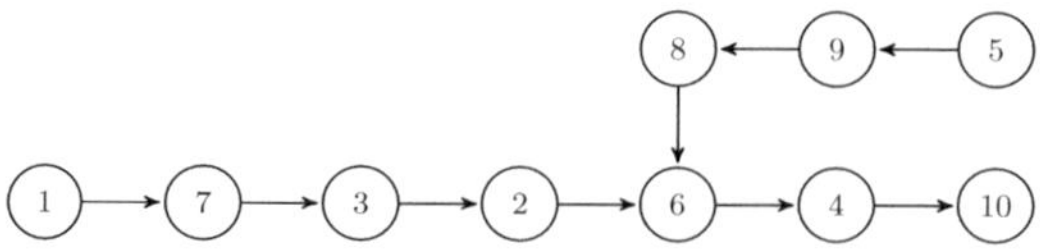

Fig. 2. Tree structure represented by the Dandelion string $\mathbf{C} = \{6, 2, 4, 9, 3, 7, 6, 8\}$.

In order to represent numerically the solutions to the problem in Expression (3) (i.e. $\boldsymbol{\zeta}$) the Dandelion code will be used to represent the tree-like network topology that jointly represents nodes connected to the backhaul under the three connection modes $\{\mathtt{D}, \mathtt{O}, \mathtt{T}\}$ considered in this work. It is important to observe that by evolving this tree the connection mode is determined depending on which nodes result to be connected to each other. The Dandelion code is a bijective mapping between a tree network topology of $M + N + 1$ nodes and an integer string with length $M+N-1$. This code has several properties in terms of inheritance and locality that make it suitable for tackling tree optimization problems by means of Evolutionary Algorithms [19], particularly in telecommunications [20–22]. A brief explanation of the encoding and decoding processes is now given:

- Dandelion encoding: Given a tree on n nodes $T \in \Gamma_n$, usually in the form of an adjacency list or connectivity matrix, with Γ_n denoting the set of possible trees interconnecting n nodes:
  - Step 1: list intermediate nodes on the path from 1 to n in T. Referring to the example tree given in Fig. 2, these are nodes 7, 3, 2, 6 and 4.
  - Step 2: find cycles in the list π by searching for limit elements, namely, elements larger than any other to their right. Such elements for π in the example list are 7, 6 and 4 and thus cycles are (7), $(3, 2, 6)$ and 4.
  - Step 3: the array A_C for tree T is constructed by filling its first row with elements 2, 3, 4, ..., $n-1$ and adding cycle-related information into its second row, i.e.

$$\mathbf{A}_c = \begin{bmatrix} 2 & 3 & 4 & 5 & 6 & 7 & 8 & 9 \\ 6 & 2 & 4 & - & 3 & 7 & - & - \end{bmatrix}.$$

  - Step 4: $C_{i+1} = \succ_i$ for every $i \in [2, n-1]$ and $\succ_x$ denoting x's successor relationship. Then, the Dandelion code C corresponding to tree T is given by the contents of bottom row of A_C.

$$\mathbf{A}_c = \begin{bmatrix} 2 & 3 & 4 & 5 & 6 & 7 & 8 & 9 \\ 6 & 2 & 4 & 9 & 3 & 7 & 6 & 8 \end{bmatrix}. \tag{4}$$

    By looking at the complete A_C representation of the example, the Dandelion code of the hypothesized tree T is $\mathbf{C} = \{6, 2, 4, 9, 3, 7, 6, 8\}$.
- Dandelion decoding: this procedure produces a tree $T \in \Gamma_n$ as follows:
  - Step 1: a $2 \times n-2$ matrix $\mathbf{A}_c$ is built by inserting the integer set $\{2, 3, \ldots, n-1\}$ in the first row and the elements of C in the second row. For the exemplifying code $\mathbf{C} = \{6, 2, 4, 9, 3, 7, 6, 8\}$, $\mathbf{A}_c$ is given as per (4)
  - Step 2: define $f_C : [2, n-1] \rightarrow [1, n]$ such that $f_C(i) = C_i$ for each $i \in [2, n-1]$. Note that $f_C(i)$ corresponds to the i-th position C_i of the code.
  - Step 3: cycles associated to f_C are computed as $\{Z_1, Z_2, \ldots, Z_L\}$. In the example 3 cycles, namely (2 6 3), (4) and (7), are obtained. Provided that b_l denotes the maximum element in Z_l (with $l \in \{1, \ldots, L\}$), cycles are then reordered such that b_l is set as the rightmost element of Z_l, and that $b_l > b_{l'}$ if $l < l'$. In words, cycles are circularly shifted so that the largest element is the rightmost and sorted so that cycle maxima decreases from left to right. In the example this step yields $\{Z_2, Z_1\} = \{(7), (3\,2\,6), (4)\}$.
  - Step 4: a list π of the elements in $\{Z_1, Z_2, \ldots, Z_L\}$ is composed in the order they occur in the cycle list, from the first element of Z_1 to the last entry of Z_L, i.e. $\pi = \{(1)(7)(3\,2\,6)(4)(10)\}$.

- Step 5: the tree $T \in \Gamma_n$ corresponding to C is constructed by arranging a set of n isolated nodes labeled with the integers from 1 to n. A path from node 1 to node n will be constructed by traversing the list π from left to right. An edge will be included between nodes i and C_i for every $i \in \{2\ldots, n-1\}$ not occurring in π. The tree corresponding to the Dandelion code $\mathbf{C} = \{6, 2, 4, 9, 3, 7, 6, 8\}$ is the tree given in Fig. 2.

The compounding steps of the proposed bi-objective HS solver are as follows:

1. Initialization: a pool of φ solutions with lengths $M + N - 1$ is initialized uniformly at random from the alphabet $\{2, \ldots, M+N\}$, which are evaluated according to the objectives as per (3).
2. Harmony improvisation: a new set of solutions is created from the previous set of harmonies by applying three stochastic operators: Harmony Memory Considering Rate (HMCR), Pitch Adjustment Rate (PAR) and Random Selection Rate (RSR), each driven by probabilistic parameters P_{HMCR}, P_{PAR} and P_{RSR}, respectively. We embrace the seminal definition of these operators proposed in [13] and extended in [23].
3. Fitness evaluation and memory update: the fitness values of the newly produced solutions is obtained and compared with those of the previous ones. As we deal with a bi-objective optimization problem a non-dominated sorting criterion based on dominance rank and crowding distance (similar to the one embedded in the well-known NSGA-II evolutionary solver [24]) is selected. Only the first φ solutions in the list of harmonies ordered by front rank (first) and crowding distance (second) will be kept for the next iteration.
4. Termination: steps 2 and 3 are repeated until a number of iterations $\mathcal{I}$ set beforehand are completed.

4 Experiments and Results

In order to assess the performance of the proposed solver when addressing the bi-objective optimization problem stated in (3), two different scenarios have been created with two operators providing services over the same geographical area. The relation between both operators is defined by α as explained in Sect. 2. In all cases the area is 50×50 with a density of 20 nodes/users per operator. Values of the coverage radii $\{R(n)\}_{n=1}^{2}$ have been dynamically adjusted in order to provide service to at least 95% of the deployed users. Parameters of the HS solver are set to $\varphi = 30$, $P_{\text{HMCR}} = 0.5$, $P_{\text{PAR}} = P_{\text{RSR}} = 0.1$ and $\mathcal{I} = 500$ iterations, with results averaged over 5 Monte Carlo experiments for each scenario. Incomes from service contracts are assumed to be $\Omega_m = 1200\ \forall m \in \{1, \ldots, M\}$, whereas costs are fixed to $C_* = 1000$ (direct connection), $C_T = 600$ (tethering incentive) and $C_{\circledast} = C_*/\alpha \in \{1000, 10000\}$. (i.e. $\alpha \in \{1, 0.1\}$). Regarding the latter it is important to note that $\alpha = 1$ will emulate a coalition market agreement between operators, whereas $\alpha = 0.1$ will correspond to a competitive market environment.

The discussion focuses on Fig. 3 and the statistics summarized in Table 1. As can be inferred from these obtained results, the income for the operators is

Table 1. Statistics of the obtained Pareto front estimations.

	$\min B(n,\zeta)$	$\max B(n,\zeta)$	$\min R(n,\zeta)$	$\max R(n,\zeta)$
Coalition ($\alpha = 1$)	4000.00	8800.00	10.43	884.98
Competition ($\alpha = 0.1$)	4400.00	8400.00	10.43	724.12

lower for the case when both implement a hostile pricing policy – i.e. a low α – to provide services to *external* users. One would expect that this reduced profit would come along with a remarkable improvement in terms of quality of service for the end users (i.e. lower values of $R(n,\zeta)$). So do the obtained values for this metric, but differences with respect to the coalition case are only found in its maximum value. Therefore, from these results it can be concluded that for the simulated scenarios, a coalition scenario between service providers is favorable for increasing their average benefit whenever the degradation of the quality of service is admissible under the contractual conditions of their users.

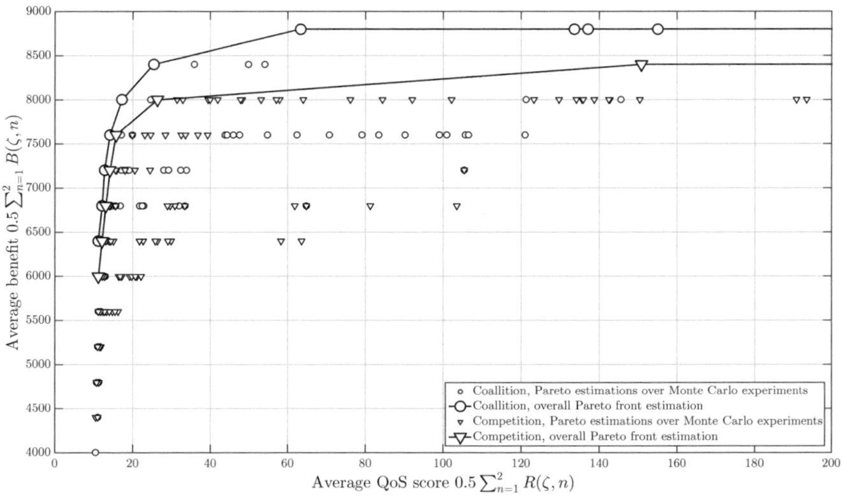

Fig. 3. Pareto Fronts trading average benefit for the quality of service offered to users under coalition and competition between operators.

5 Concluding Remarks

This manuscript has elaborated on analyzing the impact of different pricing agreements between service operators on the Pareto trade-off between their net income and the quality of service offered to their users. We have formulated this scenario as a bi-objective optimization problem, which relies on modeling the connection from users to the operators' network equipment as a tree graph that, in addition, accommodates the possibility of the operator to tether connections under incentive mechanisms. This graph is evolved towards Pareto-optimal configurations differently balancing quantitative metrics of the aforementioned optimization goals. The evolution is implemented efficiently by a heuristic solver that iteratively refines candidate solutions represented by means of the so-called Dandelion code, which possesses interesting features for evolving tree networks via crossover and mutation processes. Preliminary simulation results have evinced how such objectives behave when operators agree on a soft pricing policy for user roaming or, alternatively, impose hostile costs when processing users from any other counterpart.

Acknowledgments. This work is supported by the Basque Government through the ELKARTEK program, in particular the BID3A (ref. KK-2015/0000080) and BID3ABI projects.

References

1. Maitland, C.F., Bauer, J.M., Westerveld, R.: The European market for mobile data: evolving value chains and industry structures. Telecommun. Policy **26**(9), 485–504 (2002)
2. Kiiski, A.: Impacts of MVNOs on mobile data service market. In: 17th European Regional ITS Conference (2006)
3. Greene, W., Lancaster, B.: Over the Top Services. LTC International Inc. (2007)
4. Gruber, H.: Competition and innovation: the diffusion of mobile telecommunications in central and Eastern Europe. Inf. Econ. Policy **13**(1), 19–34 (2001)
5. Kiiski, A., Hämmäinen, H.: Mobile virtual network operator strategies: case Finland. In: ITS Conference (2004)
6. Smura, T., Kiiski, A., Hämmäinen, H.: Virtual operators in the mobile industry: a techno-economic analysis. NETNOMICS Econ. Res. Electron. Netw. **8**(1–2), 25–48 (2007)
7. Bertin, E., Noel, C., Michel, L.: A few myths about telco and OTT models. In: International Conference on Intelligence in Next Generation Networks (ICIN), pp. 6–10 (2011)
8. Xu, X., Chen, R.: Competition, cooperation, and pricing: how mobile operators respond to the challenge of over-the-top. Int. J. Market. Stud. **7**(6), 1–13 (2015)
9. Kibilda, J., Malandrino, F., DaSilva, L.A.: Incentives for infrastructure deployment by over-the-top service providers in a mobile network: a cooperative game theory model. In: IEEE International Conference on Communications, pp. 1–6 (2016)
10. Cisco visual networking index: global mobile data traffic forecast update, 20122017. Cisco, San Francisco, CA, USA. http://www.cisco.com/en/US/solutions/collateral/ns341/ns525/ns537/ns705/ns827/white_paper_c11-520862.html. Accessed Nov 2016
11. Wang, X., Chen, M., Taleb, T., Ksentini, A., Leung, V.C.: Cache in the air: exploiting content caching and delivery techniques for 5G systems. IEEE Commun. Magaz. **52**(2), 131–139 (2014)
12. Siris, V.A., Kalyvas, D.: Enhancing mobile data offloading with mobility prediction and prefetching. ACM SIGMOBILE Mobile Comput. Commun. Rev. **17**(1), 22–29 (2013)

13. Geem, Z.W., Kim, J.-H., Loganathan, G.V.: A new heuristic optimization algorithm: harmony search. Simulation **76**(2), 60–68 (2001)
14. Manjarres, D., Landa-Torres, I., Gil-Lopez, S., Del Ser, J., Bilbao, M.N., Salcedo-Sanz, S., Geem, Z.W.: A survey on applications of the harmony search algorithm. Eng. Appl. Artif. Intell. **26**(8), 1818–1831 (2013)
15. Forsati, R., Haghighat, A.T., Mahdavi, M.: Harmony search based algorithms for bandwidth-delay-constrained least-cost multicast routing. Comput. Commun. **31**(10), 2505–2519 (2008)
16. Del Ser, J., Bilbao, M.N., Gil-Lopez, S., Matinmikko, M., Salcedo-Sanz, S.: Iterative power and subcarrier allocation in rate-constrained orthogonal multicarrier downlink systems based on hybrid harmony search heuristics. Eng. Appl. Artif. Intell. **24**(5), 748–756 (2011)
17. Del Ser, J., Matinmikko, M., Gil-Lopez, S., Mustonen, M.: Centralized and distributed spectrum channel assignment in cognitive wireless networks: a harmony search approach. Appl. Soft Comput. **12**(2), 921–930 (2012)
18. Landa-Torres, I., Gil-Lopez, S., Del Ser, J., Salcedo-Sanz, S., Manjarres, D., Portilla-Figueras, J.A.: Efficient citywide planning of open wifi access networks using novel grouping harmony search heuristics. Eng. Appl. Artif. Intell. **26**(3), 1124–1130 (2013)
19. Thompson, E., Paulden, T., Smith, D.K.: The dandelion code: a new coding of spanning trees for genetic algorithms. IEEE Trans. Evol. Comput. **11**(1), 91–100 (2007)
20. Perfecto, C., Bilbao, M.N., Del Ser, J., Ferro, A., Salcedo-Sanz, S.: Dandelion-encoded harmony search heuristics for opportunistic traffic offloading in synthetically modeled mobile networks. In: Kim, J.H., Geem, Z.W. (eds.) Harmony Search Algorithm. Advances in Intelligent Systems and Computing, vol. 382. Springer, Heidelberg (2016)
21. Landa-Torres, I., Manjarres, D., Gil-Lopez, S., Del Ser, J., Salcedo-Sanz, S.: A preliminary approach to near-optimal multi-hop capacitated network design using grouping-dandelion encoded heuristics. In: IEEE International Workshop on Computer Aided Modeling and Design of Communication Links and Networks (CAMAD), pp. 85–89 (2012)
22. Perez-Bellido, A.M., Salcedo-Sanz, S., Ortiz-Garcia, E.G., Portilla-Figueras, A., Naldi, M.: A dandelion-encoded evolutionary algorithm for the delay-constrained capacitated minimum spanning tree problem. Comput. Commun. **32**(1), 154–158 (2009)
23. Mahdavi, M., Fesanghary, M., Damangir, E.: An improved harmony search algorithm for solving optimization problems. Appl. Math. Comput. **188**(2), 1567–1579 (2007)
24. Deb, K., Pratap, A., Agarwal, S., Meyarivan, T.A.M.T.: A fast and elitist multiobjective genetic algorithm: NSGA-II. IEEE Trans. Evol. Comput. **6**(2), 182–197 (2002)

Cost-Efficient Selective Network Caching in Large-Area Vehicular Networks Using Multi-objective Heuristics

Miren Nekane Bilbao[1(✉)], Cristina Perfecto[1], and Javier Del Ser[1,2,3]

[1] University of the Basque Country UPV/EHU, 48013 Bilbao, Spain
{nekane.bilbao,cristina.perfecto,javier.delser}@ehu.eus
[2] TECNALIA, 48160 Derio, Spain
[3] Basque Center for Applied Mathematics (BCAM), 48009 Bilbao, Spain

Abstract. In the last decade the interest around network caching techniques has augmented notably for alleviating the ever-growing demand of resources by end users in mobile networks. This gained momentum stems from the fact that even though the overall volume of traffic retrieved from Internet has increased at an exponential pace over the last years, several studies have unveiled that a large fraction of this traffic is usually accessed by multiple end users at nearby locations, i.e. content demands are often local and redundant across terminals close to each other, even in mobility. In this context this manuscript explores the application of multi-objective heuristics to optimally allocate cache profiles over urban scenarios with mobile receivers (e.g. vehicles). To this end we formulate two conflicting objectives: the utility of the cache allocation strategy, which roughly depends on the traffic offloaded from the network and the number of users demanding contents; and its cost, given by an cost per unit of stored data and the rate demanded by the cached profile. Simulations are performed and discussed over a realistic vehicular scenario modeled over the city of Cologne (Germany), from which it is concluded that the proposed heuristic solver excels at finding caching solutions differently balancing the aforementioned objectives.

Keywords: Network caching · Vehicular networks · Heuristics

1 Introduction

Content caching and delivery leverage the storage capabilities and wireless interfaces of user terminals to store concurrently and redundantly accessed contents. These functionalities are applied on the realistic assumption that a large percentage of such contents will be requested by nearby users in the near future. By eliminating the need for retrieving such contents from the core network the traffic flow demand from the telecommunication operator can be contained, which is crucial in light of recent forecasts around Internet stats with annual network traffic volumes expected to score above the zettabyte in 2016 [1]. This strategy,

J. Del Ser (ed.), *Harmony Search Algorithm*, Advances in Intelligent Systems and Computing 514, DOI 10.1007/978-981-10-3728-3_17

which finds its roots in the early stages of Internet (with local HTTP caches and proxies becoming the *de facto* standard [2,3]), has reborn in the last years as a means to reduce network traffic, bottlenecks, and user access latencies in 5G networks [4–6]. In-network caching, which refers to the deployment of content-driven caches through a given network, is one of the mechanisms that lie at the core of the Information-centric networking (ICN) paradigm, which aims at changing the Internet infrastructure from its traditional host-centric, connectivity based principle [7]. ICN envisages a network composed by named information entities (i.e. contents) whose labels are irrelevant with respect to their location over the network infrastructure. This radical swift involves that users consume contents rather than resources, hence information entities embody the resource unit around which users' network operations are planned [8].

To this end, many different in-network caching strategies have been proposed in the literature, each adopting different perspectives and approaches in regard to the cooperation between caches, the criterion to cache a certain content or the network topology over which such caches are deployed. The performance of content-oriented networking has been studied under different application scenarios, with recent yet scarce contributions analyzing the performance gains that an information-centric network operation yields in vehicular environments [9–11] and in general, mobility-based adhoc networks [12]. However, most previous studies rely on very simplistic mobility assumptions such as a Manhattan grid as the road layout and constant vehicle speeds, which oversimplify the network model and do not capture properly the short-lived, spotty nature of the connectivity in such a communication environment.

In this paper we analyze how multi-objective heuristics can be applied to the allocation of cache profiles through the network on the assumption that travel patterns of the vehicles demanding such contents have been inferred a priori by predictive models. We find our motivation to address content caches at a profile level in the dynamic essence of the vehicular scenario under analysis, which makes it difficult to learn or predict user's profiles in a file basis. In particular our work will assess the Pareto trade-off between the utility of the cache allocation strategy and its economical cost. On one hand, a quantitative measure of the former optimization objective will be formulated as a function of the coverage of vehicles, the number of vehicles demanding contents from cached profiles and the effective duration of the caching session delivered to the users under coverage. On the other hand, the economical cost of the cache deployment will vary depending on the cached profile and the level of service – in terms of number of simultaneously served users. Once the problem has been formulated we will resort to multi-objective heuristics in order to determine the set of Pareto-optimal caching strategies (namely, which profiles should be cached and served in the scenario) differently balancing the aforementioned goals. The performance of the proposed approach is discussed based on simulation results obtained from a realistic vehicular environment deployed over the city of Cologne, Germany.

The rest of the manuscript is structured as follows: first Sect. 2 will pose the notation and formulate the problem studied in this research work. Next, Sect. 3

describes the heuristic solver designed to tackle the optimization problem at hand. Section 4 presents and discusses the obtained simulation results and, finally, Sect. 5 concludes the paper and describes several future research lines of interest within the scope of this work.

2 Problem Formulation

According to the schematic diagram shown in Fig. 1, we assume an urban scenario S with N nodes $\{v_n\}_{n=1}^N$ representing vehicles with V2I (Vehicular to Infrastructure) communication capabilities. It is assumed that a priori probabilistic knowledge about the routes of the vehicles has been acquired by virtue of soft-output predictive models (e.g. [13, 14] and references therein), which can be modeled as a spatial-temporal probability density distribution $f_t^n(X,Y)$ such that (X,Y) denotes geographical coordinates (e.g. latitude and longitude) and $\sum_{(x,y)\in S} f_t^n(X,Y) = 1 \; \forall n \in \{1,\ldots,N\}$ and $\forall t \in [T_{in}^n, T_{out}^n]$. Here T_{in}^n and T_{out}^n stand for the time instants in which vehicle v_n enters and leaves the scenario, respectively. The coordinates for a given vehicle v_n at time will be given by $x_n(t)$ and $y_n(t)$. A circular coverage model with radii r_n will be adopted.

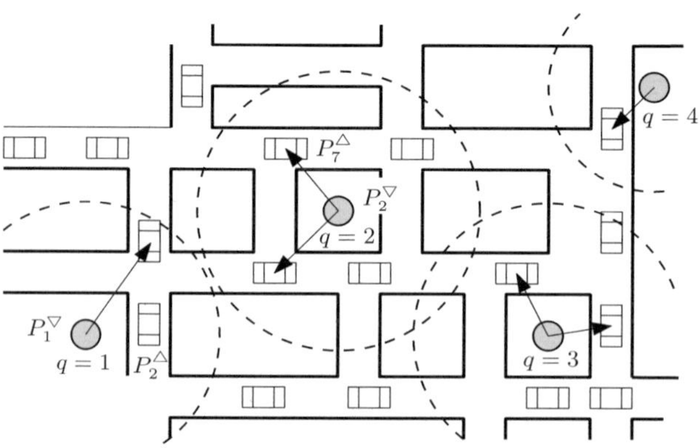

Fig. 1. Schematic diagram of the system model under consideration with $Q=4$ caches distributed over the scenario, $N=17$ vehicles and two cases highlighted: $P_1^{\triangledown} \neq P_2^{\triangle}$ (i.e. content stored at cache $q=1$ is not that demanded by vehicle $n=2$) and $P_2^{\triangledown} = P_7^{\triangle}$ (content stored at cache $q=2$ coincides with that requested by vehicle $n=7$).

The occupant at vehicle v_n is subscribed to a certain content profile $P_n^{\triangle}$ from an M-sized portfolio $\boldsymbol{\mathcal{P}} \triangleq \{\mathcal{P}_1,\ldots,\mathcal{P}_M\}$ designed and offered by the Internet Service Provider (ISP) at hand. These offered profiles are characterized by different resource levels demanded from the provider which, for the sake of simplicity, will be defined in terms of their required download rate R_m [Mb/sec]. For instance, profiles corresponding to high-definition video content will feature high values of R_m, whereas other services such as news feeds and blogs might correspond to content profiles with low values of R_m. Furthermore, it is assumed that each profile requires a minimum session establishment time T_m^{est} [sec], such that the delivery of cached content is effective only after T_m^{est} seconds since the vehicle demanding it entered the coverage area of the network cache from which the cached content is retrieved. In regards to the network cache we will assume Q locations $\{(x_q, y_q)\}_{q=1}$ over the scenario S at hand, each with coverage radii r_q [m] such that communication between vehicle n and cache q can be possible if the distance between both extremes at time t fulfills

$$d_{n,q}^t \triangleq \sqrt{(x_n(t) - x_q)^2 + (y_n(t) - y_q)^2} < \min\{r_n, r_q\}. \quad (1)$$

Due to the need for data storage over network caches a cost of the overall caching strategy will be modeled as a simple, linear cost per cached unit of data C_{unit} [monetary units/Mb]. By hereafter denoting the cached content at location $q \in \{1, \ldots, Q\}$ as $P_q^\nabla \in \boldsymbol{\mathcal{P}} \bigcup \{\emptyset\}$ – with $\emptyset$ corresponding to the case when no content is cached at location q, yielding no associated cost –, the overall cost $\Phi(\mathbf{P}^\nabla)$ of the caching strategy $\mathbf{P}^\nabla \triangleq \{P_q^\nabla\}_{q=1}^Q$ will be given by

$$\Phi(\mathbf{P}^\nabla, T) \triangleq \sum_{q=1}^{Q} T R_{P_q^\nabla} C_{unit}, \text{ [monetary units]}, \quad (2)$$

which will pose the first optimization goal of the problem tackled in this work. The second objective function will evaluate the inherent utility of the caching strategy $\mathbf{P}^\nabla$ as a function of several factors: (1) the rate of the cached profile, whose contribution to the overall utility should be higher as more data are offloaded from the mobile network; (2) the matching between the content profile of the vehicles within coverage of every network cache in the scenario and the profile cached therein; and (3) the effective session duration, which is determined by the overall time that every vehicle is within reach of the network cache penalized by the session establishment time of the cached profile to be retrieved. Mathematically, this utility $U(\mathbf{P}^\nabla, T)$ can be quantified as

$$U(\mathbf{P}^\nabla, T) \triangleq \sum_{q=1}^{Q} R_{P_q^\nabla} \sum_{n=1}^{N} I(P_n^\triangle = P_m^\nabla) \left(T_{n,q}^\triangleleft - N_{n,q}^\triangleleft T_{P_q^\nabla}^{est} \right), \quad (3)$$

where $T_{n,q}^\triangleleft$ denotes the total time during which vehicle n is within coverage of cache q, i.e.

$$T_{n,q}^\triangleleft \triangleq \sum_{t=1}^{T} I \left(d_{n,q}^t < \min\{r_n, r_q\} \right), \quad (4)$$

and $N_{n,q}^\triangleleft$ represents the number of total encounters between cache q and vehicle n, which can be quantified by

$$N_{n,q}^\triangleleft \triangleq \sum_{t=1}^{T} I \left(d_{n,q}^t \geq \min\{r_n, r_q\} \right) I \left(d_{n,q}^{t+1} < \min\{r_n, r_q\} \right), \quad (5)$$

with $I(\cdot)$ in the above formulae being an auxiliary indicator function taking value 1 if its argument is true (and 0 otherwise). By including the soft estimations of the route followed by each vehicle in the network represented by the space-temporal probability density function $f_t^n(X,Y)$, one obtains an estimation of the expected utility of the caching strategy $\mathbf{P}^{\triangledown}$ as $\mathbb{E}\{U(\mathbf{P}^{\triangledown},T)\}$, where expectation is taken over the aforementioned distribution of the vehicles' coordinates.

Based on the above notation the optimization problem undertaken in this work can be conceived as the discovery of a K-sized set of caching strategies $\{\mathbf{P}_k^{\triangledown,*}\}$ that optimally balances – in the Pareto sense of optimality – both their overall cost and utility over scenario S and a time horizon of duration T. Mathematically,

$$\{\mathbf{P}_k^{\triangledown,*}\}_{k=1}^K = \arg_{\mathbf{P}^{\triangledown}} \max \mathbb{E}\{U(\mathbf{P}^{\triangledown},T)\}, \; \min \Phi(\mathbf{P}^{\triangledown},T), \tag{6}$$

where the cardinality of the search space is given by $|\mathcal{P}\bigcup\{\emptyset\}|^Q = (M+1)^Q$, i.e. exponential with the number of network caches deployed and content profiles defined by the service provider. In other words, the goal is to determine which content profiles should be stored and served locally at caches $q \in \{1,\ldots,Q\}$ such that the expected utility of these cached profiles for the scenario at hand is maximum at a given cost level. Alternatively, $\{\mathbf{P}_k^{\triangledown,*}\}_{k=1}^K$ represent the caching strategies with minimal cost at different levels of expected utility. As explained in the next section, the exponential search space from which solutions to the above problem are drawn will be efficiently explored by a bi-objective HS solver.

3 Proposed Algorithm

The bi-objective network caching problem posed in Expression (6) can be efficiently approached using the Harmony Search (HS) algorithm, a meta-heuristic solver first proposed in [15] and applied thereafter to a plethora of optimization problems arising in diverse disciplines such as Energy [16–18], Telecommunications [19,20], Manufacturing [21,22] and Data Mining [23,24], among many others [25]. The search heuristic embodied in the HS algorithm is inspired by the music improvisation procedure and the progressive enhancement of the improvised melodies by musicians when pursuing aesthetically pleasant harmonies. From a computational standpoint the algorithm relies on a numerical solution encoding strategy, on which a set of operators is defined in a similar way to those featured by other meta-heuristic techniques from Evolutionary Computation and Swarm Intelligence [26]. In particular HS maintains a K-sized pool of candidate solutions or harmonies which are iteratively pushed towards regions of potentially higher optimality by virtue of the aforementioned operators. These operators emulate the memory-based and random pitch criteria by which musicians in practice vary their played notes (namely, optimization variables) until a stop criteria is met, e.g. a maximum number of improvisations $\mathcal{I}$ is reached.

For the problem at hand harmonies will encode a possible caching strategy as a Q-sized vector of integer numbers $\{\{X_q^k\}_{q=1}^Q\}_{k=1}^K$ such that $X_q^k \in \{0,1,\ldots,M\}$

correspond to the index of the cache profile proposed to be stored at cache $q \in \{1, \ldots, Q\}$ in harmony $k \in \{1, \ldots, K\}$. A null value (i.e. $X_q^k = 0$) reflects the case where no profile is allocated to the cache. Profile indices are sorted in increasing order of their bandwidth requirements R_m, sorting of utmost relevance for the effectiveness of the vicinity-based PAR operator explained in what follows. In regards to the HS flow diagram depicted in Fig. 2 three are the operators that are applied to the harmonies contained within the memory for each iteration:

- *Harmony Memory Considering Rate*, controlled by a probabilistic parameter HMCR $\in [0, 1]$ that sets the likelihood that the newly improvised value for a given note X_q^k is drawn from the values $\{X_q^1, \ldots, X_q^{k-1}, X_q^{k+1}, \ldots, X_q^K\}$ taken by the same note in the other $K - 1$ harmonies in the memory.
- *Pitch Adjusting Rate*, driven by the probability PAR $\in [0, 1]$, establishes the probability that the value of a given note X_q^k is replaced with any of its neighboring values in its corresponding alphabet $\{0, 1, \ldots, M\}$, i.e.

$$X_q^k \leftarrow \begin{cases} \max\{X_q^k - Z, 0\} & \text{with probability } 0.5\text{PAR}, \\ \min\{X_q^k + Z, M\} & \text{with probability } 0.5\text{PAR}, \\ X_q^k & \text{with probability1-PAR}, \end{cases} \tag{7}$$

  where Z is the realization of a uniform discrete random variable with support $\mathbb{N}[1, \lambda]$ (with λ denoting bandwidth).
- *Random Selection Rate*, whose controlling parameter RSR $\in [0, 1]$ sets the probability that the new value for a given note X_q^k will be drawn uniformly at random (i.e. without any neighborhood consideration) from its corresponding alphabet. This is equivalent to the PAR process with Z uniformly distributed over $\mathbb{Z}[-X_q^k, M - X_q^k]$.

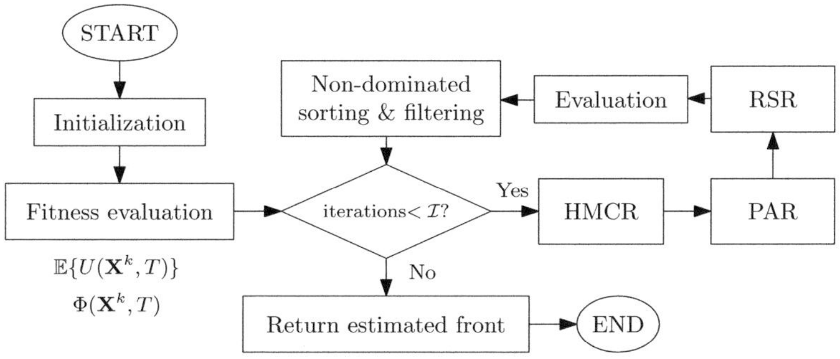

Fig. 2. Flow diagram of the proposed meta-heuristic solver.

These operators are sequentially applied to each note of the candidate solutions stored in the pool of harmonies. Once applied over the entire set of notes, the fitness functions defining the bi-objective caching criteria formulated in

Sect. 2 – namely, $\mathbb{E}\{U(\mathbf{X}^k, T)\}$ and $\Phi(\mathbf{X}^k, T)$, with $\mathbf{X}^k \triangleq \{X_1^k, \ldots, X_Q^k\} \equiv \mathbf{P}^\nabla$ – of the newly improvised solutions are evaluated. Based on the values of the objective functions for both the new harmonies and those from the previous iteration the entire set of $2K$ harmonies are ordered and filtered following the well-known Pareto dominance ranking and crowding distance criterion. To be concise, each solution is assigned a numerical score equal to its dominance level (namely, 1 for the best, non-dominated subset of solutions, 2 for the second best level, and so forth). Once all solutions have been ranked, a second criterion hinging on the sum of distances to the closest harmony along each metric drives the ordering among the solutions within a certain dominance level: as such, those solutions featuring large crowding distances are preferred rather than those with small distances to their neighboring individuals in the Pareto space. Once this second ordering has been performed only the best K solutions are kept in the pool of harmonies for the next iteration.

4 Experiments and Results

In order to assess the performance of the bi-objective heuristic solver explained in the previous section a vehicular scenario comprising realistic mobility traces over the city of Cologne (Germany) has been considered. A total of $N = 13682$ vehicles are deployed in this simulation scenario, which depart from and arrive at different spots of the city during a time horizon of $T = 3600$ s (from 8:00 to 9:00 AM). In order to shed more realism over the simulation the performed experiment utilizes the set of mobility traces released by TAPAS-Cologne [27], an initiative of the German Aerospace Center (ITS-DLR) to realistically reproduce urban vehicular traffic in this city. Caches are assumed to be located at the $Q = 247$ cellular base stations deployed in the city, which have been retrieved from public German databases as of 2012, and made available by the same initiative. Coverage radii are set to $r_n = 200$ and $r_q = 1000$ m for all vehicles and base stations, respectively. As for the cached content $M = 5$ profiles will be considered to model diverse cached services by means of different bandwidths $\{R_m\}_{m=1}^5 = \{0.01, 0.1, 0.2, 0.5, 1\}$ [Mb/sec]. The unitary cost per cached Mb is set to $C_{unit} = 100$ monetary units. The minimum session time for the content profiles will be given by $\{T_m^{est}\}_{m=1}^M = \{0, 0.05, 0.1, 0.15, 0.25\}$ [sec]. Without loss of generality a perfect estimation of the route followed by each vehicle will be assumed to reduce the computational complexity of the optimization algorithm: in this regard it should be noted that when dealing with soft estimations of the route followed by a car the caching optimization procedure would be identical, the difference being that the utility would require averaging over both space and time domains as per (6). The HS solver is configured with a pool of $K = 100$ harmonies, HMCR $= 0.7$, PAR $=$ RSR $= 0.1$, $\lambda = 1$ and a total of $\mathcal{I} = 200$ iterations. These parameters have been tuned via offline simulations over a value grid (not shown due to lack of space).

The obtained results are summarized in Figs. 3a through d, which depict the estimated Pareto front by the proposed HS solver along with three different solutions differently balancing the optimization criteria: utility and cost of the deployment. As we can observe, the caching strategies highlighted in the Pareto front span from the minimum-utility, minimum-cost region to solutions favoring high utility values at higher cost penalties. Intuitively one should expect that the former corresponds to a caching strategy where a large fraction of the caches located at the base stations do not store any content profile (especially those in the outskirts of the city to minimize the impact of the lack of caches), whereas the latter should feature a relatively higher number of high-rate cached profiles, mostly located in the city center where more vehicular traffic occurs.

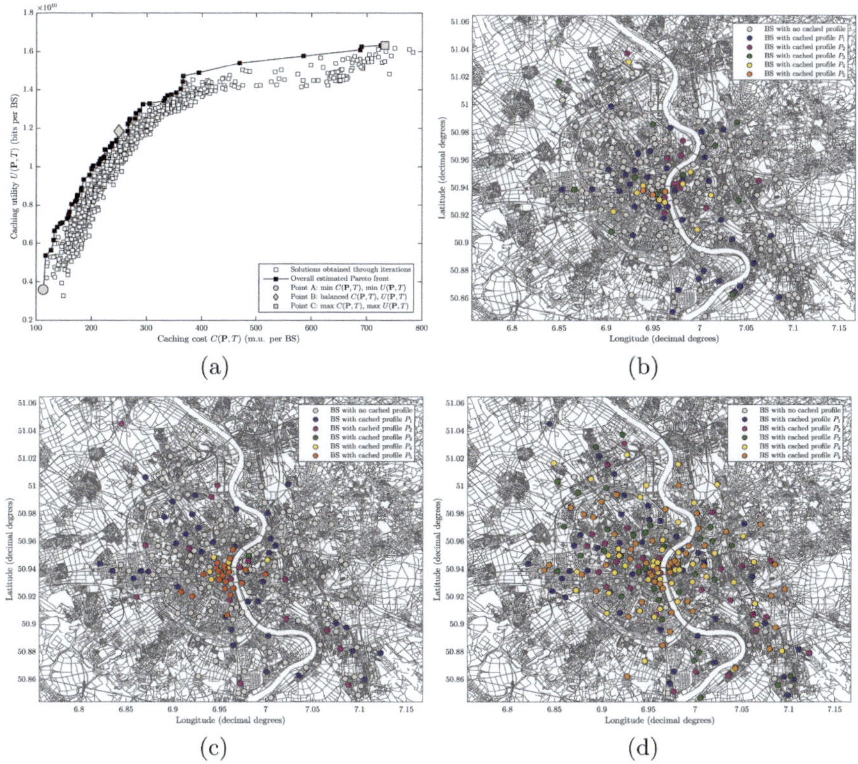

Fig. 3. (a) Estimated front by the bi-objective HS solver after $\mathcal{I} = 200$ iterations; (b) Network cache (solution) of point A in the estimated front; (c) point B; (d) point C.

This prior intuition gets confirmed through the obtained plots. To begin with, the cache deployment in Fig. 3b corresponds to Point A, namely, an strategy with minimum cost and consequently the lowest utility along the estimated Pareto front. It can be noticed that the deployment sets no contents in most caches through the urban scenario, while those activated by the caching solution concentrate around the city center and serve contents with generally low rate demands. As the strategy becomes less restrictive in terms of costs not only the set of cached contents is larger in terms of cardinality (i.e. more network caches with stored profiles are included in the strategy), but also are such strategies more diverse and sparsely distributed over the city, as Fig. 3c clearly shows for an intermediate solution in the Pareto front (Point B). Finally, Fig. 3d depicts the deployed caches associated to Point C, i.e. an strategy with no cost requirements and a maximal utility for the demands of the deployed vehicles. It can be observed that in this third example empty caches are kept to a minimum, with a large concentration of high-rate content profiles cached over the scenario.

5 Concluding Remarks and Future Research Lines

This work has capitalized on the paramount need for efficient mechanisms to optimally allocate network caches in urban environments so as to keep data traffic redundantly retrieved from nearby locations to their minimum. In this regard a novel formulation of the cost-efficient deployment of network caches over mobility-based scenarios has been proposed, which models both the coupling between the demanded content from the vehicle and the served profile by the network cache, but also the delay penalty due to the session establishment of the service at hand and the cost associated to the storage of the requested content. To efficiently solve the formulated problem in a centralized manner a heuristic technique based on a bi-objective version of the HS algorithm has been designed and put to practice over a realistic urban scenario. The Pareto-optimal caching strategies estimated by the proposed heuristic method meet the intuition and span a wide portfolio of caching options, which pave the way towards new formulations of this problem accounting for finite capacity constraints at the caches and/or mobile caches enabled at the vehicles.

Acknowledgments. This work has been supported by the Basque Government through the ELKARTEK program (ref. KK-2015/0000080) and the BID3ABI project.

References

1. Cisco: Visual networking index: forecast and methodology 2013–2018. White Paper (2014). http://www.cisco.com/go/vni
2. Luotonen, A.: Web Proxy Servers. Prentice Hall, Upper Saddle River (1997)
3. Wessels, D.: Web Caching. O'Reilly and Associates, Sebastopol (2001)
4. Wang, X., Chen, M., Taleb, T., Ksentini, A., Leung, V.C.M.: Cache in the air: exploiting content caching and delivery techniques for 5G systems. IEEE Commun. Mag. **52**(2), 131–139 (2014)
5. ElBamby, M.S., Bennis, M., Saad, W., Latva-Aho, M.: Content-aware user clustering and caching in wireless small cell networks. In: IEEE International Symposium on Wireless Communications Systems, pp. 945–949 (2014)
6. Syed, T., Bennis, M., Nardelli, P., Latva-Aho, M.: Caching in wireless small cell networks: a storage-bandwidth trade-off. IEEE Commun. Lett. (2016, to appear). http://www.sciencedirect.com/science/article/pii/S1570870516301019

7. Ahlgren, B., Dannewitz, C., Imbrenda, C., Kutscher, D., Ohlman, B.: A survey of information-centric networking. IEEE Commun. Mag. **50**(7), 26–36 (2012)
8. Xylomenos, G., Ververidis, C., Siris, V., Fotiou, N., Tsilopoulos, C., Vasilakos, X., Katsaros, K., Polyzos, G.: A survey of information-centric networking research. IEEE Commun. Surv. Tutor. **16**(2), 1024–1049 (2013)
9. Amadeo, M., Campolo, C., Molinaro, A.: CRoWN: content-centric networking in vehicular ad hoc networks. IEEE Commun. Lett. **16**(9), 1380–1383 (2012)
10. Amadeo, M., Campolo, C., Molinaro, A.: Enhancing content-centric networking for vehicular environments. Comput. Netw. **57**(16), 3222–3234 (2013)
11. TalebiFard, P., Leung, V.C.M., Amadeo, M., Campolo, C., Molinaro, A.: Information-centric networking for VANETs. In: Campolo, C., Molinaro, A., Scopigno, R. (eds.) Vehicular Ad Hoc Networks: Standards, Solutions, and Research, pp. 503–524. Springer, Cham (2015). doi:10.1007/978-3-319-15497-8_17
12. Liu, X., Li, Z., Yang, P., Dong, Y.: Information-centric mobile ad hoc networks and content routing: a survey. Ad Hoc Netw. (2016)
13. Xue, G., Li, Z., Zhu, H., Liu, Y.: Traffic-known urban vehicular route prediction based on partial mobility patterns. In: International Conference on Parallel and Distributed Systems, pp. 369–375 (2009)
14. Chen, L., Lv, M., Ye, Q., Chen, G., Woodward, J.: A personal route prediction system based on trajectory data mining. Inf. Sci. **181**(7), 1264–1284 (2011)
15. Geem, Z.W., Kim, J.-H., Loganathan, G.V.: A new heuristic optimization algorithm: harmony search. Simulation **76**(2), 60–68 (2001)
16. Ceylan, H., Ceylan, H., Haldenbilen, S., Baskan, O.: Transport energy modeling with meta-heuristic harmony search algorithm: an application to Turkey. Energy Policy **36**(7), 2527–2535 (2008)
17. Vasebi, A., Fesanghary, M., Bathaee, S.M.T.: Combined heat and power economic dispatch by harmony search algorithm. Int. J. Electr. Power Energy Syst. **29**(10), 713–719 (2007)
18. Salcedo-Sanz, S., Pastor-Sanchez, A., Del Ser, J., Prieto, L., Geem, Z.W.: A coral reefs optimization algorithm with harmony search operators for accurate wind speed prediction. Renew. Energy **75**, 93–101 (2015)
19. Del Ser, J., Bilbao, M.N., Gil-Lopez, S., Matinmikko, M., Salcedo-Sanz, S.: Iterative power and subcarrier allocation in rate-constrained orthogonal multicarrier downlink systems based on hybrid harmony search heuristics. Eng. Appl. Artif. Intell. **24**(5), 748–756 (2011)
20. Landa-Torres, I., Gil-Lopez, S., Del Ser, J., Salcedo-Sanz, S., Manjarres, D., Portilla-Figueras, J.A.: Efficient citywide planning of open WiFi access networks using novel grouping harmony search heuristics. Eng. Appl. Artif. Intell. **26**(3), 1124–1130 (2013)
21. Garcia-Santiago, C.A., Del Ser, J., Upton, C., Quilligan, F., Gil-Lopez, S., Salcedo-Sanz, S.: A random-key encoded harmony search approach for energy-efficient production scheduling with shared resources. Eng. Opt. **47**(11), 1481–1496 (2015)
22. Pan, Q.K., Suganthan, P.N., Liang, J.J., Tasgetiren, M.F.: A local-best harmony search algorithm with dynamic sub-harmony memories for lot-streaming flow shop scheduling problem. Expert Sys. Appl. **38**(4), 3252–3259 (2011)
23. Agustín-Blas, L.E., Salcedo-Sanz, S., Jiménez-Fernández, S., Carro-Calvo, L., Del Ser, J., Portilla-Figueras, J.A.: A new grouping genetic algorithm for clustering problems. Expert Syst. Appl. **39**(10), 9695–9703 (2012)
24. Karimi, Z., Abolhassani, H., Beigy, H.: A new method of mining data streams using harmony search. J. Intell. Inf. Syst. **39**(2), 491–511 (2012)

25. Manjarres, D., Landa-Torres, I., Gil-Lopez, S., Del Ser, J., Bilbao, M.N., Salcedo-Sanz, S., Geem, Z.W.: A survey on applications of the harmony search algorithm. Eng. Appl. Artif. Intell. **26**(8), 1818–1831 (2013)
26. Engelbrecht, A.P.: Fundamentals of Computational Swarm Intelligence. Wiley, Chichester (2006)
27. Uppoor, S., Trullols-Cruces, O., Fiore, M., Barcelo-Ordinas, J.M.: Generation and analysis of a large-scale urban vehicular mobility dataset. IEEE Trans. Mob. Comput. **13**(5), 1061–1075 (2014)

Harmony Search Based Algorithms for the Minimum Interference Frequency Assignment Problem

Yasmine Lahsinat[1(✉)], Dalila Boughaci[1], and Belaid Benhamou[2]

[1] LRIA-FEI/USTHB, 16111 Alger, Algeria
yasminelhsnt@gmail.com, dboughaci@usthb.dz
[2] Université Aix-Marseille, LSIS, Domaine universitaire de Saint Jérôme, Avenue Escadrille Normandie Niemen, 13397 Marseille Cedex 20, France
belaid.benhamou@univ-amu.fr

Abstract. The Minimum Interference Frequency Assignment Problem (MI-FAP) is a particularly hard combinatorial optimization problem. It consists in the assignment of a limited number of frequencies to each transceiver of the network without or at least with a low level of interference. In this work, we present an adaptation of the Harmony Search (HS) algorithm to tackle the MI-FAP problem. The results obtained by the adaptation of the classical Harmony Search algorithm are unsatisfactory. We performed a computation testing over some data sets of various sizes picked from public available benchmarks. The experimental results show that the conventional harmony search suffers from its premature convergence and therefore gets stuck in local optima. Even when it succeeds to escape from the local optimum, it does it after a long period of time. This make the process very slow. Due to these unconvincing results, we want to improve the Harmony Search algorithm's performances. To handle that, we propose some small changes and tricks that we bring to the original Harmony Search algorithm and a hybridization with a local search and the Opposition Based Learning (OPBL) principle. Here, we propose two strategies to improve the performances of the classical harmony search algorithm. We will show that both of them succeeds to enhance the performances of the harmony search in solving the MI-FAP. One of the proposed strategies gives as good results as those of the state of the art for some instances. Nevertheless, the method still need adjustment to be more competitive.

Keywords: Harmony search · MI-FAP · Optimization · Local search · OPBL

1 Introduction

The Minimum Interference Frequency Assignment Problem (MI-FAP) is a challenging combinatorial optimization problem. It arises in the modern wireless

J. Del Ser (ed.), *Harmony Search Algorithm*, Advances in Intelligent Systems and Computing 514, DOI 10.1007/978-981-10-3728-3_18

telecommunication field and has practical relevance in military and civil applications. The frequency assignment problem was first introduced by Metzeger in the sixties [20]. Since that, it has been well studied by researchers [1,7,16]. The MI-FAP is a generalization of the graph coloring problem [10]. In theory, the MI-FAP is well known to be NP-Hard [10].

Informally, we consider a number of transceivers ($TRXs$) communicating over a fixed number of frequencies. When stations in a close proximity use similar, or adjacent channels, or don't respect the predefined required frequency distance, they interfere with one another. Thus, The system designers seek to assign a frequency to each transceiver in such a way to minimize the interference level.

From a graph-theoretical point of view the MI-FAP can be defined as an undirected double weighted graph $G(V, E, S, P)$ where V is the set of vertices representing the transceivers ($TRXs$), E a set of edges that are pairs (u, v) of ($TRXs$) being constrained. Each edge (u, v) is affected with two edge weights that are elements of S respectively P. That is, S is the set of edge weights s_{uv} defining the separation constraints between the frequencies of the transceivers u and v and P is the set of edge weights (p_{uv}) corresponding to the penalty measuring the degree of interferences due to the assignment of a frequency to each of the transmitters u and v. The penalty p_{uv} occurs when the separation constraint s_{vu} of the assigned frequencies to u and v is not respected. The cost of a frequency assignment to a transmitter u is equal to the total interference caused with the other transmitters. The goal in the MI-FAP is to find an assignment of frequencies to the ($TRXs$) that avoids or minimizes the interferences and which in practice translate to allocating frequencies to the ($TRXs$) with an adequate separation between them. The formal modeling of the MI-FAP is given in Sect. 2.

Various methods and approaches have been proposed to address the MI-FAP. For instance, we can find evolutionary approaches and Ant Colony optimization algorithms [17], Tabu search and path relinking [5,15,21,22], simulated annealing algorithms [4,6], a Branch and cut algorithm [8], a Cultural algorithm [2], hybrid approaches [18,19] and hyperheuristics [14], and others methods like [3,12,13, 24]. The previous list is not exhaustive, several other works exist.

In this paper, we propose an adaptation of the Harmony Search (HS) Algorithm to solve the MI-FAP (Sect. 3). We study its ability and analyze its performances to solve this problem. Then we try to improve the performances of the Harmony Search Algorithm when dealing with the MI-FAP (Sects. 3.1 and 3.2). We make some changes to the original algorithm. First, we adopt a harmony search based algorithm that generates several new harmonies at each generation instead of one, then the best ones are selected to the next generation. Secondly, we investigate the possibility to enhance the performances through a hybridization between the HS, a Local Search (LS) to intensify and accelerate the search. To ensure a good exploration of the search space we strengthen this hybrid method with an approach based on the Opposition Based learning (OPBL). All these option methods are evaluated and compared on various publicly available benchmark problems. We analyze, compare and discuss the results and the performances obtained by the developed methods to solve the MI-FAP in Sect. 4. We conclude and give perspectives in Sect. 5.

2 Problem Description

This section gives the readers a basic understanding of the Minimum Interference Frequency Assignment Problem. A frequency assignment of the MI-FAP is a mapping f defined from $TRXs$ to F (f : TRXs$->$ F) where $TRXs$ is the set of transceivers and F the set of operational frequencies. A solution of the MI-FAP is a frequency assignment to all the $TRXs$ which satisfies all the separation constraints of the radio network. A solution X of the problem is represented by a one dimensional vector: $X = \{f_1, f_2, f_i \dots f_d\}$, where d is the number of the $TRXs$ of the network. Each element $f_i \in F$ of X corresponds to the frequency assigned to a ith transceiver. The value of f_i is an integer from the interval $[1, NF]$ /with NF being the total number of frequency. To evaluate the quality of a frequency assignment $X = \{f_1, f_2, f_i \dots f_d\}$ $X \in Sol$, we calculate the degree of interference as follows:

$$Cost(X) = \sum_{e(i,j) \in E; |f_i - f_j| <= s_{ij}} p_{i,j}. \quad (1)$$

The objective is to find an assignment of frequency that minimizes the sum of penalties $p_{i,j}$ induced by the non adequate separation between the constrained transceiver $TRX(i,j) \in E$ for which $|f_i - f_j| <= s_{ij}$.

If Sol is the set of all the solutions of the given MI-FAP, the optimal value of the objective function is $Cost^* = MIN_{X \in Sol}(Cost(X))$ and the corresponding optimal solution is noted X^*.

3 The Adapted Harmony Search Algorithm for the MI-FAP

The Harmony Search (HS) algorithm is a relatively new developed metaheuristic introduced by Geem in 2001 [9]. It is a global optimization technique that has been used to solve various optimization problem since then. The harmony Search is a population based metaheuristic inspired from the improvisation process followed by musicians playing together to create a nice and beautiful piece of music with perfect harmony. To produce a pleasing harmony the musicians composing the orchestra adjust several times the pitch of their instruments.

The designer of the HS algorithm formalize this process of improvisation into an optimization technique. Thus, the five main steps of the classical HS algorithm are as follows:

- **Step 1:**The first step initiates the algorithm parameters which consists in:
  - The harmony memory size (HMS) which represents the number of harmonies (HM) stored in the Harmony memory. Each harmony within HM is a potential solution and corresponds to a frequency plan. The Harmony Memory is the population manipulated by the Harmony Search, it is given

as a matrix $HMS \times d$: (HMS) is the number of row, each row represents a harmony. As mentioned before d is the number of (TRX) in the network.

$$HM = \begin{bmatrix} f_1^0 & \dots & f_d^0 \\ f_1^1 & \dots & f_d^1 \\ \dots & \dots & \dots \\ f_1^{HMS} & \dots & f_d^{HMS} \end{bmatrix} \quad (2)$$

- The harmony memory consideration rate: $(HMCR)$, $HMCR \in [0, 1]$.
- The pitch adjusting rate (PAR), $PAR \in [0, 1]$.

These two parameters are used during the improvisation step. Thus, they are considered as the main parameters governing the global and local search of the Harmony Search algorithm.

- **Step 2**: Initialization of the Harmony Memory (HM): in this step, the algorithm, randomly generates the sets of harmony, evaluates each harmony using the objective function and store each created harmony in the Harmony Memory.
- **Step 3**: Improvisation of a new harmony (solution): To improvise a new harmony (Frequency plan) $H = (f_1, f_2, f_i \dots f_d)$ three possible choices exist: (1) memory consideration, (2) random selection, (3) pitch adjustment. The Harmony Memory Consideration rate $HMCR$ is the probability of affecting the $i - th$ decision variable of the $i - th$ column within the HM to the $i - th$ decision variable of the new created harmony H. Otherwise, the $i - th$ decision variable of H is randomly chosen, with (1-HMCR) probability. When the value of the $i - th$ decision variable of H is chosen from HM, we use the PAR to decide whether it needs adjustment or not. To adjust the value of this variable we use a simple and basic strategy which consists in adding 1 or (with equal probability) subtracting 1 to the actual value.

  For reminder the value of a decision variable i corresponds to a frequency $f \in [1, NF]$. Thus, in case the interval bounds are crossed after the adjustment two possible choices are given:

$$f_i = \begin{cases} NF \text{ si } f_i < 1 \\ 1 \quad \text{si} f_i > NF \end{cases} \quad (3)$$

- **Step 4**: Here the algorithm updates HM if a better solution is found. It evaluates the new harmony by using the objective function, then store it in the HM if it is better than the worst solution of HM.
- **Step 5**: The algorithm repeats step 3 and 4 until the stopping criterion is met.

3.1 The First Improved Harmony Search Strategy

In the first strategy, namely the improved Harmony Search strategy one (IHS-1), designed to improve the performances of the (HS) for solving the MI-FAP, we propose to bring a small change in the HS routine. The classical HS starts with generating the HM, then creates at each generation one new harmony, evaluates its quality and store this new harmony in the HM in the case it is better than the worst one. We make some changes in the step of the creation of the new harmony. Our proposition is to generate at each iteration N_{new} new harmonies instead of one solution, then evaluate their quality. The next step consists in selecting the HMS unique best harmonies from the new created harmonies and the one stored in the HM. Further, we update the HM with these selected harmonies and use them to generate the next new (N_{new}) harmonies. We repeat these steps until the stopping criterion is met. This process allows a deeper study of the local information that is present in the HM. Indeed, creating many harmonies from the current HM allows a better exploitation and bring diversity because of the miscellaneous solutions.

3.2 The Second Improved Harmony Search Strategy

In the second strategy, namely the improved Harmony Search strategy two (IHS-2) we investigate the possibility to merge between two different techniques. Thus, we propose to enhance the process with a Local Search(LS) and use the Opposition Based learning(OPBL). We merge the HS and these last two methods as follows: The process starts with the (HSI-1), HM is then generated randomly and each harmony within is evaluated. The Best Solution is saved in $BestSol$, then, N_{new} harmonies are created and evaluated. After that, we keep the best harmonies HMS for the next generation. Now, one has to check if a better solution than $BestSol$ exists in HM following these steps:

- Sort the harmonies in HM from the best to the worst.
- ***If*** $(Cost(HM^0) < Cost(BestSol))$ ***Then*** $BestSol = (f_1^0, f_2^0, f_i^0 \dots f_d^0)$.
- ***Else if*** no new best is found ***Then*** call the Local Search (LS).

The (LS) is used in case the (IHS-1) failed in improving the actual best solution.

The Local Search, used within (IHS-2) is an iterative procedure which starts with an initial solution X picked from the Harmony Memory. This solution is improved by applying a move which consists in selecting randomly from the current solution a frequency $f_{(i)}$ assigned to one transceiver. We choose after that a new frequency $f_{(j)}$ to be assigned to the selected transceiver. We evaluate the quality of this new solution *X'* and replace X by *X'* if it is better. This process is repeated for a certain number of iterations in order to improve the current solution.

The use of the (LS) is for ensuring a better intensification and accelerating the process of convergence. Although, the introduction of the LS improve the results, but leads us to a local optimum. Further, the improvement of the actual best is not guaranteed every time. So we propose to exploit the information existing in the opposite solutions of HM. That is, when the (LS) failed to improve the actual best we call the (OPBL).

The Opposition-based learning (OPBL) is proposed in 2005 by Tizhoosh [23]. The idea behind the (OPBL) is to exploit the potential of information contained in a candidate solution and its opposite. The (OPBL) improve the ability of exploration of the global search and also avoid the premature convergence.

To find the opposite of a corresponding solution, we replace a decision variable within the solution by its opposite. The opposite of each frequency f_i within a solution X is defined as follows: Let $f_i \in [1, NF]$, the opposite f_i' of f_i is defined as:

$$f_i' = 1 + NF - f_i \quad (4)$$

The overall method (IHS-2) is as follows:

- Call (IHS-1)method.
- Sort the harmonies in HM from the best to the worst.
- ***If*** $(Cost(HM^0) < Cost(BestSol))$ ***Then*** $BestSol = HM^0$.
- ***Else if*** no new best is found ***Then*** call the Local Search (LS).
  - ***If*** (LS) does not improve the actual best ***Then*** replace each candidate solution in HM by its opposite point using the formula (2).

4 Experimental Study

All the proposed methods have been implemented using the C language programming. All the experiments are performed using a personal computer with an Intel core i5 processor and 4 GB of RAM under Windows 7. To show the effectiveness of the developed methods, we conducted our study on realistic data of various sizes taken from an available public benchmark, which can be found on-line at [11]. The best known results for these instances were obtained by an improved tabu search proposed by Montemanni et al. [21]. Further, Montemanni and Smith proposed in 2010 an algorithm named Heuristic Manipulation Technique (HMT) in [22]. Recently, Lai and al. obtained very good results with different strategies based on a path relinking algorithm [15]. We summarize in Table 1 the characteristics of the selected instances.

Due to the non-deterministic nature of the algorithms, 10 runs have been considered for each algorithm. To make a fair comparison between the proposed approaches, the running time is fixed to: 2400 s. We choose 2400 s as a stopping criterion simply because it is the time used by the reference algorithms [15,21,22]. The computational results on the tested benchmark instances are summarized in Table 2. The average and the best cost value found by each method are reported. The average value corresponds to the solution quality found by each algorithm on 10 runs. The best column corresponds to the best cost value obtained by each algorithm on 10 runs. The best results, in term of mean and best value, are highlighted in bold font. The row T corresponds to the average running time in seconds upon the data set's benchmark. The last column indicates the best known result. The first column (Instances) indicates the instance name and the second (F) column gives the number of possible frequency. Let's give some details of the parameters of the Harmony Search algorithms based methods. The values of the two main parameters of HS are fixed experimentally: $HMCR = 0.9$, and $PAR = 0.1$. We fixed $HMS = 25$ for the

Table 1. The characteristics of the instances

Instances	V (Number of $TRXs$)	E (number of constrained $TRXs$)	p_{uv} (Average penalties)
AC-45-17	45	428	1
AC-45-25	45	801	1
AC95-9	95	781	1
GSM-93	93	1073	1
P06-3	153	9193	1
GSM-246	246	7611	1
1-1-50-75-30-2-50	75	835	10.81
1-4-50-75-30-2-1	75	835	1

three methods, $n_{New} = 20$ for the (IHS-1) and (IHS-2). The number of iteration for the local search used in (IHS-2), is fixed at V which corresponds to the number of $(TRXs)$ for each instance.

As seen in Table 2, the (IHS-2) method gives better results than both HS and (IHS-1) for almost all of the instances. The quality solution of the obtained results from (IHS-2) outperforms those obtained by basic HS and (IHS-1) in all the instances except for the instance 1-1-50-75-30-2-50 where the algorithm gets lost in a local optimum and can not escape. The (IHS-2) matched the best results known for the instances AC-45-17(7), AC-45-17(9), AC-45-25 and 1- 4-50-75-30-2-1. Furthermore, (IHS-2) found results that are close to the state of the art results for the instance (AC-95-9) but failed for the others. The HS method is outperformed by the (IHS-2). The (IHS-1) shows a slight better ability on all instances than (HS), except for the AC-45-25 and GSM-246 instances. Indeed, we see that the HS finds the same results than (IHS-1) in term of average value, and a better best value for the instance GSM-246 and gives slight better results in term of average and best value for the instance AC-45-25. Although it is worth mentioning that for some instances results of HS and (IHS-1) are comparable.

Nevertheless, one can observe from Table 2 that the distance between the best solution quality and the average value of solution is slightly enlarged for the HS compared to both (IHS-1) and (IHS-2) for almost all the instances. Thus, we can say that (IHS-1) and (IHS-2) show a better stability of solution quality.

From Table 2 we can say that when comparing the average running time of the three proposed methods, we notice that on average no improvement in the quality solution is noticed after 600 s for the (IHS-2) compared to both HS and (IHS-1). For both HS and (IHS-1) the results show that there are improvements in the quality solution after a long running time. So this observation leads us to say that these two methods may give better results, but due to their slow convergence they need more time.

Table 2. The results of the HS, (IHS-1) and the (IHS-2) algorithms on the 8 benchmark instances, including the best and average values and running time over 10 independent runs.

Instances	|F|	Results	HS	(IHS-1)	(IHS-2)	Best Found
AC-45-17	7	Best	37	35	**32**	32
		Average	38.6	35.5	**32**	
		T	723.4	547.3	66.6	
AC-45-17	9	Best	18	17	**15**	15
		Average	19.2	18.4	**15**	
		T	1010.4	1052.6	922.2	
AC-45-25	11	Best	36	37	**33**	33
		Average	37.4	38	**33**	
		T	965.4	779.2	123.8	
AC-95-9	6	Best	45	45	**35**	31
		Average	49.2	46	**36**	
		T	1663	1485	562.33	
GSM-93	9	Best	47	50	**41**	32
		Average	61.5	51.75	**43.5**	
		T	1559.25	1933	142	
GSM-246	21	Best	305	309	**130**	79
		Average	312.66	312.66	**132**	
		T	1867.33	2331	340.2	
P06-3	31	Best	236	240	**135**	115
		Average	264.5	252	**142**	
		T	2022	1739	106.6	
1-1-50-75-30-2-50	5	Best	1317	**1298**	1402	1242
		Average	1378	**1327.25**	1451	
		T	962.66	1371.3	170	
1-4-50-75-30-2-1	6	Best	78	78	**70**	70
		Average	81	80	**70**	
		T	1512	1228	376.7	

Even though our methods failed to find the results obtained in [15,22] for all the checked instances, they confirm that introducing both Local Search and Opposite Based learning (in IHS-2) enhances the performances of the studied method, and reduces the gap with the best results found in the state of the art for some instances and reach a such performance for the other ones. This proves that introducing the local search and OPBL (In the IHS-2 method) has a positive impact on the improvement of the solution found. One can say that the (IHS-2) engenders a good equilibrium between the two main concepts governing the performances of any metaheuristics namely the intensification and the diversification. Since, the LS is called when the HS failed to improve the results, thus allowing to intensify the search around each solution. This ensures a better exploitation. Further, the presence of the OPBL allowed the exploration of divers point in the search space which brings diversity and then avoid being stuck in a local optimum.

The results and interpretations presented above lead to the conclusion that the (IHS-2) method is interesting and could be much more effective with a better implementation of the HS and a deeper study of its parameters.

5 Conclusion

In this work, we have treated the Minimum Interference frequency assignment problem in radio networks. To achieve this, we propose a solution using the Harmony Search Algorithm which is an interesting metaheuristic that kept our attention due to its inspiration from music. We are curious to study the specifications of this technique and investigate its ability to solve the MI-FAP. Thus we adapted, implemented and studied the Harmony Search algorithm for the MI-FAP. Further, we investigated the possibility to improve its performance on the MI-FAP. Aiming at enhancing the performance of the Harmony search when applied to the Minimum Interference frequency assignment problem in radio networks. We proposed two different variants of the latter that are: the Improved Harmony Search (1) (IHS-1) and the Improved Harmony Search (2) (IHS-2). The first one consists in creating a number of harmonies instead of one harmony at each generation. For the second strategy, we propose a collaboration between the (IHS-1), a local search and the opposite based learning.

The results obtained with the adapted HS to solve the MI-FAP are not really convincing. Nevertheless, the results presented in this study are only preliminary results. So it is clear that much work remains to be done to improve the results and ensure a good balance between intensification and exploration. Thus avoiding slow convergence and getting trap in the local optimum. The experiments show clearly that the (IHS-2) gives better results than both the HS and (IHS-1) methods. There are also many problems for (IHS-2) which need to be improved. Further work will be on the combination of LS, OPBL and HS more effectively, and at the same time investigate the possibility to add other tricks in order to improve the algorithm efficiency. It is highly imperative that our future works will be directed to find and eliminate the drawbacks that hinder the harmony search based algorithms to reach the best known results for all the instances.

References

1. Aardal, K.I., Van Hoesel, S.P., Koster, A.M., Mannino, C., Sassano, A.: Models and solution techniques for frequency assignment problems. Ann. Oper. Res. **153**(1), 79–129 (2007)

2. Alami, J., El Imrani, A.: Using cultural algorithm for the fixed-spectrum frequency assignment problem. J. Mob. Commun. **2**(1), 1–9 (2008)
3. Audhya, G.K., Sinha, K., Mandal, K., Dattagupta, R., Ghosh, S.C., Sinha, B.P.: A new approach to fast near-optimal channel assignment in cellular mobile networks. IEEE Trans. Mob. Comput. **12**(9), 1814–1827 (2013)
4. Beckmann, D., Killat, U.: Frequency planning with respect to interference minimization in cellular radio networks. Rapport technique, COST, 259 (1999)
5. Dorne, R., Hao, J.K.: Tabu search for graph coloring, T-colorings and set T-colorings. Meta-heuristics 77–92 (1999)
6. Duque-Anton, M., Kunz, D., Ruber, B.: Channel assignment for cellular radio using simulated annealing. IEEE Trans. Veh. Technol. **42**(1), 14–21 (1993)
7. Eisenblatter, A.: Frequency Assignment in GSM Networks: models, heuristics, and lower bounds. PhD Thesis, Technische Universitat Berlin (2001)
8. Fischetti, M., Lepschy, C., Minerva, G., Romanin-Jacur, G., Toto, E.: Frequency assignment in mobile radio systems using branch-and-cut techniques. Eur. J. Oper. Res. **123**(2), 241–255 (2000)
9. Geem, Z.W., Kim, J.-H., Loganathan, G.V.: A new heuristic optimization algorithm: harmony search. Simulation **76**(2), 60–68 (2001)
10. Hale, W.K.: Frequency assignment: theory and applications. Proc. IEEE **68**(12), 1497–1514 (1980)
11. http://www.idsia.ch/roberto/FAP08.zip
12. Del Ser, J., Matinmikko, M., Gil-Lopez, S., Mustonen, M.: Centralized and distributed spectrum channel assignment in cognitive wireless networks: a harmony search approach. Appl. Soft Comput. **12**(2), 921–930 (2012)
13. Del Ser, J., Bilbao, M.N., Gil-Lopez, S., Matinmikko, M., Salcedo-Sanz, S.: Iterative power and subcarrier allocation in rate-constrained orthogonal multicarrier downlink systems based on hybrid harmony search heuristics. Eng. Appl. Artif. Intell. **24**(5), 748–756 (2011)
14. Lahsinat, Y., Benhamou, B., Boughaci, D.: Trois hyper-heuristiques pour le problème d'affectation de fréquence dans un réseau cellulaire. In: 11th Journées Francophones de programmation par contraintes (JFPC), pp. 184–193 (2015)
15. Lai, X., Hao, J.K.: Path relinking for the fixed spectrum frequency assignment problem. Expert Syst. Appl. **42**(10), 4755–4767 (2015)
16. Leese, R. Hurley, S.: Methods and Algorithms for Radio Channel Assignment. Oxford Lecture Series in Mathematics and its Applications. Oxford University Press (2002)
17. Luna, F., Blum, C., Alba, E., Nebro, A.J.: ACO vs EAs for solving a real-world frequency assignment problem in GSM networks. In: Annual Conference on Genetic and Evolutionary Computation, pp. 94–101 (2007)
18. Luna, F., Estebanez, C., Coromoto, L., Chaves-Gonzalez, J.M., Nebro, A.J., Aler, R., Segura, C., Vega-Rodriguez, M.A., Alba, E., Valls, J.M., Miranda, G., Gomez Pulido, J.A.: Optimization algorithms for large-scale real-world instances of the frequency assignment problem. Soft. Comput. **15**(5), 975–990 (2011)
19. Mabed, H., Caminada, A., Hao, J.K.: Genetic tabu search for robust fixed channel assignment under dynamic traffic data. Comput. Optim. Appl. **50**(3), 483–506 (2011)
20. Metzger, B.H.: Spectrum management technique. In: National ORSA Meeting, pp. 1497–1514 (1970)
21. Montemanni, R., Moon, J.N.J., Smith, D.H.: An improved tabu search algorithm for the fixed spectrum frequency assignment problem. IEEE Trans. Veh. Technol. **52**(4), 891–901 (2003)

22. Montemanni, R., Smith, D.H.: Heuristic manipulation, tabu search and frequency assignment. Comput. Oper. Res. **37**(3), 543–551 (2010)
23. Tizhoosh, H.: Opposition-based learning: a new scheme for machine intelligence. In: International Conference on Computational Intelligence and Automation I, pp. 695–701 (2005)
24. Wu, J., Dai, Y., Zhao, Y.C.: The effective channel assignments in cognitive radio networks. Comput. Commun. **71**(1), 411–420 (2013)

A Grouping Harmony Search Algorithm for Assigning Resources to Users in WCDMA Mobile Networks

Adrian Aybar-Ruiz[1], Lucas Cuadra[1(✉)], Javier Del Ser[2,3,4], Jose Antonio Portilla-Figueras[1], and Sancho Salcedo-Sanz[1]

[1] Universidad de Alcalá, 28805 Alcalá de Henares, Madrid, Spain
lucas.cuadra@uah.es
[2] TECNALIA, 48160 Derio, Spain
javier.delser@tecnalia.com
[3] University of the Basque Country EHU/UPV, 48013 Bilbao, Spain
[4] Basque Center for Applied Mathematics (BCAM), 48009 Bilbao, Spain

Abstract. This paper explores the feasibility of a particular implementation of a Grouping Harmony Search (GHS) algorithm to assign resources (codes, aggregate capacity, power) to users in Wide-band Code Division Multiple Access (WCDMA) networks. We use a problem formulation that takes into account a detailed modeling of loads factors, including all the interference terms, which strongly depend on the assignment to be done. The GHS algorithm aims at minimizing a weighted cost function, which is composed of not only the detailed load factors but also resource utilization ratios (for aggregate capacity, codes, power), and the fraction of users without service. The proposed GHS is based on a particular encoding scheme (suitable for the problem formulation) and tailored Harmony Memory Considering Rate and Pitch Adjusting Rate processes. The experimental work shows that the proposed GHS algorithm exhibits a superior performance than that of the conventional approach, which minimizes only the load factors.

Keywords: Harmony Search · Grouping Harmony Search · Wide-band Code Division Multiple Access mobile networks

1 Introduction

Currently about 80% of mobile operators worldwide are investing to upgrade their Wide-band Code Division Multiple Access (WCDMA) networks [1], which have 1.83 billion users. High Speed Packet Access (HSPA), based on WCDMA technology, is the most widely used mobile *broadband* technology *deployed* at present. This is because HSPA allows operators to cost-efficiently *upgrade* their already deployed WCDMA networks to provide both speech and broadband data services (high speed Internet access, music-on-demand, or TV and video streaming, to name just a few). WCDMA/HSPA technology is expected to serve 90% of the world's population by 2020, with about 3.8 billion users [1].

J. Del Ser (ed.), *Harmony Search Algorithm*, Advances in Intelligent Systems and Computing 514, DOI 10.1007/978-981-10-3728-3_19

The question that motivates this work is how to assign the limited WCDMA resources to users (mobile users or users equipments). As in other mobile access systems, frequency is one of these scarce resources. This is why in WCDMA networks a number of users are allowed to use simultaneously the same frequency. To separate these communications, the network assigns a "channelization code" to each communication. However, a given amount of *interference* appears between communication links using the same frequency. To quantify the influence of interference, a parameter called "load factor" η is used. It is defined as the ratio between the interference and the total perturbation (thermal noise + interference) [2]. The most used conventional approach (CA) for dimensioning WCDMA networks is based on keeping the interference and load factor lower than some suitable empirical thresholds [2]. Other limited resources in any base station (BS) are the maximum backhaul capacity, the number of channelization codes, and the maximum power [3,4].

Regarding this, the *purpose* of this work is to explore the feasibility of a Grouping Harmony Search (GHS) algorithm [5] to near-optimally assign WCDMA resources (codes, capacity, power) of $\mathcal{N}_B$ base stations to $\mathcal{N}_U$ users, by *minimizing* a cost function composed of the following *weighted* constituents [3]: (1) "Detailed" load factors [3] (which include *all* possible interference signals), the *utilization factor* of the available resources to be used (aggregated capacity, power, codes), and the fraction of users *without service*. The latter is critical because the smaller the number of users without service, the greater service availability. High service availability help operators increase market share.

There are two recent papers [3,4] that also study this problem. The proposed work differs from [4] in the use of detailed load factors (instead of approximate ones), and also differs from [3] in the use of a GHS (instead of a Grouping Genetic Algorithm (GGA)).

The structure of the rest of this paper is as follows. While Sect. 2 states the problem along with a characterization of the resources to be assigned, Sect. 3 describes the GHS algorithm we propose. Section 4 shows the experimental work and, finally, Sect. 5 completes the paper by discussing the main findings.

2 Problem Statement

Let $\mathcal{A}$ be the service area of a WCDMA network with $\mathcal{N}_B$ base stations (BSs) and $\mathcal{N}_U$ active users. Figure 1 represents two of these $\mathcal{N}_B$ BSs for the sake of clarity. The dashed area represents the *cell* covered by a BS or "node B" (nB) in WCDMA terminology. Throughout this work, both words will be used interchangeably. $n_u^{B_k}$ is the number of users that the nB B_k is serving. In particular, a reference user u_l assigned (associated) to B_k is denoted "$u_l \in B_k$". $p_{\mathrm{R},B_k}(l)$ represents the power received at B_k emitted by user u_l ($p_e(l)$). To separate the "reference" communication link $u_l \leftrightarrow B_k$ from others using the *same* frequency f, the network assigns a different code to each communication. Although codes help ideally reduce interference, however, the remaining communications using the same frequency become interference signals. The total interference contains not only those interferences generated by the users in the "own-cell" (for instance, user u_j in Fig. 1) but also those arising from users located in *other* cells (user u_m). Note that, apart from the interferences appearing in the uplink (UL) –signals moving from the users to the BS–, there are also others in the downlink (DL). A representative example is the interference produced by the base station B_q ($q \neq k$), which interferes on the reference link $u_l \leftrightarrow B_k$. Load factors model to what extent interferences affect the network performance [2]. As explained in [3], the *detailed* UL load factor of a cell B_k with $n_u^{B_k}$ users is

$$\eta_{\mathrm{UL}}^{\mathrm{det}}(B_k) = \sum_{j=1}^{n_u^{B_k}} (1 + \xi_{u_j \to B_k}^{\mathrm{UL}}) \cdot \frac{1}{1 + \frac{1}{(e_b/n_0)_S(j)} \cdot \frac{W}{R_{b,S}^{\mathrm{UL}}(j) \cdot \nu_S^{\mathrm{UL}}(j)}}, \tag{1}$$

where $\xi_{u_j \to B_k}^{\mathrm{UL}}$, the ratio of "other-cell" to "own-cell" interference *on* the uplink $u_j \to B_k$, is defined in Table 1 along with other parameters in Eq. (1). The *key point* is that $\eta_{\mathrm{UL}}^{\mathrm{det}}(B_k)$ has very different values depending on the particular user-cell association selected [3]. Similarly, the *detailed* DL load factor is [3]:

$$\eta_{\mathrm{DL}}^{\mathrm{det}}(B_k) = \sum_{j=1}^{n_u^{B_k}} \left[(1 - \overline{\alpha}) + \xi_{B_k \to u_j}^{\mathrm{UL}} \right] \frac{(e_b/n_0)_S(j)}{\frac{W}{R_{b,S}^{\mathrm{DL}}(j) \cdot \nu_S^{\mathrm{DL}}(j)}}, \tag{2}$$

$\overline{\alpha}$ being an average orthogonality factor over cell B_k [2].

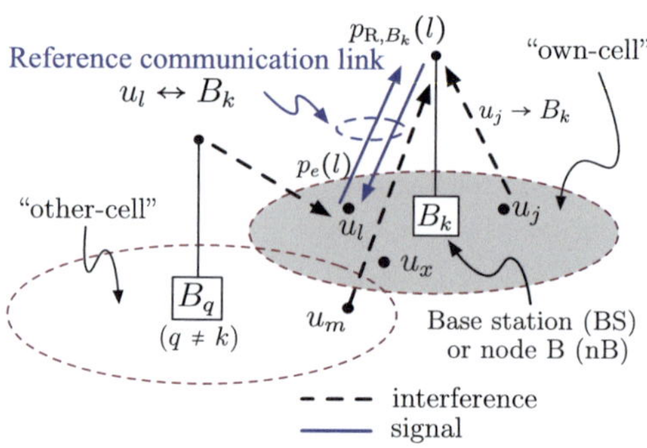

Fig. 1. Simplified representation of the communication signals (blue solid line) and interferences (black dashed lines) on the "reference" communication link $u_l \leftrightarrow B_k$.

In addition to frequency, any base station B_k has also a limited amount of each other resource, $\mathcal{R}$, which has to be shared among the $n_u^{S_k}$ users associated to B_k. For any resource, $\mathcal{R}$, we define the corresponding *utilization ratio* $\Delta_{\mathcal{R}} \doteq \mathcal{R}_{\mathrm{used}}/\mathcal{R}_{\mathrm{max}}$ as shown in Table 2. See [3] for further details.

Finally, a critical point for operators is the fraction of users *without* service, $\Delta_{n_u}^{\mathrm{WS}} \doteq n_u^{\mathrm{WS}}/\mathcal{N}_U$ (n_u^{WS} being the number of users without service), because the smaller $\Delta_{n_u}^{\mathrm{WS}}$, the higher the user satisfaction. This can help the mobile operator to increase its market share.

Table 1. Definition of parameters [3] used in this work. Uppercase UL(DL) is used to label either the uplink or downlink parameters. Subscript S stands for service.

Symbol	Definition and/or value
W	Chip rate: $W = 3.84$ Mchip/s (standardized value)
$(e_b/n_0)_S(j)$	Ratio between the mean bit energy and the noise power density (thermal noise and interference) required to achieve a given quality for service S
$R_{b,S}^{\mathrm{UL(DL)}}(j)$	Bit rate of service S in the j-th UL(DL) within cell B_k.
$\nu_S^{\mathrm{UL(DL)}}(j)$	Utilization factor: $0 < \nu_S^{\mathrm{UL(DL)}}(j) < 1$ (for voice), $\nu_S^{\mathrm{UL(DL)}}(j) = 1$ (for data services) [2]
p_{e,u_m}	Power emitted by user u_m
ℓ_{u_m,B_k}	*Total* propagation loss in the link $u_m \rightarrow B_k$
$i_{u_j \rightarrow B_k}^{\mathrm{UL},B_k}$	$= \sum_{u_m \in B_k, u_m \neq u_l} \frac{p_{e,u_m}}{\ell_{u_m,B_k}}$, UL own-cell-interference (from users on the own cell, $u_m \in B_k$)
$i_{u_j \rightarrow B_k}^{\mathrm{UL},B_q}$	$= \sum_{u_m \in B_q, B_q \neq B_k} \frac{p_{e,u_m}}{\ell_{u_m,B_q}}$, UL other-cell-interference (from users on other cell, $u_m \in B_q$)
$\xi_{u_j \rightarrow B_k}^{\mathrm{UL}}$	$= \frac{i_{u_j \rightarrow B_k}^{\mathrm{UL},B_q}}{i_{u_j \rightarrow B_k}^{\mathrm{UL},B_k}}$, ratio of other-cell to own-cell interference *on* the UL $u_j \rightarrow B_k$

With these concepts in mind, the problem consists in finding the user-cell association that assigns resources (power, codes, capacity) by *minimizing* the cost function [3]

$$\mathcal{C} = \frac{1}{\mathcal{N}_B} \sum_{k=1}^{\mathcal{N}_B} [w_\eta \cdot (\eta_{\mathrm{UL}}^{\mathrm{det}} + \eta_{\mathrm{DL}}^{\mathrm{det}}) + w_{\Delta_{C_A}} \cdot (\Delta_{C_{\mathrm{Ag}}^{\mathrm{UL}}} + \Delta_{C_{\mathrm{Ag}}^{\mathrm{DL}}} +)$$
$$+ w_{\Delta_{P_{B_k}}} \cdot \Delta_{P_{B_k}} + w_{\Delta_{\mathrm{Cod}}} \cdot \Delta_{\mathrm{Cod}} + w_{\Delta_{n_u}^{\mathrm{WS}}} \cdot \Delta_{n_u}^{\mathrm{WS}}], \tag{3}$$

constrained to the conditions that all the aforementioned components $\phi = \eta_{\mathrm{UL}}^{\mathrm{det}}$, $\eta_{\mathrm{DL}}^{\mathrm{det}}$, $\Delta_{C_{\mathrm{Ag}}^{\mathrm{UL}}}$, $\Delta_{C_{\mathrm{Ag}}^{\mathrm{DL}}}$, $\Delta_{P_{B_k}}$, Δ_{Cod}, $\Delta_{n_u}^{\mathrm{WS}}$ are real numbers fulfilling $0 \leq \phi \leq 1$ [3]. w_ϕ represents a weight factor for any of the involved components. Unlike [3], which uses a GGA, we tackle this problem by the GHS proposal that follows.

3 Proposed GHS Algorithm

A GHS [5] is a modification of the Harmony Search (HS) algorithm to deal with grouping problems. HS is a meta-heuristic, population-based algorithm, inspired by the improvisation process of an orchestra in their effort of composing the most harmonious melody. Put it simple, a candidate vector solution in HS is

Table 2. Resources ($\mathcal{R}$ = power, codes, capacity) and utilization ratios $\Delta_{\mathcal{R}}$ in BS B_k (serving $n_u^{S_k}$ users) when aiming at allocating resources to user u_l. See Fig. 1.

$\mathcal{R}$ and corresponding $\Delta_{\mathcal{R}} \doteq \mathcal{R}_{\text{used}}/\mathcal{R}_{\max}$	Definition
$p_{B_k\|\max}$	Maximum power that base station B_k can emit
$p^{\text{DL}}_{B_k \to u_j}$	Power emitted by B_k for serving user u_j
$\Delta_{P_{B_k}} \doteq \frac{1}{p_{B_k\|\max}} \sum_{j=1}^{n_u^{S_k}} p^{\text{DL}}_{B_k \to u_j}$	Power utilization ratio of B_k
$N^{S_h}_{\text{Cod}}$	Maximum no. of codes in B_k for service S_h
$n^{B_k}_{u,S_h}$	Number of users in B_k demanding service S_h
$\Delta_{\text{Cod}} \doteq \sum_{h=1}^{N_S} \frac{n^{B_k}_{u,S_h}}{N^{S_h}_{\text{Cod}}}$	Code utilization ratio in B_k
$C^{\text{UL(DL)}}_{\text{Ag}}$	Maximum aggregated capacity of B_k in UL(DL)
$R^{\text{UL(DL)}}_{b,S}(j)$	Bit rate of user $u_j \in B_k$ in UL(DL)
$\Delta_{C^{\text{UL(DL)}}_{\text{Ag}}} \doteq \frac{1}{C^{\text{UL(DL)}}_{\text{Ag}}} \sum_{j=1}^{n_u^{S_k}} R^{\text{UL(DL)}}_{b,S}(j)$	Capacity utilization ratio in B_k

called "harmony" while any of its compounding elements is named "note", the set of harmonies being commonly denoted as "Harmony Memory" (HM). The initial HM is evolved by applying optimization processes –"Harmony Memory Considering Rate" (HMCR) and "Pitch Adjusting Rate (PAR) –, producing a new improvised harmony in any iteration. The way the HS algorithm works can be summarized in four basic steps: (1) Initialization of the HM; (2) Improvisation of a new harmony; (3) Inclusion of the newly generated harmony in the HM (its fitness improves the worst fitness value in the previous HM); (4) Returning to step (2) until a termination criteria (maximum number of iterations or fitness stalls) is fulfilled. A useful survey on applications of the HS algorithm is [6].

3.1 Problem Encoding

The encoding is based on separating each harmony $\mathbf{h}$ into two parts: $\mathbf{h} = [\mathbf{e}|\mathbf{g}]$, the first one being the *element* section, while the second part, the *group* section. Since the number of base stations in our network is *constant* ($\mathcal{N}_B$), we have used the following variations of the classical grouping encoding:

1. The element part $\mathbf{e}$ is an $\mathcal{N}_U$-length vector whose elements ($u_j^{B_k}$) mean that user u_j has been assigned to base station B_k.
2. The group section $\mathbf{g}$ is an ($\mathcal{N}_B + 1$) length vector, whose elements (labeled $n_u^{B_j}$) represent the number of users assigned to each j-th base station (B_j). Subscript j ranges from -1 to $\mathcal{N}_B$, $j = -1$ being used to represent those users that are *not* connected to any node, that is, those in an "imaginary" or virtual base station that we have labelled "base station -1". As will be shown, this group part is necessary since the PAR operator acts first on the group part.

As an example, following our notation, a candidate harmony $\mathbf{h}_i$, belonging to an HM with Γ harmonies $\{\mathbf{h}_1 \cdots \mathbf{h}_\Gamma\}$, could encode a solution –i.e., an assignment of $\mathcal{N}_U$ elements (users) to $\mathcal{N}_B$ base stations, forming thus groups of users– as

$$\mathbf{h}_i = [u_{1(i)}^{B_h} \dots u_{m(i)}^{B_j} \dots u_{\mathcal{N}_U(i)}^{B_w} \mid n_{u(i)}^{B_{-1}} n_{u(i)}^{B_1} \dots n_{u(i)}^{B_k} \dots n_{u(i)}^{B_{\mathcal{N}_B}}], \tag{4}$$

where $n_{u(i)}^{B_{-1}}$ is the number of users without service (n_u^{WS}), those that have not been able to be assigned to any nB and do not have service. We represents this by assigning then to a "virtual" nB labeled B_{-1}. Figure 2(a) shows a simple example of codification of an assignment in which $\mathcal{N}_U = 10$ users have been assigned to $\mathcal{N}_B = 4$ base stations (c1).

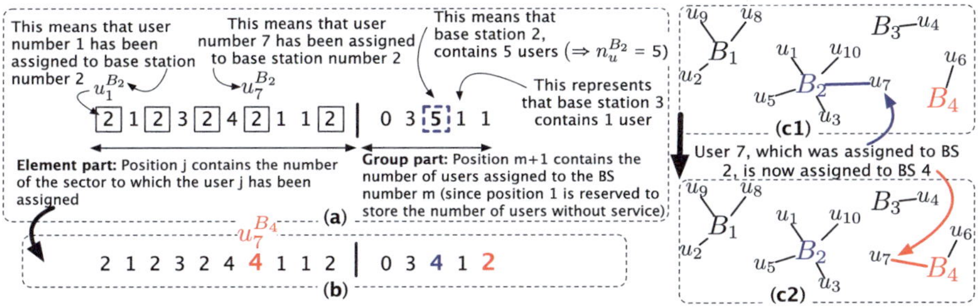

Fig. 2. (a) Harmony with 10 users and 4 base stations. The PAR process selects, in the group part, the BS B_2 (position 3 since the first one is reserved to quantify the number of users without service), which has assigned 5 users ($u_1, u_3, u_5, u_7, u_{10}$, as shown in (c1) and represented in the element part of (a)). (b) In a second step, the PAR process selects one of the elements (u_7, which was assigned to B_2 in (a)), and reassign it to B_4. (c) Representation of the reassignment process driven by PAR.

3.2 Algorithm Implementation

(a) The *initialization* of the notes' values of all harmonies included in the HM is only executed at the first iteration.

(b) The *improvisation* process is sequentially applied to each note of the complete set of Γ harmonies. Two processes are used for improvising the new refined set of harmonies:

(b.1) The Harmony Memory Considering Rate (HMCR) establishes the probability that the new value for a note $u_{m(i)}^{B_j}$ (in the *element* part $\mathbf{e}_i$ of $\mathbf{h}_i$) is drawn from the values of the same note taken in all the other $\Gamma - 1$ harmonies existing in the HM, ($u_{m(\gamma)}^{B_p}, \gamma = 1 \cdots \Gamma$, $\gamma \neq i$). Note that the smaller HMCR is, the less the use of partial knowledge acquired during the iterative process will be, and hence the more explorative the algorithm will behave. The new note will be chosen at random if it is not drawn from the HM.

(b.2) The Pitch Adjusting Rate (PAR) process works as a fine adjusting rate of the note vocabulary. In our implementation, it first selects at random a *group* in $\mathbf{g}_i$ (for instance, BS $B_k = B_2$ (blue dashed squared) in Fig. 2(a), which has assigned $n_u^{B_k} = 5$ users). In a second step, it selects one of the users assigned to B_k (for instance, element $u_{m(i)}^{B_k} = u_{7(i)}^{B_2}$ (user 7) in $\mathbf{e}_i$) and assigns it to another BS (B_4 in Fig. 2 (b), (c2)) with a given probability, $\mathcal{P}$. This probabilistic process defines the new value $u_m^{B_k^*}(i)$ for a certain note $u_m^{B_k}(i)$ (after HMCR processing) as

$$u_{m(i)}^{B_k^*} = \begin{cases} u_{m(i)}^{B_q}, & \text{with } \mathcal{P} = \text{PAR}, \\ u_{m(i)}^{B_k}, & \text{with } \mathcal{P} = 1 - \text{PAR}, \end{cases} \tag{5}$$

where B_q ($q \neq k$) is another BS (selected at random) at which user u_m will be assigned only if the distance between user u_m and BS B_q fulfills $d(u_m, B_q) < d_{\text{MAX}}$.

Analogously to the HMCR process, a high value of PAR jointly with a increased value of d_{MAX} sets a highly explorative behavior of the algorithm around the iteratively-identified potential candidates or harmonies, while narrower bandwidths (i.e. lower values of d_{MAX}) the PAR process leads to a restricted local search procedure (the user will be assigned to an adjacent cell instead of farther cells).

c) At each iteration the quality of the improvised harmonies is evaluated by means of the cost function $\mathcal{C}$ stated by Eq. (3). A harmony $\mathbf{h}_p$ "sound best" than another $\mathbf{h}_g$ if $\mathcal{C}(\mathbf{h}_p) < \mathcal{C}(\mathbf{h}_g)$. Then, based on these metric evaluations and their comparison with the cost of harmonies remaining from the previous iteration, the Γ best harmonies are kept and the HM is hence updated by excluding the worst harmonies.

d) The stopping criterion is selected based on a fixed number of iterations $\mathcal{T}$.

4 Experimental Work

4.1 Experimental Set up and Comparative Framework

We have considered the three different services listed in Table 3 along with their characteristic parameters. Other network parameters used are [2]: $\overline{\alpha} = 0.65$, $\overline{\xi} = 0.55$, $p_{B_k|\max} = 36$ W, and $C_{\text{Ag}}^{\text{UL}} = C_{\text{Ag}}^{\text{DL}} = 1536$ kbps. With these services, we have considered the following service profiles: 90% of users with service S_1, 9% with S_2, and 1% with S_3. We have carried out 20 runs of each GHS algorithm, with $\mathcal{T} = 300$ iterations each. This number has been found to be large enough for the algorithm to converge.

For comparative purposes to the conventional approach (CA) we have implemented a combination of two CAs: the "Best-Server Cell Selection" (BSCS) [2] and the "Radio Prioritized Cell Selection" (RPCS) [2] algorithms. For any user u_j (with $j = 1, 2, \cdots, \mathcal{N}_U$), we compute the SINR between u_j and all the base stations B_k (with $k = 1, 2, \cdots, \mathcal{N}_B$): $\Upsilon_{j,k}$. This leads to a $\mathcal{N}_U \times \mathcal{N}_B$ matrix of SINR

Table 3. Values of service parameters. ARM means Adaptive Multi-Rate.

S_h	$(E_b/N_0)_i$ (dB)	$R_{b,i}^{\text{UL}}$ (dB)	$R_{b,i}^{\text{DL}}$ (kbps)	$\nu_i^{\text{UL}} = \nu_i^{\text{DL}}$	$N_{\text{Cod}}^{S_h}$ (codes)
S_1 (ARM)	5	12.2	12.2	0.58	256
S_2 (data)	1.5	64	64	1	32
S_3 (data)	1	64	384	1	4

ratios. For any user u_j, we compute an "assignment vector", $\mathbf{A}_j$, which contains a list of BSs, sorted from the one that provides the best SINR down to that giving the worst one. Initially, each user u_j is assigned to the nB with the corresponding best SINR ("best base station" (BBS) in the BSCS algorithm [2]), that is, to the first one of the assignment vector $\mathbf{A}_j$. In any cell, the algorithm checks whether or not the assignment leads to a load factor higher that the threshold (overload). In each overloaded cell (let say, for instance, B_g), the user with the worst SINR with respect to B_g (let say, for instance, u_f) is detached from B_g and assigned to the next non-overloaded BS of its assignment vector $\mathbf{A}_f$. The algorithm iterates until either the cells are no longer overloaded, which may cause some users fail to be assigned to any station.

4.2 Comparison and Discussion

Figures 3(a) and (b), which represent respectively the different assignments that the GHS and CA algorithms have found, will help us compare both approaches.

Each BS in Fig. 3 has been represented by a square box containing a different symbol $(+, \times, \Diamond, \triangleright, \cdots)$ so that any user attached, for instance, to base station B_3 ($\Diamond$-symbol inside the box), will be represented with that symbol ($\Diamond$). They correspond to $\mathcal{N}_U = 500$ users, which leads to a user density $D_{\text{U}} \approx 31.25$ users/km^2. Note that both figures have *identical* user locations, but *differ in the*

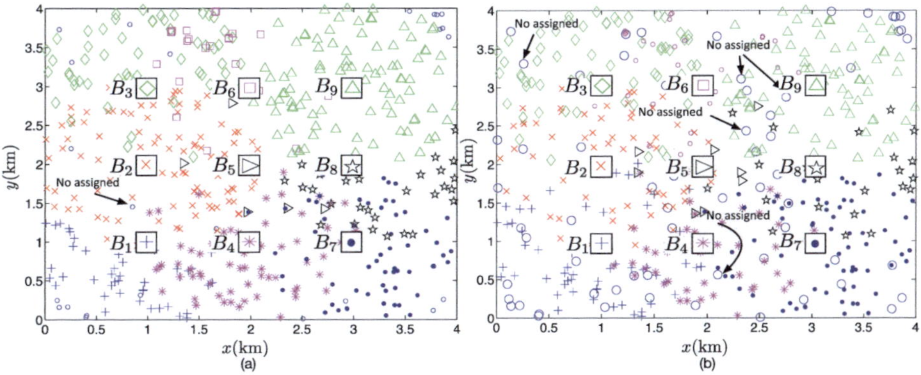

Fig. 3. Assignments of $\mathcal{N}_U = 500$ uniformly distributed users to $\mathcal{N}_B = 9$ BSs found, respectively, by the GHS algorithm (a) and by the CA (b).

way they are assigned to different stations. This can be easily seen by taking a look at those users located in-between base stations B_6 and B_9 in both figures. While in Fig. 3(a) the users are mostly labeled with green $\triangle$-symbols (what means that they have been assigned to B_9 ($\triangle$ symbol)), however, in Fig. 3(b), many of these users located between stations B_6 and B_9 (which in Fig. 3(a) were mostly assigned to B_9) are now however *without service* (represented with blue $\bigcirc$ symbols) since they have not been assigned to any nB. The CA assignment (Fig. 3(b)) works worse in the sense that it leaves more customers unserved.

To proceed further in this regard, it is convenient to focus on Fig. 4(a). It compares, respectively, the fraction of the different constituents ($\phi = \eta_{\text{UL}}, \eta_{\text{DL}}, \Delta_{C_{\text{Ag}}^{\text{UL}}}, \Delta_{C_{\text{Ag}}^{\text{DL}}}, \Delta_{P_{B_k}}, \Delta_{\text{Cod}}, \Delta_{n_u}^{\text{WS}}$) of the *minimized* cost function, computed by the CA (grey bars), the GGA approach [3] (blue bars), and the proposed GHS method (red bars). The most relevant aspect is that the proposed GHS method assigns resources to *many more users* than the CA: the fraction of users *without service* in the GHS assignment is only $\Delta_{n_u}^{\text{WS}} \mid_{\text{GHS}} = 3\%$ (mean value over 20 runs). This represents only 15 users in absolute terms, which is much smaller than that achieved by the conventional assignment, which is $\Delta_{n_u}^{\text{WS}} \mid_{\text{CA}} = 18\%$ (i.e., 90 users). Note that $\Delta_{n_u}^{\text{WS}} \mid_{\text{GHS}}$ is 6 times smaller than $\Delta_{n_u}^{\text{WS}} \mid_{\text{CA}}$. In this respect, the GHS strategy is more practical for the operator's economical strategy since it helps increase the number of active users without having to draw upon new, expensive deployments. Note also that the GHS method works slightly better than the GGA approach [3].

The true potential of the proposed GHS algorithm can be seen much more clearly in Fig. 4(b), which represents the fraction of resources assigned *per user*. The fraction of resources used per user in GHS is lower than that of CA. On average, this is ≈85% of those of the CA. In this sense, the use of resources is *more efficient* because the proposed method leads to an assignment in which there are *more users* with the required service ($500 - 15 = 485 = \mathcal{N}_{\text{U}}^{\text{GHS}} > 500 - 90 = 410 = \mathcal{N}_{\text{U}}^{\text{CA}}$) along with a *lower consumption-per-user* than that achieved by the CA and that by the GGA [3].

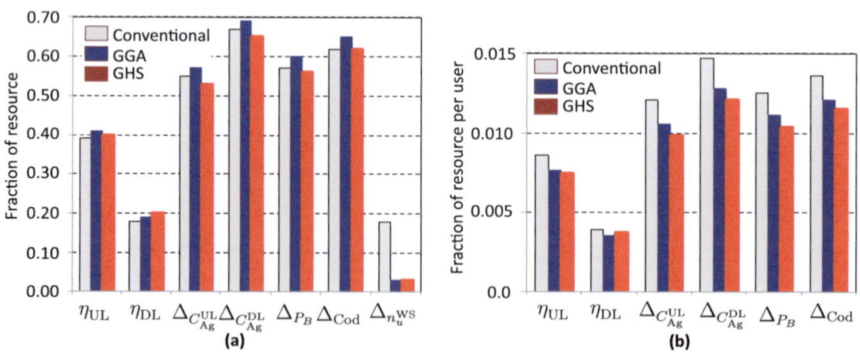

Fig. 4. (a) Fraction of used resources corresponding to the assignment computed by the CA (grey bars), the GGA method (blue bars) and the proposed GHS (red bars). (b) Fraction of resources *per* user (same color convention as (a)).

5 Summary and Conclusions

In this work we have proposed a novel implementation of a Grouping Harmony Search (GHS) algorithm to assign resources (codes, capacity, power) to users in Wide-band Code Division Multiple Access (WCDMA) networks. The GHS algorithm aims at *minimizing* a cost function composed of not only the *detailed* load factors (including *all* interferences) but also *resource utilization ratios* and the fraction of users without service. We have proposed an encoding scheme, which is novel in GHS, and based on this, also tailored Harmony Memory Considering Rate (HMCR) and Pitch Adjusting Rate (PAR) processes. In particular, the proposed PAR process acts on the group part (by selecting a base station) and assigns one of its users to another base station with a given probability. The explored GHS exhibits a *superior* performance than that of the conventional approach (CA) –which minimizes only the load factors–, and is slightly better than that of the Grouping Genetic Algorithm (GGA) approach. The GHS not only assigns resources to more users (97% of users –in a scenario with 31.25 users/km^2, uniformly distributed–, higher than 82% of users assigned by the CA) but also it does it *more efficiently*, since the mean value of the *resources used per user* in the GHS assignment is 85% of that of the CA one.

Acknowledgments. This work has been partially supported by projects TIN2014-54583-C2-2-R (Spanish Ministerial Commission of Science and Technology) and S2013ICE-2933_02 (Comunidad Autónoma de Madrid).

References

1. Ericsson Company: Ericsson Mobility Report. On the Pulse of the Networked Society. Technical Paper (2016). https://www.ericsson.com/
2. Holma, H., Toskala, A.: WCDMA for UMTS: HSPA Evolution and LTE. Wiley, New York (2010)
3. Cuadra, L., Aybar-Ruiz, A., Del Arco, M.A., Navío-Marco, J., Portilla-Figueras, J.A., Salcedo-Sanz, S.: A Lamarckian hybrid grouping genetic algorithm with repair heuristics for resource assignment in WCDMA networks. Appl. Soft Comput. **43**, 619–632 (2016)
4. Cuadra, L., Salcedo-Sanz, S., Carnicer, A.D., Del Arco, M.A., Portilla-Figueras, J.A.: A novel grouping genetic algorithm for assigning resources to users in WCDMA networks. In: Mora, A.M., Squillero, G. (eds.) EvoApplications 2015. LNCS, vol. 9028, pp. 42–53. Springer, Heidelberg (2015). doi:10.1007/978-3-319-16549-3_4
5. Gil-Lopez, S., Del Ser, J., Landa-Torres, I., Garcia-Padrones, L., Salcedo-Sanz, S., Portilla-Figueras, J.A.: On the application of a novel grouping harmony search algorithm to the switch location problem. In: Chatzimisios, P., Verikoukis, C., Santamaría, I., Laddomada, M., Hoffmann, O. (eds.) Mobilight 2010. LNICSSITE, vol. 45, pp. 662–672. Springer, Heidelberg (2010). doi:10.1007/978-3-642-16644-0_57
6. Manjarres, D., Landa-Torres, I., Gil-Lopez, S., Del Ser, J., Bilbao, M.N., Salcedo-Sanz, S., Geem, Z.W.: A survey on applications of the harmony search algorithm. Eng. Appl. Artif. Intell. **26**(8), 1818–1831 (2013)

Wireless Network Optimization for Massive V2I Data Collection Using Multiobjective Harmony Search Heuristics

Iker Sobron[1(✉)], Borja Alonso[1], Manuel Vélez[1], and Javier Del Ser[1,2,3]

[1] University of the Basque Country UPV/EHU, 48013 Bilbao, Spain
{iker.sobron,manuel.velez,javier.delser}@ehu.eus,
balonso024@ikasle.ehu.eus
[2] TECNALIA, 48160 Derio, Spain
javier.delser@tecnalia.com
[3] Basque Center for Applied Mathematics (BCAM), 48009 Bilbao, Spain

Abstract. This paper proposes to improve the efficiency of the deployment of wireless network infrastructure for massive data collection from vehicles over regional areas. The increase in the devices that are carried by vehicles makes it especially interesting being able to gain access to that data. From a decisional point of view, this collection strategy requires defining a wireless Vehicular-to-Infrastructure (V2I) network that jointly optimizes the level of service and overall CAPEX/OPEX costs of its deployment. Unfortunately, it can be intuitively noted that both optimization objectives are connecting with one another: adding more equipment will certainly increase the level of service (i.e. coverage) of the network, but costs of the deployment will rise accordingly. A decision making tool blending together both objectives and inferring therefrom a set of Pareto-optimal deployments would be of utmost utility for stakeholders in their process of provisioning budgetary resources for the deployment. This work will explore the extent to which a multi-objective Harmony Search algorithm can be used to compute the aforementioned Pareto-optimal set of deployment by operating on two different optimization variables: the geographical position on which wireless receivers are to be deployed and their type, which determines not only their coverage range but also their bandwidth and cost. In particular we will utilize a non-dominated sorting strategy criterion to select the harmonies (solution vectors) evolved by Harmony Search heuristics.

Keywords: Vehicular networks · Cost-efficient deployment · Harmony search

1 Introduction

The latest technological advances in the design of wireless sensors have lately given rise to lower fabrication costs, which have in turn favored their widespread integration in vehicles. Indeed the automotive sector is a rich substrate

J. Del Ser (ed.), *Harmony Search Algorithm*, Advances in Intelligent Systems and Computing 514, DOI 10.1007/978-981-10-3728-3_20

for mobility-based applications leveraging the geo-located information seamlessly collected, processed and stored by vehicles. These data, which is usually sensed locally at each vehicle, would certainly open new horizons if further processing is done to aggregate, fuse and infer knowledge from the data collected by all vehicles [1,2].

To this end communications infrastructure must be deployed over the area to compile all the information captured by vehicles so that it can be subsequently processed, stored and managed by the applications at hand [3]. The scope of this research work is framed around the search of solutions that improve the efficiency of the Vehicular-to-Infrastructure (V2I) wireless network for the collection of massive data from vehicles over regional areas. From a decisional viewpoint this collection strategy requires defining a wireless V2I network that jointly optimizes the level of service and overall CAPEX/OPEX costs of its deployment. Unfortunately, it can be intuitively noted that both optimization objectives are conflicting with one another: adding more equipment will certainly increase the level of service (i.e. coverage) of the network, but costs of the deployment will rise accordingly. A decision making tool blending together both objectives and inferring therefrom a set of Pareto-optimal deployments would be of utmost utility for stakeholders in their process of provisioning budgetary resources for the deployment.

This work will explore to which extent a multi-objective Harmony Search (HS) algorithm can be used to compute the aforementioned Pareto-optimal set of deployment by operating on two different optimization variables: the geographical position on which wireless receivers are to be deployed and their type, which determines not only their coverage range but also their bandwidth and cost. In particular we will utilize a non-dominated sorting strategy criterion to select the harmonies (solution vectors) evolved by the Harmony Search heuristic. Experiments with the designed solver will be run over a massive case study comprising more than $7 \cdot 10^5$ vehicles deployed over the region of the Basque Country [4], as well as real target locations for the wireless receivers.

2 Vehicular and Infrastructure Databases

As a starting point we have created a geolocated vehicles database from the Basque Country region in Spain that allows simulating a realistic wireless network scenario of the entire region. For this purpose we will exploit information on the main cities such as the quantity of vehicles on each geographical location. In order to distribute the vehicles over the region, we assume that most of the traffic will be located around the most populated spots of the region. As a result we define vehicle distribution areas with a maximum radius, r_{max}, around the main populated cities in the region. The circular area of each village is divided in several concentric rings with inner and outer radius r_{int} and r_{ext}, respectively. The ratio of distributed vehicles within a m-th ring is thus given by

$$P_m = \frac{\int_{r_{int,m}}^{r_{ext,m}} e^{-cr} dr}{\int_0^{r_{max}} e^{-cr} dr}, \tag{1}$$

where r the radius and c the coefficient that modifies the gradient according to the size of the city. Note that P_m is the ratio of vehicles deployed on roads of the m-ring area with respect to its total registered in a given village. Thus, the sum of all the concentric rings around the village equals 1. The vehicles are distributed over the existing motorways, which will be classified in three categories according to the traffic. Therefore the most crowded type of motorway will have more chance to have vehicles assigned than the least crowded ones. Due to the fact that the quantity of running vehicles is variable depending on the time of the day, we can apply a correcting factor F which controls such a variability. The number of vehicles V_m to distribute in a random manner on the motorways of the m-th ring will be

$$V_m = FP_mTV, \tag{2}$$

where TV the total of vehicles recorded on each village.

It will be also necessary to create a database of wireless communications infrastructures to which the vehicles can be connected to send the data collected by their sensors. Bearing in mind that low-power wide-area systems (LPWA) such as Sigfox are focused on low-rate communications over wide areas, we assume that base stations (BSs) could be located covering wide areas. For this rationale we have built two databases of already deployed wireless networks by the Spanish operator for Internet of Things (IoT) – Cellnex Telecom – [5] and the Basque operator of TDT, Itelazpi [6]. We assume that IoT BSs can be located at the same point of the TDT repeaters since the operators may employ part of the already deployed infrastructure for wireless transmission-reception purposes.

3 Deployment Cost Profiles

For the definition of the deployment costs we consider two criteria: power and throughput of the BSs. An increase of the transmission power and the throughput corresponds to a greater range and capacity, respectively. Since we assume that infrastructure is already deployed we do not consider a location-based optimization process. The power level will be measured over the coverage radius that every transmitter reaches and the throughput will be measured on the amount of users/cars that can be simultaneously connected at the same transmitter. As a result, the cost function depends on these two parameters according to

$$C = \alpha BW + \beta \pi R^2, \tag{3}$$

where C is the cost, BW the capacity measured in number of cars/users, R the range radius and weights α and β permit to adjust the cost. We have defined the series of profiles listed in Table 1, which combine different capacities and coverages for a given transmitter. We have chosen $\alpha = 0.5$ and $\beta = 10$. The capacity values are 1000, 2000 and 5000 vehicles, and ranges are given by 5, 10 and 30 km of radius. We will also consider a profile where both values are zero, incurring no cost. This profile will emulate the case when the BS at hand is off (not activated).

4 Computing the Deployment Coverage and Cost

After establishing the optimization criteria we define the algorithm that calculates the deployment coverage level for the users on roads according to the combination of profiles for the deployed network. Following the previously defined vehicle distribution, each profile combination – that determines the range radius and capacity for each BS within the network – yields the total amount of vehicles that will be covered at certain time. To compute this value we use the flowchart depicted in Fig. 1. Using the combination of profiles as an input, the algorithm calculates the coverage level and the corresponding cost to each specific case.

In the region of the Basque Country 413 BSs are available as shown in Fig. 2, each capable of operating under any of the profiles defined in Table 1. Firstly, we arrange the vehicles of the database according to the amount of BSs that it will be able to be connected to from the lowest to the highest number of BSs. Therefore, vehicles located inside the coverage area of a unique BS will be the first ones to be connected. This is due to the fact that those vehicles will be only able to connect to one BS (in this case, there is no overlapped coverage among BSs). In this manner vehicles, which can be connected to more than one BS, do not limit the access to the ones that can be only connected to a single BS. We start, car by car, with the decision criterion shown in Fig. 1. Vehicles that fall only in range of one BS will be connected if it is possible, because BSs are limited by the capacity defined in their profiles. As can be seen in Fig. 1, vehicles located within the range of two or more BSs try to connect first with their nearest BS. If this connection is not possible (due to e.g. lack of bandwidth resources of the selected profile for the BS at hand), they will attempt at connecting to their second closest BS. If the vehicle cannot connect because BSs have reached their maximum number of vehicles, the vehicle has no coverage. When the algorithm ends, we shall compute the deployment coverage level and the overall cost of the deployed profiles.

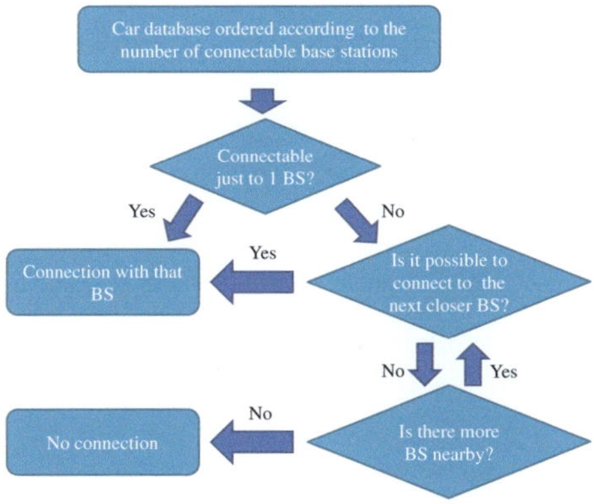

Fig. 1. Algorithmic scheme used to compute the deployment coverage ratio according to the BS profiles.

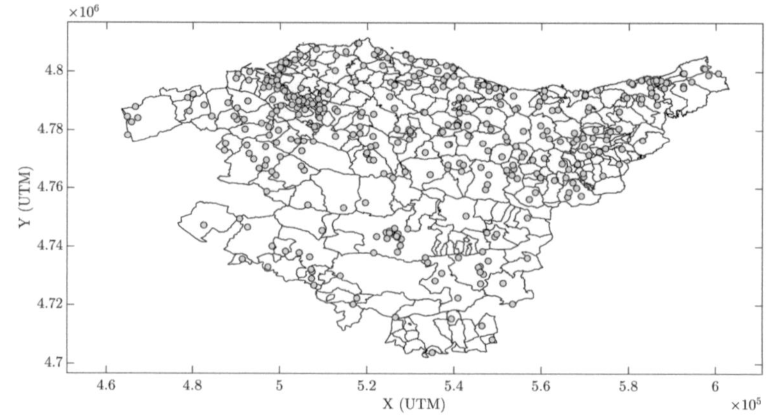

Fig. 2. Geo-location of the deployed BSs in the basque country in Spain.

Table 1. BS profiles considered in this work.

Profile type	Radius R (km)	Cars per cell (BW)	Cost (C)
1	0	0	0.00
2	5	1000	1285.40
3	5	2000	1785.40
4	5	5000	3285.40
5	10	1000	3641.59
6	10	2000	4141.59
7	10	5000	5641.59
8	30	1000	28774.33
9	30	2000	29274.33
10	30	5000	30774.33

5 Review of Multi-objective Harmony Search Algorithm

The bi-objective problem addressed in this paper – i.e. the simultaneous maximization of the coverage ratio of the deployed receivers and the minimization of their cost – will be tackled by resorting to a multi-objective version of the HS algorithm, a heuristic solver proposed in [7] that emulates the music composition process observed in jazz bands when their members jointly improvise harmonies driven by an aesthetic measure of musical quality. In essence HS is a population-based optimization algorithm that resembles other techniques from Swarm Intelligence and Evolutionary Computation, with subtle yet relevant differences in the operators that iteratively refine the solutions contained in the K-sized population of the algorithm (commonly referred to as Harmony Memory). Such operators iteratively enhance the candidate solutions or harmonies by permuting and mutating intelligently their constituent variables or notes until a convergence criterion set beforehand is met (e.g. a maximum number of iterations $\mathcal{I}$ are completed). The outperforming behavior of HS has been evinced in several application scenarios [8], with scarce albeit interesting practices regarding the optimization of cost-efficient deployments [9–11].

The HS solver proposed in this work utilizes three improvisation operators that are applied to the harmonies within the harmony memory at every iteration:

- *Harmony Memory Considering Rate* HMCR $\in [0, 1]$, which establishes the probability that the improvised value for a note is drawn from the values taken by the same note in the other $K-1$ harmonies in the harmony memory.
- *Pitch Adjusting Rate* PAR $\in [0, 1]$, which tunes the probability that the value of a note is replaced with any of its neighboring values in its alphabet. In order to establish a proper relation of vicinity BS profiles are sorted in increasing order of their bandwidth, in such a way that they are equally ordered in terms of cost.
- *Random Selection Rate* RSR $\in [0, 1]$, which is similar to PAR: it sets the probability that the new value for a given note is taken uniformly at random (i.e. without any neighborhood consideration) from its alphabet.

The above operators are applied to each note of the candidate solutions, after which the fitness functions of the newly improvised solutions (namely, coverage ratio and deployment cost) are evaluated. Based on the values of the objective functions for both the new harmonies and those from the previous iteration a Pareto dominance ranking and crowding distance criterion is applied to filter out the prevailing population for the next iteration. In short each solution is ranked according to its dominance level (namely, `rank` = 1 for the best, non-dominated subset of solutions, `rank` = 2 for the second best front, etc.). After ranking all solutions the sum of distances to the closest harmony along each metric (crowding distance) permits to prioritize solutions within a certain dominance level: those solutions featuring large crowding distances are preferential.

6 Results

In Fig. 3, we can observe the Pareto efficiency curve achieved by HS for the defined profile combinations. The HS algorithm has been configured with $K = 30$, HMCR $= 0.5$, PAR $= 0.15$, RSR $= 0.1$, $\mathcal{I} = 200$ iterations. The overall cost has been normalized by the most expensive scenario: the overall number of transmitters at the most expensive cost (i.e. 1.2710e+07). The amount of 10^6 vehicles have been distributed according to the radial distribution in (1) using $c = 0.01$ and $r_{max} = 25$ km around the 30 most populated villages in the region. The temporal factor F has been chosen as 0.7 in the rush hour. We have also distributed in a uniform random manner 10% of the running vehicles on roads out of the 30 circular areas.

Solutions A, B and C have been highlighted in Fig. 3 with 205, 150 and 70 transmitters switched off out of the overall 413, respectively. One can observe that an acceptable 55% of coverage level can be achieved with the minimum Pareto efficiency value (208 transmitters on). This result arises from the effect of a radial traffic distribution model centered in cities, most of cars will be distributed close to the city where a higher number of transmitters are usually deployed. As a result, those BSs can provide coverage to more amount of users compared with the rural BSs which are deployed far away from urban areas.

In Fig. 4 the deployment profiles of the marked scenarios are depicted. In Fig. 4a it can be observed the high amount of BS in low profiles (type 1 (205 BSs) and 2) and that a few BSs are set with profiles with high cost (type 9 and 10). This deployment provides a coverage level of 55%. In Fig. 4b we show an intermediate point where the trade-off between cost and coverage level is interesting: 82% of users on roads are covered at rush hour with a reduced cost since 150 transmitters can be off and rest of profiles are mainly low-cost. Finally, we can see the C deployment in Fig. 4c where the highest coverage level can be achieved. One can notice that deployments at the rightmost part of the Pareto curve with higher cost can not achieve that coverage level. These scenarios have shown the importance of deployment analysis for operator in terms of cost and coverage since adequate deployments with lower cost can obtain relevant results in terms of coverage.

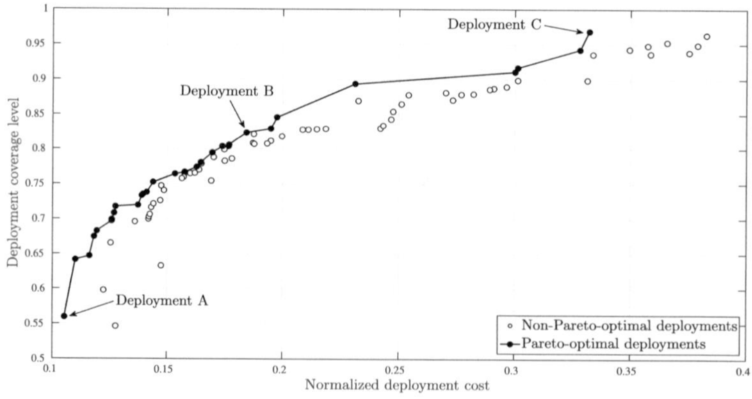

Fig. 3. Pareto efficiency curve of V2I deployment coverage level for the basque region in Spain.

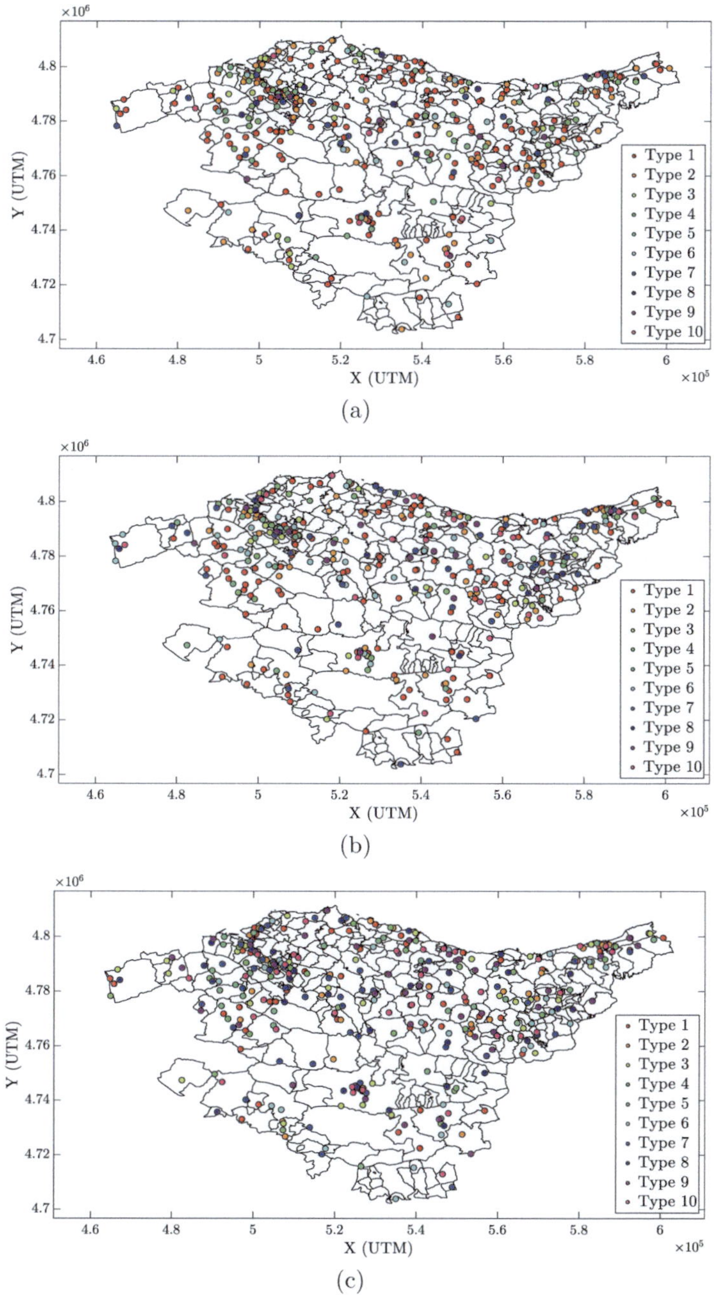

Fig. 4. Deployment profiles of the A, B and C pareto solutions.

7 Conclusions

This paper have presented a multi-objective HS-based strategy to improve the usage of already deployed wireless network infrastructure for massive data collection from vehicles over regional areas. This HS strategy is based on a non-dominated sorting criterion in terms of two conflicting fitness metrics: the level of service (number of connected cars) and maintenance cost of the network. Realistic results have been obtained by leveraging open data from the Basque Country region in Spain. The set of Pareto-optimal solutions have demonstrated the utility of this tool for local operators in their process of providing cost-efficient resources for the deployment of a wireless network aimed at collecting data from vehicles. An important reduction of the budget can be achieved by carefully selecting the operation profiles of the already deployed wireless infrastructure.

Acknowledgments. This work has been financially supported in part by the University of the Basque Country (UFI 11/30), in part by the Basque Government (IT-683-13 and ELKARTEK program under BID3A and BID3ABI projects) and in part by the Spanish Ministry of Economy and Competitiveness under Project 5G-NewBROs (TEC2015-66153-P MINECO/FEDER) and the European Regional Development Fund, ERDF.

References

1. Papadimitratos, P., Fortelle, A.D.L., Evenssen, K., Brignolo, R., Cosenza, S.: Vehicular communication systems: enabling technologies, applications, and future outlook on intelligent transportation. IEEE Commun. Mag. **47**(11), 84–95 (2009)
2. Reininger, M., Miller, S., Zhuang, Y., Cappos, J.: A first look at vehicle data collection via smartphone sensors. In: IEEE Sensors Applications Symposium (SAS), pp. 1–6 (2015)
3. Paulin, T., Ruehrup, S., Fuxjaeger, P., Paier, A.: Improving V2I edge communication by performance maps. In: IFIP Wireless and Mobile Networking Conference (WMNC), pp. 200–207 (2015)
4. OTEUS: Basque Transport Observatory. http://www.garraioak.ejgv.euskadi.eus/r41-4833/es/
5. Cellnex Telecom. http://www.cellnextelecom.com/
6. Itelazpi: Basque Public Telecommunications Company. http://www.itelazpi.eus/
7. Geem, Z.W., Kim, J.-H., Loganathan, G.V.: A new heuristic optimization algorithm: harmony search. Simulation **76**(2), 60–68 (2001)
8. Manjarres, D., Landa-Torres, I., Gil-Lopez, S., Del Ser, J., Bilbao, M.N., Salcedo-Sanz, S., Geem, Z.W.: A survey on applications of the harmony search algorithm. Eng. Appl. Artif. Intell. **26**(8), 1818–1831 (2013)
9. Bilbao, M.N., Del Ser, J., Salcedo-Sanz, S., Gil-López, S., Portilla-Figueras, J.A.: A Bi-objective harmony search approach for deploying cost-effective multi-hop communications over large-area wildfires. In: Puerta, J.G., Ferreira, I.G., Bringas, P.G., Klett, F., Abraham, A., Carvalho, A.C.P.L.F., Herrero, Á., Baruque, B., Quintián, H., Corchado, E. (eds.) International Joint Conference SOCO'14-CISIS'14-ICEUTE'14. AISC, vol. 299, pp. 93–103. Springer, Heidelberg (2014). doi:10.1007/978-3-319-07995-0_10

10. Bilbao, M.N., Gil-Lopez, S., Del Ser, J., Salcedo-Sanz, S., Sanchez-Ponte, M., Arana-Castro, A.: Novel hybrid heuristics for an extension of the dynamic relay deployment problem over disaster areas. TOP **22**, 997–1016 (2014)
11. Bilbao, M.N., Del Ser, J., Salcedo-Sanz, S., Casanova-Mateo, C.: On the application of multi-objective harmony search heuristics to the predictive deployment of firefighting aircrafts: a realistic case study. Int. J. Bio-Inspired Comput. (IJBIC) **7**(5), 270–284 (2015)

Harmony Search and Engineering Optimization

Modified Harmony Search for Optimization of Reinforced Concrete Frames

Gebrail Bekdaş(✉) and Sinan Melih Nigdeli

Department of Civil Engineering, Istanbul University, Avcilar, 34320 Istanbul, Turkey
{bekdas,melihnig}@istanbul.edu.tr

Abstract. In this study, harmony search algorithm is modified with several random search stages for optimum design of reinforced concrete frame structures. The structure is subjected to both static and dynamic forces. A detailed optimum design was proposed without grouping the design variables of the frame structure. The dynamic forces calculated according to the time history analyses. The objective is to find the most economical design supporting the design requirements of ACI-318 (Building code requirements for reinforced concrete structure). The method was applied to a non-symmetric structure and the proposal is feasible.

Keywords: Modified harmony search · Reinforced concrete structures · Optimization · Frames · Metaheuristic methods · Structural optimization

1 Introduction

Cross-sectional dimensions of reinforced concrete (RC) elements are assumed by a designer and these dimensions are checked according to design requirements given in design codes. These codes propose us the minimum and maximum limits in order to ensure security but the design of engineer may not be the optimum one with minimum cost. Since reinforced concrete is a composite of concrete and steel bars, the dimensions of structural elements are depended to steel reinforcements. Additionally, the costs of concrete and steel are very different. For that reason, an optimization approach is a mandatory for the most economical design of reinforced concrete structures.

Several approaches have been proposed for the optimization of different RC members. A state of the art review, including the pioneering studies about the cost optimization of concrete structures was developed by Sarma and Adeli [1]. Also, the cost optimization of structures with fuzzy logic, genetic algorithm and parallel computing are described in the book developed by Adeli and Sarma [2].

Genetic algorithm (GA) was employed for the optimization of RC beams and the approach was compared with conventional trials of a designer [3]. Genetic algorithm was also used for RC biaxial columns in the optimization approach proposed by Rafiq and Southcombe [4]. Koumousis and Arsenis optimized several RC members by using the GA employed optimization approach [5]. Rajeev and

J. Del Ser (ed.), *Harmony Search Algorithm*, Advances in Intelligent Systems and Computing 514, DOI 10.1007/978-981-10-3728-3_21

Krishnamoorthy proposed a methodology based on GA for RC frames, including detailed reinforcement design [6]. Rath et al. optimized the shape of RC members by using sequential quadratic programming (SQP) technique [7]. The SQP technique was used to modify the rectangular shape of the cross-section and GA was employed for the cost optimization. Camp et al. optimally designed RC frames by considering slenderness of columns. In the study employing GA, two-bay six-story frame was optimized by grouping cross sections of beams and columns for a constant vertical distributed load and lateral loads [8]. According to various design codes, T-shaped RC beams were optimized by Ferreira et al. [9]. Leps and Sejnoha used the combination two metaheuristic methods (GA and simulated annealing (SA) method) in their approach developed for the optimum design of continuous beams [10]. Lee and Ahn employed GA for RC frames by using a database for possible design of members. Earthquake loads were considered as lateral equivalent static earthquake load applied as joint loads [11]. Balling and Yao proposed an optimization method for three-dimensional RC frames subjected to the several loads such as dead, live, snow and earthquake [12]. Ahmadkhanlou and Adeli [13] optimized RC slabs by using a methodology with two stages, including the use of the neural dynamics model developed by Adeli and Park [14,15] for optimum solution of continuous variables and perturbation technique to find practical values of the design variables. Optimum expressions of the bending moment, steel area and the ratio for singly and doubly RC beams were proposed by Barros et al. [16]. In order to minimize the total cost of the pre-stressed concrete bridges, an optimization procedure considering the cost of concrete, reinforcement, formwork and fabrication was developed by Sirca Jr and Adeli [17]. Govindaraj and Ramasamy optimized continuous beams by using GA. The possible optimum results were chosen from a typical database with reinforcement template [18]. Sahab et al. optimized RC flat slab building by using a hybrid optimization algorithm which is the combination of GA and discretized form of the Hook and Jeeves method [19]. Zou and Chan proposed an optimization approach for reinforced concrete buildings using nonlinear pushover analyses in order to conduct the performance-based seismic design [20]. Govindaraj and Ramasamy were optimized statically load RC frames by using GA [21]. Guerra and Kiousis optimized several frame structures such as single-bay multi-story and multi-bay single story RC frames [22]. Paya et al. proposed a multi-objective optimization approach for RC frames using simulated annealing [23]. Perea et al. proposed two heuristic methods (random walk and the descent local search) and two metaheuristic methods (the threshold accepting and the simulated annealing) for the optimum design of RC bridges frames [24]. RC frames were optimized by considering minimum embedded CO2 emissions and cost by employing simulated annealing [25] and big bang-big crunch [26]. The objectives are the cost of the constructability, the environmental impact and the overall safety in these studies. Gil-Martin et al. proposed the reinforcement sizing diagram (RSD) approach for RC beams and biaxial columns [27]. Optimum depth and reinforcement of rectangular RC elements in bending were investigated by Barros et al. [28]. Fedghouche and Tiliouine optimized T-shaped beams according to Eurocode2 by using GA [29]. Several metaheuristic methods such as simulated annealing [30,31], big bang-big crunch method [32], harmony search [33] and charged system search [34] were used for optimum design of RC retaining walls. The harmony search algorithm was used for optimum design of RC continuous beams [35], RC frames [36,37] and T-shaped RC beams [38]. The combination of several metaheuristic algorithms was employed for the optimum design of RC frames by Kaveh and Sabzi [39]. For optimum design of RC beams, a methodology using big bang-big crunch was proposed by Kaveh and Sabzi [40]. A metaheuristic method for optimization of hybrid fiber-reinforced composite plates is proposed by combining the best features of classical local search algorithms such as SA and tabu search (TS) [41].

By using a classical form of an algorithm, a detailed optimization cannot be found since the number of the types of design variables is too many. Only, particular optimum designs can be found for several members of the structure. The other members of the structure may randomly assigned with local optimum solutions or a solution with constraint violation. In that case, the classical algorithms can be modified with additional random stages.

In this study, harmony search algorithm is modified with random stages for different types of design steps. RC frame structures are excited by static and dynamic forces. Earthquake effects are defined according to time history analyses of three recorded earthquakes. In the optimization methodology, a detailed reinforcement design considering the positioning of the bars is done without grouping the members. The optimum design of the RC elements was done according to ACI-318 building code requirements for structural concrete [42] in order to minimize the material cost. The method was presented with a multi-bay multi-story non-symmetric RC frame structure.

2 Optimum Design Methodology

In the design methodology, the music inspired harmony search algorithm [43] is proposed with random search stages in order to find effective global solutions for cost minimization of RC frame structures. In Fig. 1, the flowchart of the methodology is presented.

In the first step, the design constants of RC frame, loading condition (live-L and Dead-D loads), earthquake records and design variable ranges are defined.

In the next step, random search stages are conducted. In step 2a, the cross-section dimensions of the beams are generated and ACI-318 constraints are checked. In the step 2b, the cross-section dimensions of the columns are generated and the constraints are checked. After the dimensions of the structure is defined, static and dynamic analyses are done in step 2c. Slenderness is an important factor in RC design. For that reason, the design constraints about preventing brittle fracture are checked in step 2d and the process continue from the step 2a if a violation occurs. In steps 2e and 2f, the reinforcement designs of beams and columns are respectively done. The steps 2a, 2b, 2e and 2f are repeated until the corresponding constraints are provided.

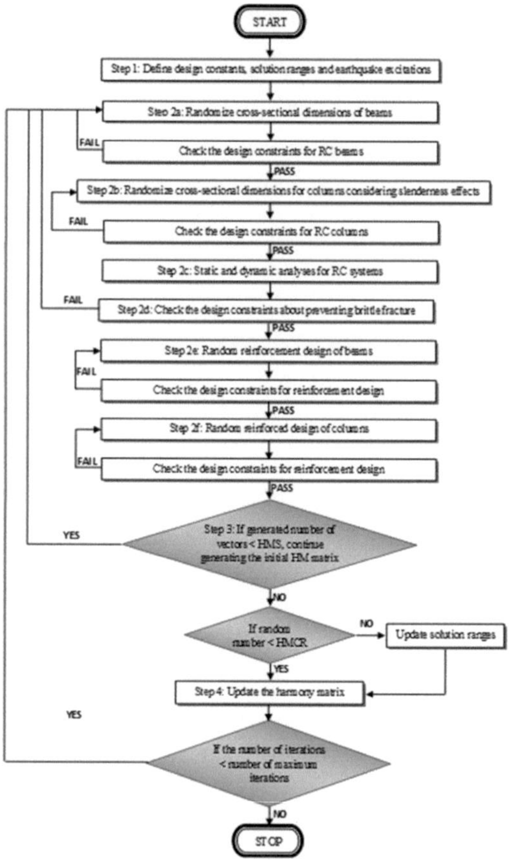

Fig. 1. Flowchart of the optimization process

Table 1. Earthquake records

Earthquake	Date	Station	Component	PGA(g)
Imperial Valley	1940	117 El Centro	I-ELC180	0.313
Northridge	1994	24514 Sylmar	SYL360	0.843
Loma Prieta	1989	16 LGPC	LGP000	0.563

Note: Earthquake records were taken from PEER NGA Database [44]

In the step 3, the number of generations is checked and the step 2 is done for harmony memory size (HMS). Thus, the initial harmony matrix is generated.

In step 4, the solutions are updated according to the rules of the HS algorithm. The objective of the problem is to minimize the total material cost of the frame including concrete cross sections and steel reinforcement bars. Elimination of the existing set of solutions is done according to the total cost of the frames.

Table 2. Design constants and ranges of design variables

Definition	Symbol	Unit	Value
Range of web width	b_w	mm	250–400
Range of height	h	mm	300–600
Clear cover	c_c	mm	30
Range of reinforcement	ϕ	mm	16–30
Range of shear reinforcement	ϕ_v	mm	8–14
Max. aggregate diameter	D_{max}	mm	16
Yield strength of steel	f_y	MPa	420
Comp. strength of concrete	$f_c^{'}$	MPa	25
Elasticity modulus of steel	E_s	MPa	200000
Specific gravity of steel	γ_s	t/m^3	7.86
Specific gravity of concrete	γ_c	t/m^3	2.5
elastic response parameter	R	-	8.5
Cost of the concrete per m^3	C_c	$	40
Cost of the steel per ton	C_s	$	400

3 Numerical Examples

The numerical example is a three-span three-story RC frame. The optimization was conducted under three different earthquake records given in Table 1. The value of design constants and the ranges are shown in Table 2. The dimensions

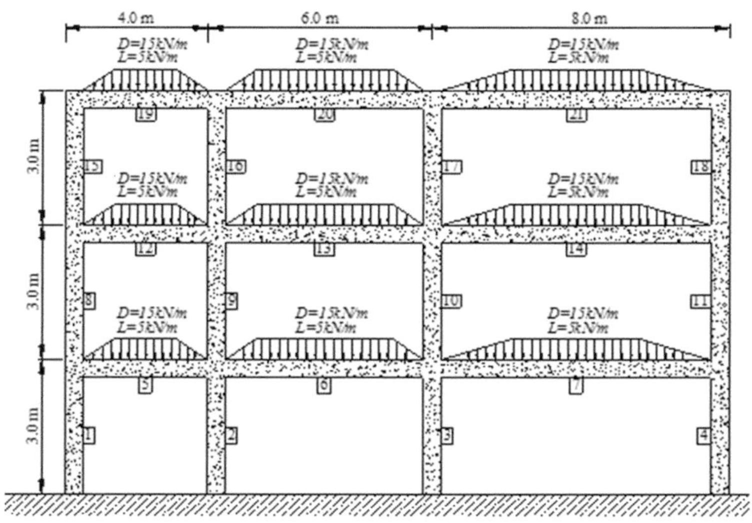

Fig. 2. Model of the numerical example

Table 3. Optimum design of columns

Element Number	b_w (mm)	h (mm)	Bars in each face	Shear reinforcement diameter/distance (mm)
1	250	300	2ϕ10+2ϕ12	ϕ8/120
2	250	350	3ϕ10+1ϕ12	ϕ8/150
3	250	350	1ϕ12+1ϕ18+1ϕ10	ϕ8/150
4	250	300	2ϕ10+2ϕ12	ϕ8/120
8	250	300	2ϕ10+2ϕ12	ϕ8/120
9	250	300	2ϕ10+2ϕ12	ϕ8/120
10	250	300	4ϕ10+1ϕ12	ϕ8/120
11	250	300	2ϕ10+2ϕ12	ϕ8/120
15	250	300	2ϕ10+2ϕ12	ϕ8/120
16	250	300	2ϕ10+2ϕ12	ϕ8/120
17	250	300	1ϕ14+2ϕ12	ϕ8/120
18	250	300	3ϕ18	ϕ8/120

Table 4. Optimum design of beams

Element Number	bw (mm)	h (mm)	Bars in comp. section	Bars in tensile section	Shear reinforcement diameter/distance (mm)
LJ5	250	350	1ϕ10+2ϕ14+1ϕ16	1ϕ24+1ϕ28	ϕ8/150
5			2ϕ10+1ϕ12	1ϕ14+1ϕ16+1ϕ12	
RJ5-LJ6	250	350	1ϕ10+1ϕ26	1ϕ16+ 2ϕ14+1ϕ28	ϕ8/150
6			2ϕ10+ 1ϕ12	1ϕ20+ 1ϕ14	
RJ6-LJ7	250	450	1ϕ22+1ϕ14+1ϕ18	1ϕ20+1ϕ18+1ϕ26+1ϕ22	ϕ8/200
7			1ϕ10+1ϕ12+1ϕ14	1ϕ24+ 1ϕ20	
RJ7			1ϕ20+ 1ϕ22	1ϕ14+1ϕ16+1ϕ28+1ϕ20	
LJ12	250	450	1ϕ20+ 1ϕ12	1ϕ30+1ϕ12	ϕ8/200
12			1ϕ16+ 1ϕ14	1ϕ12+ 1ϕ22	
RJ12-LJ13	250	400	4ϕ12	1ϕ28+ 1ϕ18	ϕ8/170
13			4ϕ10	1ϕ14+ 1ϕ12+1ϕ16	
RJ13-LJ14	250	400	1ϕ28+ 1ϕ14	2ϕ28+ 1ϕ18	ϕ8/170
14			1ϕ16+ 1ϕ12	1ϕ10+ 2ϕ12+1ϕ26	
RJ14			1ϕ16+1ϕ22+1ϕ12	1ϕ10+1ϕ22+1ϕ28+1ϕ16	
LJ19	250	400	1ϕ16+ 1ϕ12	1ϕ10+ 1ϕ18+1ϕ14	ϕ8/170
19			2ϕ14	2ϕ14	
RJ19-LJ20	250	300	1ϕ12+ 1ϕ22	1ϕ10+ 2ϕ22+1ϕ12	ϕ8/120
20			2ϕ12	1ϕ10+1ϕ16+1ϕ12+1ϕ14	
RJ20-LJ21	250	400	1ϕ24+1ϕ18	1ϕ18+1ϕ26+1ϕ24+1ϕ12	ϕ8/170
21			1ϕ12+ 1ϕ16	1ϕ24+ 1ϕ26	
LJ21			1ϕ20+ 1ϕ12	1ϕ26+ 1ϕ20	

are assigned to the values with are the multiples of 50 mm. The diameters of reinforcement bars are assigned to the values which are the multiples of 2 mm. Thus, the final results are practical for construction. The three-span three-story structure is given in Fig. 2. For the spans, the ratio of a and l is 0.25.

4 Results and Conclusions

The optimum results are given in Tables 3 and 4 for column and beam elements, respectively. The total cost of the optimum design is 724.54 \$. The optimum value for the first five attempts (1148 \$) is nearly 58% higher than the optimum cost. These five attempts may be the classical solution of an engineer. For that reason, the modified HS with random search stages is a feasible approach for the investigated civil engineering problem.

References

1. Sarma, K.C., Adeli, H.: Cost optimization of concrete structures. J. Struct. Eng. **124**(5), 570–578 (1998)
2. Adeli, H., Sarma, K.: Cost Optimization of Structures - Fuzzy Logic, Genetic Algorithms, and Parallel Computing. Wiley (2006)
3. Coello, C.C., Hernandez, F.S., Farrera, F.A.: Optimal design of reinforced concrete beams using genetic algorithms. Expert Syst. Appl. **12**, 101–108 (1997)
4. Rafiq, M.Y., Southcombe, C.: Genetic algorithms in optimal design and detailing of reinforced concrete biaxial columns supported by a declarative approach for capacity checking. Comput. Struct. **69**, 443–457 (1998)
5. Koumousis, V.K., Arsenis, S.J.: Genetic algorithms in optimal detailed design of reinforced concrete members. Comput-Aided Civ. Inf. **13**, 43–52 (1998)
6. Rajeev, S., Krishnamoorthy, C.S.: Genetic algorithm-based methodology for design optimization of reinforced concrete frames. Comput-Aided Civ. Inf. **13**, 63–74 (1998)
7. Rath, D.P., Ahlawat, A.S., Ramaswamy, A.: Shape optimization of RC flexural members. J. Struct. Eng.-ASCE **125**, 1439–1446 (1999)
8. Camp, C.V., Pezeshk, S., Hansson, H.: Flexural design of reinforced concrete frames using a genetic algorithm. J. Struct. Eng.-ASCE **129**, 105–111 (2003)
9. Ferreira, C.C., Barros, M.H.F.M., Barros, A.F.M.: Optimal design of reinforced concrete T-sections in bending. Eng. Struct. **25**, 951–964 (2003)
10. Leps, M., Sejnoha, M.: New approach to optimization of reinforced concrete beams. Comput. Struct. **81**, 1957–1966 (2003)
11. Lee, C., Ahn, J.: Flexural design of reinforced concrete frames by genetic algorithm. J. Struct. Eng.-ASCE **129**(6), 762–774 (2003)
12. Balling, R., Yao, X.: Optimization of reinforced concrete frames. J. Struct. Eng. ASCE **123**, 193–202 (1997)
13. Ahmadkhanlou, F., Adeli, H.: Optimum cost design of reinforced concrete slabs using neural dynamics model. Eng. Appl. Artif. Intell. **18**(1), 65–72 (2005)
14. Adeli, H., Park, H.S.: Optimization of space structures by neural dynamics. Neural Networks **8**(5), 769–781 (1995)
15. Adeli, H., Park, H.S.: Neurocomputing for Design Automation. CRC Press (1998)
16. Barros, M.H.F.M., Martins, R.A.F., Barros, A.F.M.: Cost optimization of singly and doubly reinforced concrete beams with EC2-2001. Struct. Multidisciplinary Optim. **30**(3), 236–242 (2005)
17. Sirca, G.F., Adeli, H.: Cost optimization of prestressed concrete bridges. J. Struct. Eng. **131**(3), 380–388 (2005)
18. Govindaraj, V., Ramasamy, J.V.: Optimum detailed design of reinforced concrete continuous beams using genetic algorithms. Comput. Struct. **84**(1), 34–48 (2005)

19. Sahab, M.G., Ashour, A.F., Toropov, V.V.: Cost optimisation of reinforced concrete flat slab buildings. Eng. Struct. **27**(3), 313–322 (2005)
20. Zou, X.K., Chan, C.M.: Optimal seismic performance-based design of reinforced concrete buildings using nonlinear pushover analysis. Eng. Struct. **27**(8), 1289–1302 (2005)
21. Govindaraj, V., Ramasamy, J.V.: Optimum detailed design of reinforced concrete frames using genetic algorithms. Eng. Optim. **39**(4), 471–494 (2007)
22. Guerra, A., Kiousis, P.D.: Design optimization of reinforced concrete structures. Comput. Concrete **3**(5), 313–334 (2006)
23. Paya, I., Yepes, V., Gonzalez-Vidosa, F., Hospitaler, A.: Multiobjective optimization of concrete frames by simulated annealing. Comput.-Aided Civil Infrastruct. Eng. **23**(8), 596–610 (2008)
24. Perea, C., Alcala, J., Yepes, V., Gonzalez-Vidosa, F., Hospitaler, A.: Design of reinforced concrete bridge frames by heuristic optimization. Adv. Eng. Softw. **39**(8), 676–688 (2008)
25. Paya-Zaforteza, I., Yepes, V., Hospitaler, A., Gonzalez-Vidosa, F.: CO_2 optimization of reinforced concrete frames by simulated annealing. Eng. Struct. **31**(7), 1501–1508 (2009)
26. Camp, C.V., Huq, F.: CO_2 and cost optimization of reinforced concrete frames using a big bang-big crunch algorithm. Eng. Struct. **48**, 363–372 (2013)
27. Gil-Martín, L.M., Hernandez-Montes, E., Aschheim, M.: Optimal reinforcement of RC columns for biaxial bending. Mater. Struct. **43**(9), 1245–1256 (2010)
28. Barros, A.F.M., Barros, M.H.F.M., Ferreira, C.C.: Optimal design of rectangular RC sections for ultimate bending strength. Struct. Multidisciplinary Optim. **45**(6), 845–860 (2012)
29. Fedghouche, F., Tiliouine, B.: Minimum cost design of reinforced concrete T-beams at ultimate loads using Eurocode2. Eng. Struct. **42**, 43–50 (2012)
30. Ceranic, B., Fryer, C., Baines, R.W.: An application of simulated annealing to the optimum design of reinforced concrete retaining structures. Comput. Struct. **79**(17), 1569–1581 (2001)
31. Yepes, V., Alcala, J., Perea, C., Gonzalez-Vidosa, F.: A parametric study of optimum earth-retaining walls by simulated annealing. Eng. Struct. **30**(3), 821–830 (2008)
32. Camp, C.V., Akin, A.: Design of retaining walls using big bang–big crunch optimization. J. Struct. Eng. **138**(3), 438–448 (2011)
33. Kaveh, A., Abadi, A.S.M.: Harmony search based algorithms for the optimum cost design of reinforced concrete cantilever retaining walls. Int. J. Civil Eng. **9**(1), 1–8 (2011)
34. Talatahari, S., Sheikholeslami, R., Shadfaran, M., Pourbaba, M.: Optimum design of gravity retaining walls using charged system search algorithm. Mathematical Problems in Engineering (2012)
35. Akin, A., Saka, M.P.: Optimum detailed design of reinforced concrete continuous beams using the harmony search algorithm. In: Proceedings of the Tenth International Conference on Computational Structures Technology, vol. 131, pp. 109–119 (2010)
36. Akin, A., Saka, M.P.: Optimum detailing design of reinforced concrete plane frames to ACI 318–05 using the harmony search algorithm. In: Proceedings of the Eleventh International Conference on Computational Structures Technology 72 (2012)
37. Akin, A., Saka, M.P.: Harmony search algorithm based optimum detailed design of reinforced concrete plane frames subject to ACI 318–05 provisions. Comput. Struct. **147**, 79–95 (2015)

38. Bekdaş, G., Nigdeli, S.M.: Cost optimization of T-shaped reinforced concrete beams under flexural effect according to ACI 318. In: 3rd European Conference of Civil Engineering (2012)
39. Kaveh, A., Sabzi, O.: A comparative study of two meta-heuristic algorithms for optimum design of reinforced concrete frames. Int. J. Civil Eng. **9**(3), 193–206 (2011)
40. Kaveh, A., Sabzi, O.: Optimal design of reinforced concrete frames using big bang-big crunch algorithm. Int. J. Civil Eng. **10**(3), 189–200 (2012)
41. Rama Mohan Rao, A.R., Shyju, P.P.: A.R., Shyju, P. P.: A meta-heuristic algorithm for multi-objective optimal design of Hybrid Laminate Composite Structures. Comput.-Aided Civil Infrastruct. Eng. **25**(3), 149–170 (2010)
42. ACI 318M–05, Building code requirements for structural concrete and commentary, American Concrete Institute (2005)
43. Geem, Z.W., Kim, J.-H., Loganathan, G.V.: A new heuristic optimization algorithm: harmony search. Simulation **76**(2), 60–68 (2001)
44. PEER Pacific earthquake engineering resource center: NGA database. University of California, Berkeley (2005). http://peer.berkeley.edu/nga. Accessed Nov 2011

Optimum Tuning of Mass Dampers by Using a Hybrid Method Using Harmony Search and Flower Pollination Algorithm

Sinan Melih Nigdeli[1(✉)], Gebrail Bekdaş[1], and Xin-She Yang[2]

[1] Department of Civil Engineering, Istanbul University, Avcılar, 34320 Istanbul, Turkey
{melihnig,bekdas}@istanbul.edu.tr

[2] Design Engineering and Mathematics, Middlesex University London, The Burroughs, London NW4 4BT, UK
x.yang@mdx.ac.uk

Abstract. In this study, a new approach is proposed for optimization of tuned mass damper positioned on the top of seismic structures. The usage of metaheuristic algorithms is a well-known and effective technique for optimum tuning of parameters such as mass, period and damping ratio. The aim of the study is to generate a new methodology in order to improve the computation capacity and precision of the final results. For that reason, harmony search (HS) and flower pollination algorithm (FPA) are hybridized by proposing a probability based approach. In the methodology, global and local search processes of HS are used together with global and local pollination stages of FPA. In that case, four different types of generation are used. In the methodology, these four types of generation have the same chance at the start of the optimization process and probabilities are reduced when the corresponding type of the generation is chosen. If an improvement is provided for the objective of the optimization, the probability of the effective type is increased. The proposed method has an effective convergence by providing improvement of the optimization objective comparing to classical FPA.

Keywords: Tuned mass damper · Optimization · Earthquake · Harmony search · OptFlower pollination algorithm · Structures

1 Introduction

In passive seismic control of structures, the properties of the control systems must be optimized for a feasible reduction of structural vibrations. Tuned mass dampers (TMDs) are mechanical devices which are also used in structure subjected to undesired excitations. In that case, the properties of TMDs such as mass, period and damping ratio must be tuned and the problem is nonlinear because of inherent damping and vibrations with random frequency. For that reason, numerical algorithms have been used. Most of the proposed methods employ metaheuristic algorithms.

J. Del Ser (ed.), *Harmony Search Algorithm*, Advances in Intelligent Systems and Computing 514, DOI 10.1007/978-981-10-3728-3_22

The oldest and well-known method in tuning of TMDs was derived by Den Hartog. The formulations of Den Hartog are for single degree of freedom systems subjected to harmonic loading. Also, the inherent damping of the system is not considered. For that reason, this method is not effective for multiple degree of freedom structures subjected to earthquake excitation [1]. For frequency and damping ratio, Warburton proposed closed form expression for harmonic and random excitations [2]. Since the formulations of Den Hartog [1] and Warburton [2] are lack of consideration of inherent damping, Sadek et al. [3] proposed to use numerical trials and obtained the formulations by using curve fitting. In that case, the inherent damping of the main structure was considered and a proposal is done for multiple degree of freedom structures by considering a single vibration mode. In order to consider all vibration modes of multiple degree of freedom structures, the use of numerical algorithms is needed for a precise optimum in order to consider the limits of design parameters of TMD and design constraints such as stroke capacity of TMD. For that reason, numerical optimization techniques have been developed for tuning of TMDs [4–7].

In recent years, metaheuristic methods inspired from nature or a process are the most popular employed methods in the optimization of TMDs. The employed metaheuristic methods contain evolutionary algorithms, swarm intelligence, biology inspired algorithms. Genetic algorithms are the oldest method used for the TMD optimization problem [8–12]. Swarm intelligence was used in the optimum TMD design using particle swarm optimization [13,14]. The examples of other metaheuristics used for TMD design are bionic optimization [15], music inspired harmony search [16–19], ant colony optimization [20], artificial bee optimization [21], shuffled complex evolution [22] and education inspired teaching learning based optimization [23].

In this study, the flower pollination algorithm (FPA) used in the development of a methodology for TMD design. In order to increase the performance of the algorithm, FPA is hybridized with harmony search algorithm. A probability based strategy is proposed for the decision of the type of the generation of new solutions. In the numerical example, a 40 story structure is investigated for optimum TMD with a limited stroke capacity. A global optimum was searched for 44 different earthquake records. The example is optimized for the classical FPA and the developed hybrid method (HS-FPA).

2 Equations of Motion for TMD Implemented Earthquake Excited Structures

As seen in Fig. 1, the N-story structure is modeled as a shear building. The structure has N degree of freedom for all lateral motion of the stories and the number of degrees of freedom of the TMD controlled structure is $N{+}1$. The equations of motion of a shear building is written as follows:

$$M\ddot{x}(t) + C\dot{x}(t) + Kx(t) = -M1\ddot{x}_g(t) \tag{1}$$

The M, C and K matrices are shown as Eqs. (2)–(4) and these are diagonal lumped mass, damping and stiffness matrices, respectively. In the equations of motion, x(t), $\ddot{x}_g(t)$ and 1 are the vector containing structural displacements of all stories and TMD (given as Eq. (5)), ground acceleration and a vector of ones with a dimension of (N + 1, 1), respectively.

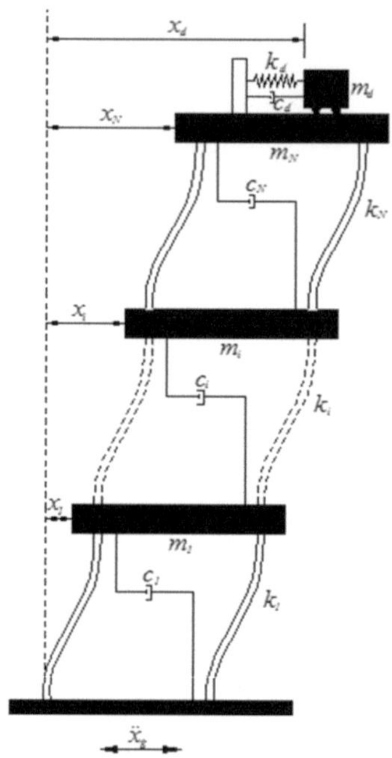

Fig. 1. Model of N-story shear building including a TMD on the top

$$M = diag[m_1 \quad m_2 \quad \ldots \quad m_N \quad m_d] \tag{2}$$

$$\begin{bmatrix} (c_1+c_2) & c_2 & & & & \\ -c_2 & (c_2+c_3) & -c_3 & & & \\ & & \vdots & \vdots\vdots & & \\ & & \vdots & \vdots\vdots & & \\ & & & c_N & (c_N+c_d) & -c_d \\ & & & & -c_d & c_d \end{bmatrix} \tag{3}$$

$$\begin{bmatrix} (k_1+k_2) & k_2 & & & & \\ -k_2 & (k_2+k_3) & -k_3 & & & \\ & & \vdots & \vdots\vdots & & \\ & & \vdots & \vdots\vdots & & \\ & & & k_N & (k_N+k_d) & -k_d \\ & & & & -k_d & k_d \end{bmatrix} \quad (4)$$

$$x(t) = [x_1 \quad x_2 \ldots x_N \quad x_d]^T \quad (5)$$

In the matrices and vectors given in the paper, m_i, mc_i, k_i and x_i are mass, damping coefficient, stiffness coefficient and displacement of ith story of structure for i = 1:N. The parameters which need to be optimized, are mass (m_d), damping coefficient (c_d) and stiffness coefficient (k_d) of TMD. The displacement of the TMD is represented with x_d. The damping and stiffness coefficients can be found from the design variables such as period (T_d) and damping ratio (ξ_d) of TMD as shown in Eqs. (6) and (7).

$$T_d = 2\pi\sqrt{\frac{m_d}{k_d}}, \quad (6)$$

$$\xi_d = \frac{c_d}{2m_d\sqrt{\frac{k_d}{m_d}}} \quad (7)$$

3 Flower Pollination Algorithm and the Optimization Methodology

The flower pollination developed by Yang [24] is a nature inspired metaheuristic algorithm. The main source of the algorithm is the pollination process of flowering plants. The pollination process can be explained in two types. These types are cross-pollination done by the pollen transfer of pollinators or self-pollination done by same flower or plant. The inspired rules of flower pollination are as follows:

1. The pollinators such as birds, insects or bees obey the rules of Lévy distribution used in the global pollination process in order to formulize the cross-pollination process.
2. In self-pollination, the reproduction process occurs from the pollens of the same flower or plant. This type of pollination is the inspiration of the local pollination process of the algorithm.
3. In order to consider the similarity of two flowers involved in the pollination, a probability of reproduction called flower constancy is used.
4. A switch between the global and local pollination process is used and the name of the parameter is the switch probability (p).

In methodology, structural properties, external excitations and ranges of design variables are defined as constants. Then, the structure without TMD is analyzed in order to compare the effectiveness of the TMD. After that, initial solutions are generated for TMD parameters such as mass, period and damping ratio. For all set of variables, the dynamic analyses are done for the structure. Then, the essential optimization process starts.

In the global pollination, the generation of a new i_{th} solution (x_i^{t+1}) is done by using Eq. (8). In this equation, L is a Lévy distribution, (x_i^t) is the existing solution and (g^*) is the best solution of the population.

$$x_i^{t+1} = x_i^t + L(x_i^t - g^*) \tag{8}$$

The local pollination is formulized as seen in Eq. (9). In this equation, two random solutions (j^{th} and k^{th}) are used with a random number (ϵ) between 0 and 1.

$$x_i^{t+1} = x_i^t + \epsilon(x_j^t - x_k^t) \tag{9}$$

In this paper, FPA is combined with the music inspired harmony search (HS) algorithm developed by Geem et al. [25]. The global search of the music inspired HS is formulized in Eq. (10). The lower and upper bounds of the design variables are defined with x_L and x_U, respectively.

$$x_i^{t+1} = x_L + \epsilon(x_U - x_L) \tag{10}$$

The local search is done in a smaller range around an existing solution. It is formulated as follows.

$$x_i^{t+1} = x_j + (\epsilon - 0.5)PAR(x_U - x_L) \tag{11}$$

The parameter PAR is the pitch adjusting rate. It is used for reducing the solution range. For all generations defined in Eqs. (8)–(11), new design variables must be limited with x_L and x_U.

In HS, harmony memory considering rate (HMCR) is used control the type of the search. In the proposed approach, HMCR and p parameters of the both algorithms are not used. Instead of these parameters, a probabilistic approach is proposed and the two algorithms are combined.

At the start of the optimization, all four generation types (global and local optimizations of the two algorithms) have same possibility. The type of the generation is chosen according to a random number and possibility is reduced according to the maximum iteration number. If the solution of the objective is better than the existing ones, the possibility is increased. In that way, the possibility of the chosen type is reduced, but the reduction is not effective for the generation type with better results. Thus, the best type of generation will gain more chance.

The new solutions are updated according to the objective function. The maximum stroke capacity of the TMD is also considered. The main objective is to reduce the maximum top story displacement of the structure under a user defined value (x_{max}) as defined in Eq. (12). The other objective is about the limitation of the stroke of the TMD. This objective is formulated as Eq. (13). First, the objective defined as Eq. (13) is considered. If this objective function is lower than the user defined limit: st_max, the main objective function given in Eq. (12) is considered. This iterative optimization continue for a user defined maximum iterations.

$$max|x_N| \leq x_{max}, \tag{12}$$

$$\frac{max[|x_{N+1} - x_N|]_{withTMD}}{max[|x_N|]_{withoutTMD}} \leq st\_max. \tag{13}$$

4 Numerical Examples

The proposed method was applied to the optimization of a TMD on the 40-story structure [26]. The properties of the structure as shown in Table 1. The example structure is a shear building with different stiffness and damping coefficient for stories. These values are linearly decreased by increase of the number of stories. For limitation of the stroke capacity, st_max is taken as 1.5 and the optimization is done for 200 maximum iterations. The value of PAR is taken as 0.5 while the population number is 5 for the numerical example.

Table 1. Properties of the example building

m_i(t)	980
k_1-k_{40}(MN/m)	2130-998
c_1-c_{40}	42.6-20

For global optimization, 44 earthquake records (2 components of 22 stations of historical earthquakes) were used in optimum tuning of TMD. The information about the excitations are shown in Table 2 and these records are the far-field ground motion records presented in FEMA P-695 [27].

The design variable ranges and the optimum parameters for the FPA and the proposed HS-FPA approaches are shown in Table 3. According to the results, the maximum displacement under critical excitation (1.9278 m) is reduced to 1.7957 m and 1.72047 m for FPA and HS-FPA methods, respectively. The critical

Table 2. FEMA P-695 far-field ground motion records [27]

Earthquake number	Date	Name	Component 1	Component 2
1	1994	Northridge	NORTHR/MUL009	NORTHR/MUL279
2	1994	Northridge	NORTHR/LOS000	NORTHR/LOS270
3	1999	Duzce, Turkey	DUZCE/BOL000	DUZCE/BOL090
4	1999	Hector Mine	HECTOR/HEC000	HECTOR/HEC090
5	1979	Imperial Valley	IMPVALL/H-DLT262	IMPVALL/H-DLT352
6	1979	Imperial Valley	IMPVALL/H-E11140	IMPVALL/H-E11230
7	1995	Kobe, Japan	KOBE/NIS000	KOBE/NIS090
8	1995	Kobe, Japan	KOBE/SHI000	KOBE/SHI090
9	1999	Kocaeli, Turkey	KOCAELI/DZC180	KOCAELI/DZC270
10	1999	Kocaeli, Turkey	KOCAELI/ARC000	KOCAELI/ARC090
11	1992	Landers	LANDERS/YER270	LANDERS/YER360
12	1992	Landers	LANDERS/CLW-LN	LANDERS/CLW-TR
13	1989	Loma Prieta	LOMAP/CAP000	LOMAP/CAP090
14	1989	Loma Prieta	LOMAP/G03000	LOMAP/G03090
15	1990	Manjil, Iran	MANJIL/ABBAR–L	MANJIL/ABBAR–T
16	1987	Superstition Hills	SUPERST/B-ICC000	SUPERST/B-ICC090
17	1987	Superstition Hills	SUPERST/B-POE270	SUPERST/B-POE360
18	1992	Cape Mendocino	CAPEMEND/RIO270	CAPEMEND/RIO360
19	1999	Chi-Chi, Taiwan	CHICHI/CHY101-E	CHICHI/CHY101-N
20	1999	Chi-Chi, Taiwan	CHICHI/TCU045-E	CHICHI/TCU045-N
21	1971	San Fernando	SFERN/PEL090	SFERN/PEL180
22	1976	Friuli, Italy	FRIULI/A-TMZ000	FRIULI/A-TMZ270

Table 3. The ranges of design variables and optimum values

Design variable	Range definition	FPA	HS-FPA
Mass (t)	Between 1% and 5% total mass of structure	935.55	1203
Period (s)	Between 0.5 and 1.5 times of the critical period of structure	3.8652	4.0753
Damping ratio (%)	Between 0.1% and 30%	28.04	30

excitation is the CHY101-N records of the Chi-Chi earthquake. The displacement time history plot for the top story of the structure under the critical excitation is shown as Fig. 2.

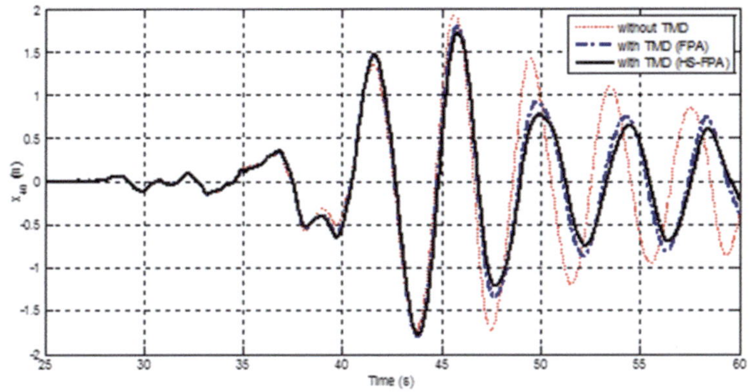

Fig. 2. Time history plot of the top story of the structure

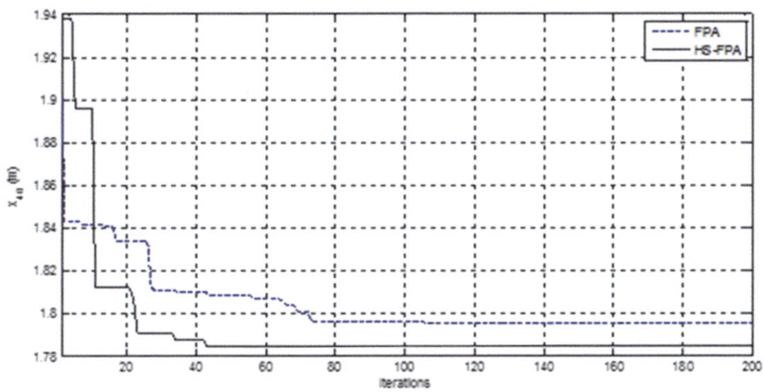

Fig. 3. Convergence plot of the methods

5 Conclusions

A new hybrid method combining HS and FPA method is proposed for the optimum tuning of mass damper. According to the results of the numerical example, the proposed method is more effective than the classical FPA because a better reduction of the maximum displacement is provided. Additionally, the convergence of the proposed method is better than the classical algorithm. This situation can be clearly seen from the Fig. 3.

In the numerical example, the global pollination process is more effective than the other generation types because the probability of the global pollination is not zero while the others are zeroized. The new proposal is effective in shortening the computation time and improvement of the optimum results.

References

1. Den Hartog, J.P.: Mechanical Vibrations, 3rd edn. McGraw-Hill, New York (1947)
2. Warburton, G.B.: Optimum absorber parameters for various combinations of response and excitation parameters. Earthq. Eng. Struct. Dyn. **10**(3), 381–401 (1982)
3. Sadek, F., Mohraz, B., Taylor, A.W., Chung, R.M.: A method of estimating the parameters of tuned mass dampers for seismic applications. Earthq. Eng. Struct. Dyn. **26**(6), 617–636 (1997)
4. Rana, R., Soong, T.T.: Parametric study and simplified design of tuned mass dampers. Eng. Struct. **20**(3), 193–204 (1998)
5. Chang, C.C.: Mass dampers and their optimal designs for building vibration control. Eng. Struct. **21**(5), 454–463 (1999)
6. Lee, C.L., Chen, Y.T., Chung, L.L., Wang, Y.P.: Optimal design theories and applications of tuned mass dampers. Eng. Struct. **28**(1), 43–53 (2006)
7. Bakre, S.V., Jangid, R.S.: Optimum parameters of tuned mass damper for damped main system. Struct. Control Health Monit. **14**(3), 448–470 (2007)
8. Hadi, M.N., Arfiadi, Y.: Optimum design of absorber for MDOF structures. J. Struct. Eng. **124**(11), 1272–1280 (1998)
9. Marano, G.C., Greco, R., Chiaia, B.: A comparison between different optimization criteria for tuned mass dampers design. J. Sound Vib. **329**(23), 4880–4890 (2010)
10. Singh, M.P., Singh, S., Moreschi, L.M.: Tuned mass dampers for response control of torsional buildings. Earthq. Eng. Struct. Dyn. **31**(4), 749–769 (2002)
11. Desu, N.B., Deb, S.K., Dutta, A.: Coupled tuned mass dampers for control of coupled vibrations in asymmetric buildings. Struct. Control Health Monit. **13**(5), 897–916 (2006)
12. Pourzeynali, S., Lavasani, H.H., Modarayi, A.H.: Active control of high rise building structures using fuzzy logic and genetic algorithms. Eng. Struct. **29**(3), 346–357 (2007)
13. Leung, A.Y.T., Zhang, H.: Particle swarm optimization of tuned mass dampers. Eng. Struct. **31**(3), 715–728 (2009)
14. Leung, A.Y., Zhang, H., Cheng, C.C., Lee, Y.Y.: Particle swarm optimization of TMD by non-stationary base excitation during earthquake. Earthq. Eng. Struct. Dyn. **37**(9), 1223–1246 (2008)
15. Steinbuch, R.: Bionic optimisation of the earthquake resistance of high buildings by tuned mass dampers. J. Bionic Eng. **8**(3), 335–344 (2011)
16. Bekdaş, G., Nigdeli, S.M.: Estimating optimum parameters of tuned mass dampers using harmony search. Eng. Struct. **33**(9), 2716–2723 (2011)
17. Bekdaş, G., Nigdeli, S.M.: Mass ratio factor for optimum tuned mass damper strategies. Int. J. Mech. Sci. **71**, 68–84 (2013)
18. Nigdeli, S.M., Bekdaş, G.: Optimum tuned mass damper design for preventing brittle fracture of RC buildings. Smart Struct. Syst. **12**(2), 137–155 (2013)
19. Yang, X.-S., Bekdaş, G., Nigdeli, S.M.: Review and applications of metaheuristic algorithms in civil engineering. In: Yang, X.-S., Bekdaş, G., Nigdeli, S.M. (eds.) Metaheuristics and Optimization in Civil Engineering, vol. 7, pp. 1–24. Springer, Cham (2016). doi:10.1007/978-3-319-26245-1_1
20. Farshidianfar, A., Soheili, S.: Ant colony optimization of tuned mass dampers for earthquake oscillations of high-rise structures including soil-structure interaction. Soil Dyn. Earthq. Eng. **51**, 14–22 (2013)

21. Farshidianfar, A., Soheili, S.: ABC optimization of TMD parameters for tall buildings with soil structure interaction. Interact. Multiscale Mech. **6**(4), 339–356 (2013)
22. Farshidianfar, A., Soheili, S.: Optimization of TMD parameters for earthquake vibrations of tall buildings including soil structure interaction. Int. J. Optim. Civ. Eng. **3**, 409–429 (2013)
23. Nigdeli, S.M., Bekdaş, G.: Teaching-learning-based optimization for estimating tuned mass damper parameters. In: International Conference on Optimization Techniques in Engineering (2015)
24. Yang, X.-S.: Flower pollination algorithm for global optimization. In: Durand-Lose, J., Jonoska, N. (eds.) UCNC 2012. LNCS, vol. 7445, pp. 240–249. Springer, Heidelberg (2012). doi:10.1007/978-3-642-32894-7_27
25. Geem, Z.W., Kim, J.-H., Loganathan, G.V.: A new heuristic optimization algorithm: harmony search. Simulation **76**(2), 60–68 (2001)
26. Liu, M.Y., Chiang, W.L., Hwang, J.H., Chu, C.R.: Wind-induced vibration of high-rise building with tuned mass damper including soil-structure interaction. J. Wind Eng. Ind. Aerodyn. **96**(6), 1092–1102 (2008)
27. FEMA P-695: Quantification of Building Seismic Performance Factors, Federal Emergency Management Agency (2009)

Tuning and Position Optimization of Mass Dampers for Seismic Structures

Sinan Melih Nigdeli(✉) and Gebrail Bekdaş

Department of Civil Engineering, Istanbul University, 34320 AvcıLar, Istanbul, Turkey
{melihnig,bekdas}@istanbul.edu.tr

Abstract. Tuned mass dampers (TMDs) used in seismic vibration control are positioned on the top of the structures and it is the optimum position for regular structures. For structure with different stiffness values for each story, the optimum place of a TMD may be different. In this study, the story position of TMD is also included as a design variable. In methodology, harmony search algorithm is employed. The stroke capacity of the TMD is also considered in the methodology. As a numerical example, a six story structure is investigated. The stiffness of the first two stories are lower than the following two stories. The last two stories have higher stiffness than the other ones. In that case, the optimum position is not the top story. For that reason, the position of TMD is an important design variable for irregular structures in rigidity.

Keywords: Tuned mass damper · Optimization · Earthquake · Harmony search · Optimum position · Irregular structures

1 Introduction

In the reduction of seismic responses, tuned mass dampers (TMDs) can be used on structures, but the parameters of TMDs must be optimally tuned for an important seismic control. In that case; numerical optimization is needed since mathematical methods cannot be effectively used because of several limitations related to inherent damping, multiple vibration modes and the characteristic of earthquake excitation with random frequency. In addition to classical methods proposing close-form expressions [1–3], numerical methods [3–7] and metaheuristic methods are effectively used in finding optimum parameters of TMD such as mass (in several studies), period and damping ratio. Genetic algorithm (GA) is the one of the oldest metaheuristic methods and it is the reason of usage of GA in several approaches for TMD optimization [8–12]. Particle swarm optimization [13,14], bionic optimization [15], harmony search (HS) algorithms [16–19], ant colony optimization [20], artificial bee optimization [21], shuffled complex evolution [22] and teaching learning based optimization [23] are the other successfully

J. Del Ser (ed.), *Harmony Search Algorithm*, Advances in Intelligent Systems and Computing 514, DOI 10.1007/978-981-10-3728-3_23

employed metaheuristic algorithms. Differently from the TMD properties, the optimum position of TMD can be taken as a design variable in the optimization processes. If the rigidities of stories of the structure are regular (the same or higher stiffness in lower stories), the optimum position of TMD is the top of the structures. Because of architectural reasons, irregular stiffness values of structures can be seen. In that case, the optimum place of a TMD may not be the top story because the relative displacement of a story may be bigger than the top story. For that reason, the optimization of the position of the TMD may be important. In this study, additional to the TMD properties such as mass, period and damping ratio, the story containing a TMD is also taken as a design variable. In the optimization methodology, harmony search algorithm is employed together with dynamic analyses of the structure. As a constraint, the stroke capacity of the TMD is also investigated. Numerical investigation is done for a six story structure with irregular rigidity.

2 Equations of Motion and Optimization Methodology

The physical model of an N-story structure is shown as seen in Fig. 1. The TMD is positioned in the middle, but the optimum position can also be the top story. The shear building has N degree of freedom for all lateral motion of the stories. By including a TMD, the total number of degrees of freedom is N+1. The equation of motion is written as seen in Eq. (1).

$$M\ddot{x}(t) + C\dot{x}(t) + Kx(t) = -M1\ddot{x}_g(t) \tag{1}$$

The diagonal lumped mass (M), damping (C) and stiffness (K) matrices are presented as Eqs. (2), (3) and (4), respectively. Also, x(t), $\ddot{x}_g(t)$ and 1 are the vector containing structural displacements, ground acceleration and a vector of ones with a dimension of (N+1, 1), respectively. x(t) is shown in Eq. (5).

$$M = diag[m_1 \quad m_2 ... m_i \quad m_d \quad m_{i+1} ... m_{N-1} \quad m_N] \tag{2}$$

The mass, damping coefficient, stiffness coefficient and displacements are symbolized with m, c, k and x. The subscripts show the story number between 1 and N. The TMD parameters are mass (m_d), damping coefficient (c_d) and stiffness coefficient (k_d) while the displacement of the TMD is x_d.

The damping and stiffness coefficients can be written as period (T_d) and damping ratio (ξ_d) of TMD as shown in Eqs. (6) and (7). In the numerical example, these two values are taken as design variables together with mass of

TMD (m_d) and TMD positioned story (i).

$$\begin{bmatrix} (c_1+c_2) & -c_2 & & & & & & & \\ -c_2 & (c_2+c_3) & -c_3 & & & & & & \\ & \vdots & \vdots\ \vdots & & & & & & \\ & \vdots & \vdots\ \vdots & & & & & & \\ & & c_i & (c_i+c_{i+1}+c_d) & -c_d & -c_{i+1} & & & \\ & & & -c_d & c_d & & & & \\ & & & -c_d & c_d & & & & \\ & & & -c_{i+1} & & (c_{i+1}+c_{i+2}) & -c_{i+2} & & \\ & & & & & \vdots & \vdots & \vdots & \\ & & & & & \vdots & \vdots & \vdots & \\ & & & & & & c_{N-1} & (c_{N-1}+c_N) & -c_N \\ & & & & & & & -c_N & c_N \end{bmatrix} \quad (3)$$

$$\begin{bmatrix} (k_1+k_2) & -k_2 & & & & & & & \\ -k_2 & (k_2+k_3) & -k_3 & & & & & & \\ & \vdots & \vdots\ \vdots & & & & & & \\ & \vdots & \vdots\ \vdots & & & & & & \\ & & k_i & (k_i+k_{i+1}+k_d) & -k_d & -k_{i+1} & & & \\ & & & -k_d & k_d & & & & \\ & & & -k_d & k_d & & & & \\ & & & -k_{i+1} & & (k_{i+1}+k_{i+2}) & -k_{i+2} & & \\ & & & & & \vdots & \vdots & \vdots & \\ & & & & & \vdots & \vdots & \vdots & \\ & & & & & & k_{N-1} & (k_{N-1}+k_N) & -k_N \\ & & & & & & & -k_N & k_N \end{bmatrix} \quad (4)$$

$$x(t) = [x_1 \quad x_2 \ldots x_i \quad x_d \quad x_{i+1} \ldots x_{N-1} \quad x_N]^T \quad (5)$$

$$T_d = 2\pi\sqrt{\frac{m_d}{k_d}} \quad (6)$$

$$\xi_d = \frac{c_d}{2m_d\sqrt{\frac{k_d}{m_d}}} \quad (7)$$

Music inspired harmony search (HS) algorithm [24] is employed in the proposed optimization methodology. Firstly, structural properties, external excitations and ranges of design variables are defined. These are the design constants and it is possible to constraint the mass and other parameters by using an upper bound values. As the second step of the methodology, the structure without TMD is analyzed in time domain by using a wide range of earthquake records. After these two primary steps, the initial harmony memory (HM) matrix is constructed. This matrix contains harmony vectors (HVs) and the number of

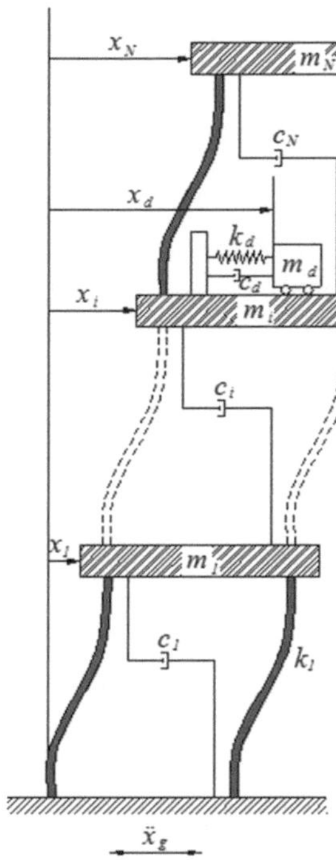

Fig. 1. Model of N-story shear building including a TMD on the ith storey

these vectors are equal to an algorithm parameter called Harmony Memory Size (HMS). In Eq. (8), a HV is shown including the design variables. The parameters are named with dv and the subscript shows the design variable number. The possible design variables are randomly assigned within the solution range in the first generations. Also, the numerical analyses are done for all set of solutions and the required objective functions are calculated.

$$HV = \begin{bmatrix} m_d \\ T_d \\ \xi_d \\ i \end{bmatrix} = \begin{bmatrix} dv_1 \\ dv_2 \\ dv_3 \\ dv_4 \end{bmatrix} \tag{8}$$

After the generation of the initial matrix, a new vector is generated in two ways. In the global search, a design variable (dv) is generated from the whole range as shown in Eq. (9). The lower and upper bounds of the solution range are shown with dvL and dvU, respectively. ϵ is random number between 0 and 1.

$$dv = dv_L + \epsilon(dv_u - dv_L) \tag{9}$$

The local search is done by using a smaller range around an existing solution (dv*) as seen below.

$$dv = dv^* + (\epsilon - 0.5)PAR(dv_u - dv_L) \quad (10)$$

The parameter called the pitch adjusting rate is symbolized with PAR. By using this parameter, a smaller range than the initial range is used, but the generated solution must be also limited with the initial ranges. Harmony memory considering rate (HMCR) is used to control the search type. It is used as a probability of using the local search in the iterative process. The newly generated solutions are compared and updated according to the objective functions. The essential objective is to minimize the maximum top story displacement of the structure (Eq. (11)). The iterations continue until it is reduced under a user defined value (x_{max}). The other objective represents the limitation of the stroke of the TMD and it can be formulated as Eq. (12). First, the objective about the stroke capacity is considered. If it is lower than the user defined limit: st_max, the essential objective function is considered. The user defined value; x_{max} is iteratively increased if the solution is not physical.

$$max|x_N| \leq x_{max} \quad (11)$$

$$\frac{max[|x_d - x_i|]_{withTMD}}{max[|x_N|]_{withoutTMD}} \leq st\_max \quad (12)$$

3 Numerical Examples

The numerical example is a six story structure. The stiffness of the first two stories are lower than the following two stories. The last two stories have higher stiffness than the other ones. The structural properties are shown in Table 1.

Table 1. Properties of the structure

Story	m(t)	k(MN/m)	c(MNs/m)
1	360	450	4.5
2	360	450	4.5
3	360	550	5.5
4	360	550	5.5
5	360	650	6.5
6	360	650	6.5

Table 2. The optimum results

	Case 1	Case 2
m_d(t)	23.17	22.38
T_d(s)	0.6590	0.5365
ξ_d	0.2912	0.2057
Story of TMD	3	6
Maximum displacement (m)	0.1947	0.1965

In the study, the mass of TMD (md) was assigned with the value between 1% and 5%. The period of TMD (Td) was searched between 0.5 and 1.5 times of the fundamental period of the structure. The range of damping ratio of TMD (ξ_d) is between 1% and 30%. The optimum results are shown in Table 2 for two cases. In case 1, the position of the TMD is also optimized. The position is fixed

Table 3. FEMA P-695 far-field ground motion records [25]

Earthquake number	Date	Name	Component 1	Component 2
1	1994	Northridge	NORTHR/MUL009	NORTHR/MUL279
2	1994	Northridge	NORTHR/LOS000	NORTHR/LOS270
3	1999	Duzce, Turkey	DUZCE/BOL000	DUZCE/BOL090
4	1999	Hector Mine	HECTOR/HEC000	HECTOR/HEC090
5	1979	Imperial Valley	IMPVALL/H-DLT262	IMPVALL/H-DLT352
6	1979	Imperial Valley	IMPVALL/H-E11140	IMPVALL/H-E11230
7	1995	Kobe, Japan	KOBE/NIS000	KOBE/NIS090
8	1995	Kobe, Japan	KOBE/SHI000	KOBE/SHI090
9	1999	Kocaeli, Turkey	KOCAELI/DZC180	KOCAELI/DZC270
10	1999	Kocaeli, Turkey	KOCAELI/ARC000	KOCAELI/ARC090
11	1992	Landers	LANDERS/YER270	LANDERS/YER360
12	1992	Landers	LANDERS/CLW-LN	LANDERS/CLW-TR
13	1989	Loma Prieta	LOMAP/CAP000	LOMAP/CAP090
14	1989	Loma Prieta	LOMAP/G03000	LOMAP/G03090
15	1990	Manjil, Iran	MANJIL/ABBAR–L	MANJIL/ABBAR–T
16	1987	Superstition Hills	SUPERST/B-ICC000	SUPERST/B-ICC090
17	1987	Superstition Hills	SUPERST/B-POE270	SUPERST/B-POE360
18	1992	Cape Mendocino	CAPEMEND/RIO270	CAPEMEND/RIO360
19	1999	Chi-Chi, Taiwan	CHICHI/CHY101-E	CHICHI/CHY101-N
20	1999	Chi-Chi, Taiwan	CHICHI/TCU045-E	CHICHI/TCU045-N
21	1971	San Fernando	SFERN/PEL090	SFERN/PEL180
22	1976	Friuli, Italy	FRIULI/A-TMZ000	FRIULI/A-TMZ270

and it is the top story in the second case. The optimization was conducted under earthquake excitations grouped as far-fault records in FEMA P-695 [25]. These records are shown in Table 3.

4 Conclusions

In the study, a global solution was searched and the usage of a wide range of earthquake records is the reason of it. The most critical excitation is the subject of the optimization and the critical one can be change according to structural properties and randomly assigned TMD parameters. In the final optimum result, the critical excitation is the NORTHR/LOS270 record of 1994 Northridge Earthquake. The plot of the top story displacement is shown as Fig. 2 for Case 1.

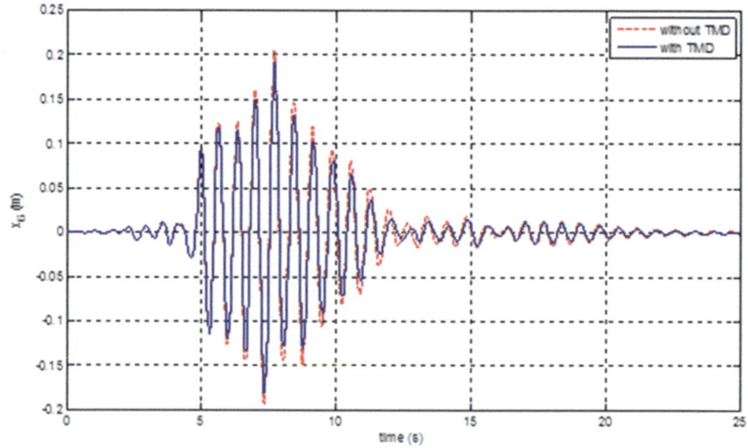

Fig. 2. Time history plot of the top story of the structure (Case 1)

For Case 1, the optimization objective defined as the maximum displacement under the recorded excitation is lower than Case 2. In that case, the optimum place of the TMD is on the third story. In that situation, the position of TMD is important if the structures are irregular.

References

1. Den Hartog, J.P.: Mechanical Vibrations, 3rd edn. Mc Graw-Hill, New York (1947)
2. Warburton, G.B.: Optimum absorber parameters for various combinations of response and excitation parameters. Earthq. Eng. Struct. D. **10**, 381–401 (1982)
3. Sadek, F., Mohraz, B., Taylor, A.W., Chung, R.M.: A method of estimating the parameters of tuned mass dampers for seismic applications. Earthq. Eng. Struct. D. **26**, 617–635 (1997)
4. Rana, R., Soong, T.T.: Parametric study and simplified design of tuned mass dampers. Eng. Struct. **20**, 193–204 (1998)

5. Chang, C.C.: Mass dampers and their optimal designs for building vibration control. Eng. Struct. **21**, 454–463 (1999)
6. Lee, C.L., Chen, Y.T., Chung, L.L., Wang, Y.P.: Optimal design theories and applications of tuned mass dampers. Eng. Struct. **28**, 43–53 (2006)
7. Bakre, S.V., Jangid, R.S.: Optimal parameters of tuned mass damper for damped main system. Struct. Control Health **14**, 448–470 (2007)
8. Hadi, M.N.S., Arfiadi, Y.: Optimum design of absorber for MDOF structures. J. Struct. Eng. ASCE **124**, 1272–1280 (1998)
9. Marano, G.C., Greco, R., Chiaia, B.: A comparison between different optimization criteria for tuned mass dampers design. J. Sound Vib. **329**, 4880–4890 (2010)
10. Singh, M.P., Singh, S., Moreschi, L.M.: Tuned mass dampers for response control of torsional buildings. Earthq. Eng. Struct. D. **31**, 749–769 (2002)
11. Desu, N.B., Deb, S.K., Dutta, A.: Coupled tuned mass dampers for control of coupled vibrations in asymmetric buildings. Struct. Control Health **13**, 897–916 (2006)
12. Pourzeynali, S., Lavasani, H.H., Modarayi, A.H.: Active control of high rise building structures using fuzzy logic and genetic algorithms. Eng. Struct. **29**, 346–357 (2007)
13. Leung, A.Y.T., Zhang, H.: Particle swarm optimization of tuned mass dampers. Eng. Struct. **31**, 715–728 (2009)
14. Leung, A.Y.T., Zhang, H., Cheng, C.C., Lee, Y.Y.: Particle swarm optimization of TMD by non-stationary base excitation during earthquake. Earthq. Eng. Struct. D. **37**, 1223–1246 (2008)
15. Steinbuch, R.: Bionic optimisation of the earthquake resistance of high buildings by tuned mass dampers. J. Bionic Eng. **8**, 335–344 (2011)
16. Bekdaş, G., Nigdeli, S.M.: Estimating optimum parameters of tuned mass dampers using harmony search. Eng. Struct. **33**, 2716–2723 (2011)
17. Bekdaş, G., Nigdeli, S.M.: Mass ratio factor for optimum tuned mass damper strategies. Int. J. Mech. Sci. **71**, 68–84 (2013)
18. Nigdeli, S.M., Bekdaş, G.: Optimum tuned mass damper design for preventing brittle fracture of RC buildings. Smart Struct. Syst. **12**(2), 137–155 (2013)
19. Yang, X.-S., Bekdaş, G., Nigdeli, S.M.: Review and applications of metaheuristic algorithms in civil engineering. Metaheuristics and Optimization in Civil Engineering. Modeling and Optimization in Science and Technologies, vol. 7, pp. 1–24. Springer, Heidelberg (2016). Chap. 1
20. Farshidianfar, A., Soheili, S.: Ant colony optimization of tuned mass dampers for earthquake oscillations of high-rise structures including soil-structure interaction. Soil Dyn. Earthq. Eng. **51**, 14–22 (2013)
21. Farshidianfar, A., Soheili, S.: ABC optimization of TMD parameters for tall buildings with soil structure interaction. Interact. Multiscale Mech. **6**, 339–356 (2013)
22. Farshidianfar, A., Soheili, S.: Optimization of TMD parameters for earthquake vibrations of tall buildings including soil structure interaction. Int. J. Optim. Civ. Eng. **3**, 409–429 (2013)
23. Nigdeli, S.M., Bekdaş, G.: Teaching-learning-based optimization for estimating tuned mass damper parameters. In: 3rd International Conference on Optimization Techniques in Engineering (OTENG 2015), pp. 7–9, Rome, Italy, November 2015
24. Geem, Z.W., Kim, J.H., Loganathan, G.V.: A new heuristic optimization algorithm: harmony search. Simulation **76**(2), 60–68 (2001)
25. FEMA P-695, Quantification of Building Seismic Performance Factors, Federal Emergency Management Agency, Washington DC (2009)

Utilization of Harmony Search Algorithm in Optimal Structural Design of Cold-Formed Steel Structures

Serdar Carbas[1(✉)] and Ibrahim Aydogdu[2]

[1] Civil Engineering Department, Karamanoglu Mehmetbey University, Karaman, Turkey
scarbas@kmu.edu.tr

[2] Civil Engineering Department, Akdeniz University, Antalya, Turkey
aydogdu@akdeniz.edu.tr

Abstract. The most important concern for structural design engineers is, nowadays, how to design and build a structure which is really sustainable. The course to design and construct of buildings has to be urgently changed if the overall carbon dioxide emission would like to be reduced. Otherwise, the increase in global warming arising out of building construction will continue in a great majority. The application of cold-formed steel skeleton frames increasingly in building trade makes possible sustainable structures. In this study a harmony search algorithm (HSA) and an improved version, called as adaptive harmony search (AHSA) algorithm to obtain optimum design of cold-formed steel frames. These algorithms choose the cold-formed thin-walled C-sections treated as design variables from a list in AISI-LRFD (American Iron and Steel Institution, Load and Resistance Factor Design). This selection minimize the weight of the cold-formed steel frame while the design constraints specified by the code are satisfied.

Keywords: Structural optimization · Harmony search algorithm · Cold-formed steel frame · Discrete sizing · AISI-LRFD specification

1 Introduction

The greenhouse gas emission, especially carbon dioxide emission, has great impact contribution to global warming. The construction industry with over a third of gas emissions is a main actor of carbon dioxide emission and act as a catalyzer to other environmental impacts. The structural engineers have an opportunity to make an important support to sustainable design by implementing the cold-formed steel framing [1]. Cold-formed steel framing has earned a great popularity by a growing implementation, especially in low-rise steel frames from four to nine stories. This more extensive utilization increases the importance of cold-formed steel frame design due to the nonlinear characterization of the thin-walled steel structural members. Due to having very thin wall thickness the member can buckle under axial load, shear, bending or bearing before the stresses attain to yield stress [2, 3]. On this account, the local buckling occurs at the member walls constitute the principal design criteria. To hurdle

J. Del Ser (ed.), *Harmony Search Algorithm*, Advances in Intelligent Systems and Computing 514, DOI 10.1007/978-981-10-3728-3_24

this, it is necessary to apply a suitable optimization technique. Further to that, since the design variables treated as the cross-section areas of the steel profiles which are generally discrete values and are selected from a list of available steel sections provided by manufacturers, the optimization problem becomes more complicated. As it is obvious, to reach the solution of this discrete optimization problem described is not in an easy way. The appearance of the metaheuristics brings about open a new gate to find optimal solutions for this type of complex programming problems [4, 5].

Harmony search algorithm (HSA) [6] gains a great popularity among metaheuristics during last decades due to having very wide application fields of engineering research and practice. To put it differently, procurement pertinence between pitches to reach a better state of harmony in music is resembled trying to yield the optimum solution of an optimization problem in HSA [7]. In the standard implementation of the technique appropriate constant values are assigned to two main parameters, harmony memory considering rate (HMCR) and pitch adjusting rate (PAR), at the beginning of the optimization process and they stay unchanged during search. But the selection of appropriate values to these parameters has direct effect on the accomplishment of the algorithm. So-called adaptive harmony search algorithm (AHSA) come in sight in this study embodies an authentic perspective for adjusting these parameters automatically during the search for the most efficient optimization process [8].

In current work, the algorithm developed for optimal design is to obtain the minimum weight of cold-formed steel frames made out of thin-walled open steel sections. The constraints that are the design limitations are conducted in sight of AISI-LRFD (American Iron and Steel Institute, Load and Resistance Factor Design) [9, 10]. The displacement limitations, inter-story drift restrictions, effective slenderness ratio, strength requirements for beams and combined axial and bending strength requirements which include the elastic torsional lateral buckling for beam-columns as well as the additional restrictions are regarded as practical design requirements.

2 Discrete Design Optimization of Cold-Formed Steel Structures to AISI-LRFD

The constraints are implemented from AISI-LRFD [9] in the formulation of the design problem the following discrete programming problem is obtained.

Find a vector of integer values **I** (Eq. 1) representing the sequence numbers of C-sections assigned to ng member groups

$$I^T = [I_1, I_2, \ldots, I_{ng}] \tag{1}$$

to minimize the weight (W) of the frame

$$\text{Minimize} \quad W = \sum_{k=1}^{ng} m_k \sum_{i=1}^{nk} L_i \tag{2a}$$

Subjected to

2.1 Serviceability Constraints

$$\frac{\delta_{jl}}{L/Ratio} - 1.0 \leq 0 \quad j = 1, 2, \ldots, nsm, l = 1, 2, \ldots, nlc \tag{2b}$$

$$\frac{\Delta_{jl}^{top}}{H/Ratio} - 1.0 \leq 0 \,, \quad j = 1, 2, \ldots, nj_{top}, l = 1, 2, \ldots, nlc \tag{2c}$$

$$\frac{\Delta_{jl}^{oh}}{h_{sx}/Ratio} - 1.0 \leq 0 \,, \quad j = 1, 2, \ldots, n_{st}, l = 1, 2, \ldots, nlc \tag{2d}$$

where, δ_{jl} is the maximum deflection of jth member under the lth load case, L is the length of member, *nsm* is the total number of members where deflections limitations are to be imposed, *nlc* is the number of load cases, H is the height of the frame, nj_{top} is the number of joints on the top story, Δ_{jl}^{top} is the top story displacement of the jth joint under l^{th} load case, n_{st} is the number of story, *nlc* is the number of load cases and Δ_{jl}^{oh} is the story drift of the jth story under l^{th} load case, h_{sx} is the story height and *Ratio* is limitation ratio for lateral displacements described in ASCE Ad Hoc Committee report [11].

2.2 Strength Constraints: Combined Tensile Axial Load and Bending

It is stated in AISI-LRFD that when a cold-formed members are subject to concurrent bending and tensile axial load, the member shall satisfy the interaction equations given in section C5.1 of reference [9] which is repeated below.

$$\frac{M_{ux}}{\phi_b M_{nxt}} + \frac{M_{uy}}{\phi_b M_{nyt}} + \frac{T_u}{\phi_t T_n} \leq 1.0 \tag{2e}$$

$$\frac{M_{ux}}{\phi_b M_{nx}} + \frac{M_{uy}}{\phi_b M_{ny}} - \frac{T_u}{\phi_t T_n} \leq 1.0 \tag{2f}$$

where,

M_{ux}, M_{uy} = the required flexural strengths [factored moments] with respect to centroidal axes.

$Ø_b$ = for flexural strength [moment resistance] equals 0.90 or 0.95.

M_{nxt}, M_{nyt} = $S_{ft}F_y$ (where, S_{ft} is the section modulus of full unreduced section relative to extreme tension fiber about appropriate axis and F_y is the design yield stress).

T_u = required tensile axial strength [factored tension].

$Ø_t$ = 0.95.

T_n = nominal tensile axial strength [resistance].

M_{nx}, M_{ny} = nominal flexural strengths [moment resistances] about centroidal axes.

2.3 Strength Constraints: Combined Compressive Axial Load and Bending

It is stated in AISI-LRFD that when a cold-formed steel members are subject to concurrent bending and compressive axial load, the member shall satisfy the interaction equations given in section C5.2 of reference [9] which is repeated below.

For $\frac{P_u}{\phi_c P_n} > 0.15$,

$$\frac{P_u}{\phi_c P_n} + \frac{C_{mx} M_{ux}}{\phi_b M_{nx} \alpha_x} + \frac{C_{my} M_{uy}}{\phi_b M_{ny} \alpha_y} \leq 1.0 \tag{2g}$$

$$\frac{P_u}{\phi_c P_{no}} + \frac{M_{ux}}{\phi_b M_{nx}} + \frac{M_{uy}}{\phi_b M_{ny}} \leq 1.0 \tag{2h}$$

For $\frac{P_u}{\phi_c P_n} \leq 0.15$,

$$\frac{P_u}{\phi_c P_n} + \frac{M_{ux}}{\phi_b M_{nx}} + \frac{M_{uy}}{\phi_b M_{ny}} \leq 1.0 \tag{2i}$$

where,

P_u = required compressive axial strength [factored compressive force].
$\varnothing_c$ = 0.85.

and

$$\alpha_x = 1 - \frac{P_u}{P_{E_x}} > 0.0, \quad \alpha_y = 1 - \frac{P_u}{P_{E_y}} > 0.0 \tag{2j}$$

where,

$$P_{E_x} = \frac{\pi^2 EI_x}{(K_x L_x)^2}, \quad P_{E_y} = \frac{\pi^2 EI_y}{(K_y L_y)^2} \tag{2k}$$

where,

I_x = moment of inertia of full unreduced cross section about x axis.
K_x = effective length factor for buckling about x axis.
L_x = unbraced length for bending about x axis.
I_y = moment of inertia of full unreduced cross section about y axis.
K_y = effective length factor for buckling about y axis.
L_y = unbraced length for bending about y axis.
P_{no} = nominal axial strength [resistance] determined in accordance with section C4 of AISI [9], with $F_n = F_y$.
C_{mx}, C_{my} = coefficients taken as 0.85 or 1.0.

2.4 Allowable Slenderness Ratio Constraints

The maximum allowable slenderness ratio of cold-formed compression members has been limited to 200.

$$\frac{K_x{*}L_x}{r_x} \text{ or } \frac{K_y{*}L_y}{r_y} < 200 \tag{2l}$$

where,

K_x = effective length factor for buckling about x axis.
L_x = unbraced length for bending about x axis.
K_y = effective length factor for buckling about y axis.
L_y = unbraced length for bending about y axis.
r_x, r_y = radius of gyration of cross section about x and y axes.

2.5 Geometric Constraints

Geometric constraints are required to make sure that steel C-section selected for the columns of two consecutive stories are either equal to each other or the one above storey is smaller than the one in the below storey. Similarly when a beam is connected to flange of a column, the flange width of the beam is less than or equal to the flange width of the column in the connection. Furthermore when a beam is connected to the web of a column, the flange width of the beam is less than or equal to $(D - 2t_b)$ of the column web dimensions in the connections where D and t_b are the depth and the flange thickness of C-section as shown in Fig. 1.

$$\frac{D_i^a}{D_i^b} - 1 \le 0 \quad \text{and} \quad \frac{m_i^a}{m_i^b} - 1 \le 0, \quad i = 1,\ldots, n_{ccj} \tag{2m}$$

$$\frac{B_i^{Bi}}{D_i^{Ci} - 2t_b^{Ci}} - 1 \le 0, \quad i = 1,\ldots, n_{j1} \tag{2n}$$

$$\frac{B_f^{Bi}}{B_f^{Ci}} - 1 \le 0, \quad i = 1,\ldots, n_{j2} \tag{2o}$$

where n_{ccj} is the number of column-to-column geometric constraints defined in the problem, m_i^a is the unit weight of C-section selected for above story, m_i^b is the unit weight of C-section selected for below story, D_i^a is the depth of C-section selected for above story, D_i^b is the depth of C-section selected for below story, n_{j1} is the number of joints where beams are connected to the web of a column, n_{j2} is the number of joints where beams connected to the flange of a column. D_i^{Ci} is the depth of C-section selected for the column, t_b^{Ci} is the flange thickness of C-section selected for the column,

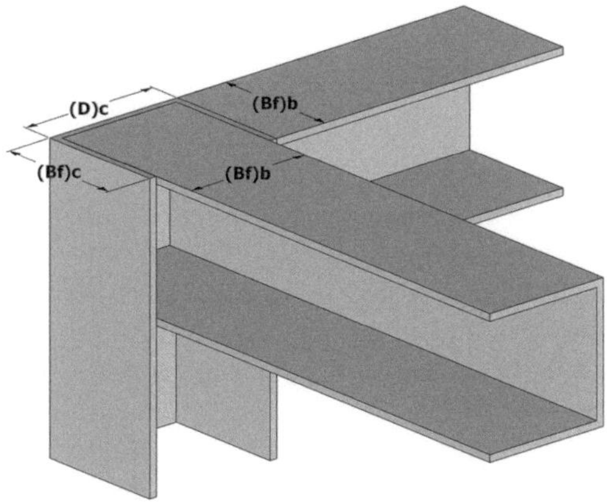

Fig. 1. Typical beam-column connection of steel C-section

B_f^{Ci} is the flange width of *C*-section selected for the column and B_f^{Bi} is the flange width of *C*-section selected for the beam, at joint *i*.

3 Harmony Search Algorithm (HSA)

The harmony search algorithm (HSA) is originated by Geem and Kim [12]. The algorithm was inspired by using the musical performance processes that emerge when a musician searches for a perfect state of harmony, such as during jazz improvisation. A musician always intends to bring out a piece of music with perfect harmony. On the other hand, the optimal solution of an optimization problem should be the best solution available to the problem under given objective and limited by constraints. Both processes aim at reaching the best solution that is the optimum. The main steps of a standard HSA are summarized as follows. The detail explanations of each step can be found in reference [12]:

Step 1: Assign the algorithm parameters (HMCR and PAR).
Step 2: Initialize the harmony memory (HM) matrix.
Step 3: Improvise a new solution from the HM matrix.
Step 4: Update the HM matrix.
Step 5: Repeat step 3 and step 4 until the stopping criterion is satisfied.

3.1 Adaptive Harmony Search Algorithm (AHSA)

In standard harmony search method the HMCR and PAR are assigned to constant values that are arbitrarily chosen within their recommended ranges [12] based on the observed efficiency of the technique in different problem fields. It is observed through the application of the standard HSA that the selection of these values is problem dependent in full. While a certain set of values yields a good performance of the technique in one type of design problem, the same set may not demonstrate the same performance in another type of design problem. Hence, it is not possible to come up with a certain set of values that can be used in every optimal design problem. In each problem, a sensitivity analysis should be conducted to identify of parameter values. AHSA eliminates the necessity of finding the best set of parameter values by adopting the values of these parameters automatically during the optimization process. The HMCR and PAR are set to initial values for all the solution vectors in the initial HM matrix. After filling this matrix randomly, adaptive algorithm is initialized by a new set of values is sampled for HMCR and PAR parameters each time prior to improvisation (generation) of a new harmony vector, which in fact forms the basis for the algorithm to gain adaptation to varying features of the design space. Accordingly, to generate a new harmony vector in the proposed algorithm, a main sampling of control parameters is activated as formulated in Eqs. 3a and 3b.

$$(\mathrm{HMCR})^{c} = \left(1 + \frac{1-(\mathrm{HMCR})^{ave}}{(\mathrm{HMCR})^{ave}} * e^{-\gamma N(0,1)}\right)^{-1} \tag{3a}$$

$$(\mathrm{PAR})^{c} = \left(1 + \frac{1-(\mathrm{PAR})^{ave}}{(\mathrm{PAR})^{ave}} * e^{-\gamma N(0,1)}\right)^{-1} \tag{3b}$$

where $(\mathrm{HMCR})^{c}$ and $(\mathrm{PAR})^{c}$ represent the sampled values of the control parameters for a new harmony vector. The notation N(0,1) designates a normally distributed random number having expectation 0 and standard deviation 1. The symbols $(\mathrm{HMCR})^{ave}$ and $(\mathrm{PAR})^{ave}$ signify the average values of control parameters within the HM matrix, obtained by averaging the corresponding values of all the solution vectors within the HM matrix. The γ refers to the learning rate of control parameters, which is recommended to be selected within a range of [0.25, 0.50] and here this parameter is set to 0.35. The detail information for AHSA is given in reference [8]. Repetition of each expression of AHSA is not possible due to lack of space in the article; hence readers are referred to reference [8].

4 Constraint Handling

To obtain the solution of constrained optimization problems, penalty function is utilized (Eq. 4). In this study the following function is used in this transformation.

$$W_p = W(1 + C)^{\varepsilon} \tag{4}$$

where W is the value of objective function of optimum design problem given in Eq. 2a. W_p is the penalized weight of structure, C is the value of total constraint violations which is calculated by summing the violation of each individual constraint. ε is penalty coefficient which is taken as 2.0 in this study.

5 Design Optimization Algorithms with Discrete Variables

The solution of the discrete optimum design problem given in Eqs. 2a–2o is obtained using HS and AHS algorithms. In both algorithms the sequence number of the steel C-sections in the available profile list is treated as design variable. For this purpose complete set of 85 C-sections starting from 4CS2 × 059 to 12CS4 × 105 as given in AISI is taken into account as a design pool from which the optimum design algorithms select steel C-sections for cold-formed steel frame members. Once a sequence number is selected, then the sectional designation and properties of that section becomes available from the section table for the algorithms. The optimization algorithms proposed assume continuous design variables. However the design problem considered requires discrete design variables. This necessity is resolved by rounding the numbers to a discrete value.

The analysis of cold-formed steel structures is performed using finite element method. Noticing the fact that steel structures made out of cold-formed thin-walled steel sections are quite slender structures, large deformations compare to their initial dimensions may take place under external loads. In structures with large displacements, although the material behaves linear elastic, the response of the structure becomes nonlinear [13]. In such structures, it is necessary to take into account the effect of axial forces to member stiffness. This is achieved by carrying out P-δ analysis in the application of the stiffness method. The details of the derivation of the nonlinear stiffness matrix and consideration of geometric nonlinearity in the analysis of steel frames made out of thin-walled sections are given in [14].

6 Design Example

Two-storey, 1211-member lightweight cold-formed steel frame shown in Fig. 2 is selected as design problem [15]. 3-D, plan and floor views of the frame are shown in the same figure respectively. The spacing between columns is decided to be 0.6 m span and each floor has 2.8 m height. The total height of the building is 5.6 m. The frame consists of 708 joints (including supports) and 1211 members that are grouped into 14 independent member groups which are treated as design variables. The member grouping of the frame is illustrated in Table 1. The frame is subjected to gravity and lateral loads, which are computed as per given in ASCE 7-05 [16]. The loading consists of a design dead load of 2.89 kN/m^2, a design live load of 2.39 kN/m^2, a ground snow load of 0.755 kN/m^2. Unfactored wind load values are taken as 0.6 kN/m^2. The load and combination factors are applied according to code specifications of LRFD-AISC [10] as; Load Case 1: 1.2D + 1.6L + 0.5S, Load Case 2: 1.2D + 0.5L + 1.6S and Load Case 3: 1.2D + 1.6WX + 1.0L + 0.5S where D represents dead load, L is live load, S is snow load and WX is the wind load applied on X global direction respectively. The top story drift in both X and Y directions are restricted to 14 mm and inter-story drift limitation is specified to 7 mm. The complete single C-section with lips list given in AISI Design Manual 2007 [17] which consists of 85 section designations is considered as a design pool for design variables.

The cold-formed steel frame is designed by using HSA and AHSA. The size of harmony memory matrix HMS = 30, a maximum search number Itmax = 20000,

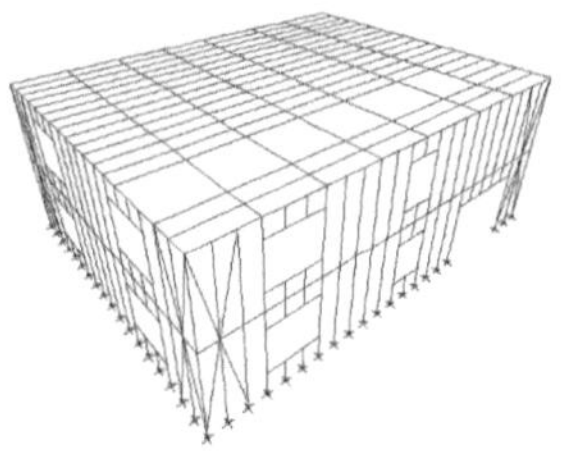

a front and left shot

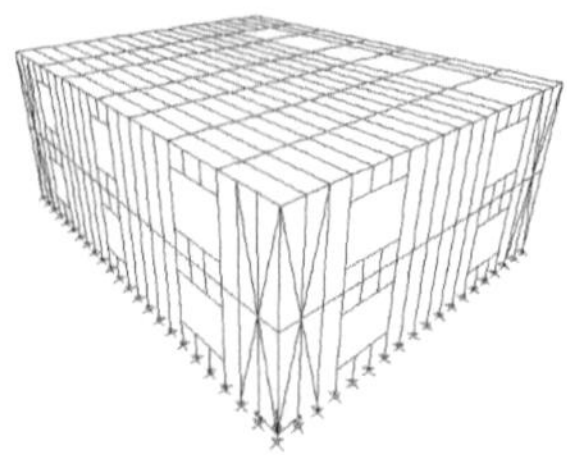

b back and right shot

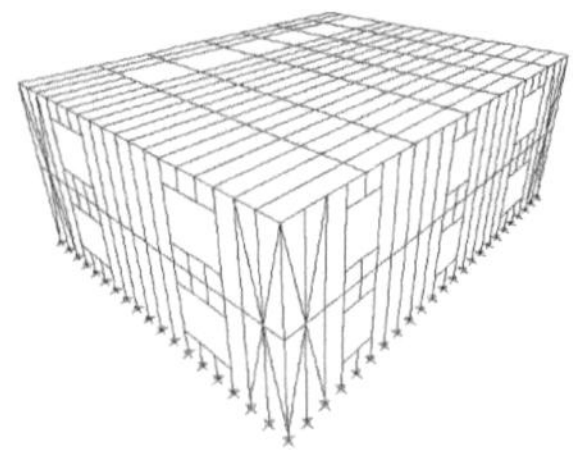

c back and left shot

a 3-D views from different shots

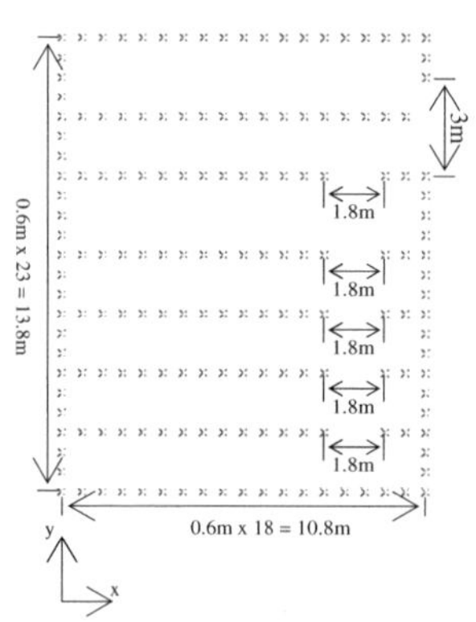

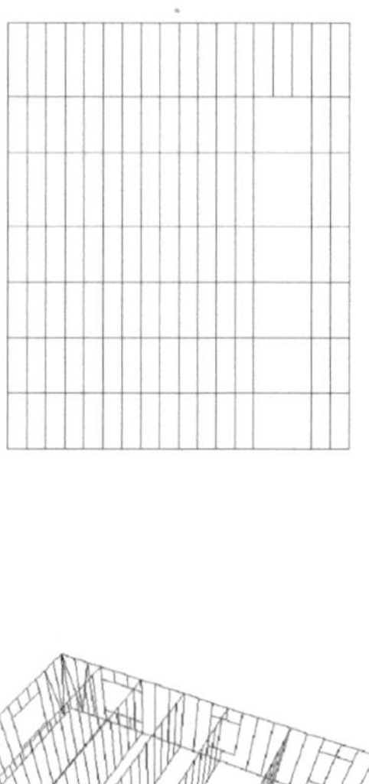

b Plan views

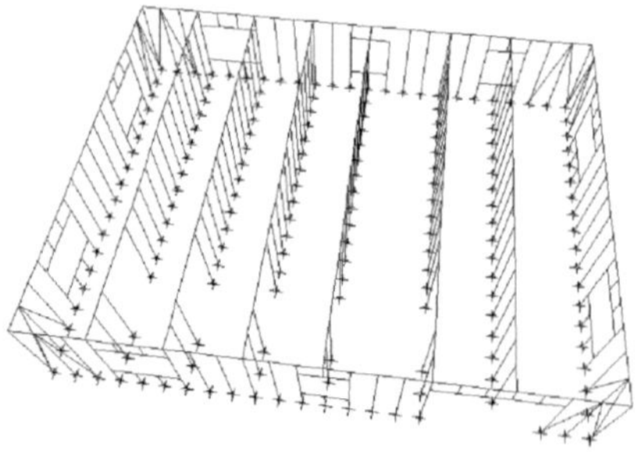

a First floor top view without slabs

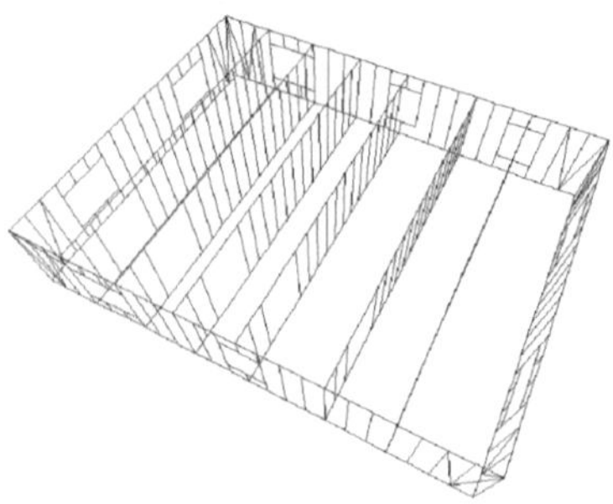

b Second floor top view without slabs

c First and second floors top views without slabs, **a** First floor top view without slabs, **b** Second floor top view without slabs

Fig. 2. 1211-member three dimensional lightweight cold-formed steel frame, a 3-D views from different shots, b Plan views, c First and second floors top views without slabs.

Table 1. The member grouping of 1211-member lightweight cold-formed steel frame

Storey	Beams outer short	Beams inner short	Beams inner gates	Beams windows	Beams outer gate
1	1	2	3	4	5
2	1	2	3	4	–
Storey	Columns connected short beams	Columns connected long beams	Columns near inner gates	Columns windows	Braces
1	6	7	8	9	14
2	10	11	12	13	14

Table 2. Optimum design results of 1211-member lightweight steel frame

Group no.	Group type	Sections selected by AHS algorithm	Sections selected by HS algorithm
1	1st and 2nd floors outer short beams	4CS2 × 105	4CS2 × 085
2	1st and 2nd floors inner short beams	4CS2 × 059	4CS2 × 065
3	1st and 2nd floors inner gates' beams	6CS2.5 × 059	6CS2.5 × 059
4	1st and 2nd floors windows' beams	4CS2.5 × 059	4CS2.5 × 059
5	1st floor outer gate beams	4CS2 × 059	4CS2 × 059
6	1st floor columns connected short beams	4CS2 × 059	4CS2 × 059
7	1st floor columns connected long beams	4CS4 × 059	4CS2 × 059
8	1st floor columns near inner gates	4CS2 × 059	4CS2 × 059
9	1st floor windows' columns	4CS2 × 059	4CS2 × 059
10	2nd floor columns connected short beams	4CS2 × 059	4CS2 × 059
11	2nd floor columns connected long beams	4CS4 × 059	4CS4 × 059
12	2nd floor columns near inner gates	12CS2.5 × 070	8CS2 × 059
13	2nd floor windows' columns	4CS2 × 059	4CS2 × 059
14	1st and 2nd floors braces	4CS2 × 059	4CS2 × 059
Minimum weight (kN (kg))		52.525 (5356.059)	54.162 (5522.987)
Maximum top storey drift (mm)		8.768	8.624
Maximum inter-storey drift (mm)		2.491	2.484
Maximum deflection (mm)		0.195	0.254
Maximum strength ratio		0.998	0.992
Maximum number of iterations		20000	20000
No. of structural analysis to reach optimum design		3459	6648

HMCR = 0.95, and PAR = 0.30 are taken as parameter set. Although the values of control parameters for HMCR and PAR remain unchanged in the standard HSA, they are only assigned to initial values of these parameters in the AHSA, that is, $HMCR^0 = 0.95$ and $PAR^0 = 0.30$. The optimum designs determined by HS and AHS algorithms are listed in Table 2.

It is interesting to notice that both of the algorithms have almost found optimum designs that are close to each other. The AHSA has attained the best global optimum design with the minimum weight of 52.525 kN (5356.059 kg). The HSA is determined the optimum weight of the frame is 54.162 kN (5522.987 kg) which is only 3.117% heavier than the optimum design attained by AHSA. This indicates the fact that HSA and AHSA are robust algorithms used in confidence to optimal design of cold-formed steel structures. It is quite apparent from Table 2 that the strength constraints are dominant in the design optimization problem. For both of the algorithms, the maximum strength ratio is very close to 1.0 while displacement and inter-story drift constraints are much less than their upper bounds. The convergence history of each algorithm is shown in Fig. 3. It is apparent from this figure that AHS algorithm has much better convergence rate than HS algorithm.

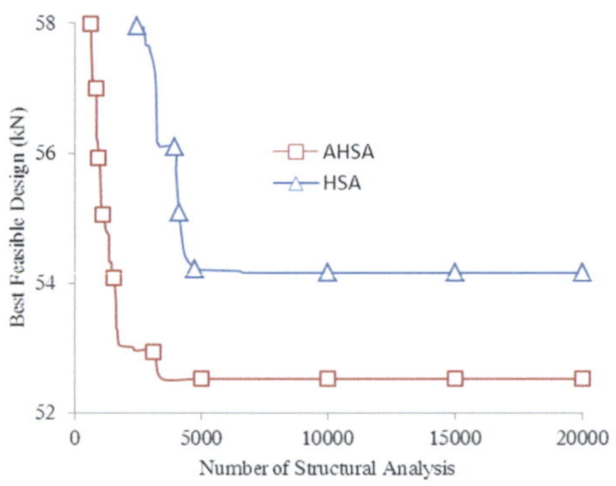

Fig. 3. Search histories of 1211-member lightweight cold-formed steel frame

7 Conclusions

The cold-formed thin-walled steel framing becomes prominent, due to upsurge in gas emissions causing global warming, the sustainable building notion has gained a great reputation recently. A harmony search (HS) algorithm and an enhanced version of which, so-called adaptive harmony search (AHS) algorithm are improved for attaining the design optimization of cold-formed thin-walled steel structures by which systems it is possible to reduce the required amount of material aiding the sustainability of the construction. In the standard HS algorithm, two fundamental managing parameters, harmony memory considering rate and pitch adjusting rate, are allocated as changeless from beginning to the end of the search process. By contrast with this, in adaptive HS algorithm those parameters are attuned in a dynamic manner and varying each time when new harmony vector is generated. With the help of this state-of-the-art characteristic the design space can exploitatively be searched. The algorithms proposed choose the optimum cold-formed thin-walled steel C-section notations from the profile list in a way that design constraints depicted in AISI-LRFD are met and so the frame has the minimum weight. In consideration of the attained results it can be rendered a verdict that the HS and AHS algorithms are efficient and mighty design tools to optimize the light weight cold-formed steel frames in which the influence of geometric nonlinearity is reckoned into. The containing of geometric nonlinearity in the response of cold-formed steel structures is an essentiality if the acquired designs are to be requisitioned as more realistic.

References

1. Yudelson, J.: Green Building Through Integrated Design. McGrow-Hill Professionals, New York (2009)
2. Ghersi, A., Landolfo, R., Mazzolani, F.M.: Design of Metallic Cold-Formed Thin-Walled Members. Spon Press, London (2005)
3. Yu, W.-W., LaBoube, R.A.: Cold-Formed Steel Design, 4th edn. Wiley, New Jersey (2010)
4. Paton, R.: Computing with Biological Metaphors. Chapman and Hall, London (1994)
5. Adami, C.: An Introduction to Artificial Life. Springer, Heidelberg (1998)
6. Lee, K.S., Geem, Z.W.: A new structural optimization method based on harmony search algorithm. Comput. Struct. **82**, 781–798 (2004)
7. Lee, K.S., Geem, Z.W.: A new meta-heuristic algorithm for continuous engineering optimization: harmony search theory and practice. Comput. Methods Appl. Mech. Eng. **194**, 3902–3933 (2005)
8. Hasancebi, O., Erdal, F., Saka, M.P.: An adaptive harmony search method for structural optimization. J. Struct. Eng. ASCE **136**(4), 419–431 (2010)
9. AISI (American Iron and Steel Institute) S100-07: North American Specification for the Design of Cold-Formed Steel Structural Members (2007)
10. AISC (American Institute of Steel Construction): LRFD, Vol. 1, Structural Members, Specifications & Code, Manual of Steel Construction (1991)
11. Ad Hoc Committee on Serviceability: Structural serviceability: a critical appraisal and research needs. J. Struct. Eng. ASCE **112**(12), 2646–2664 (1986)
12. Geem, Z.W., Kim, J.H.: A new heuristic optimization algorithm: harmony search. Simulation **76**, 60–68 (2001)
13. Majid, K.I.: Nonlinear Structures. Butterworth, London (1972)
14. Carbas S.: Optimum design of low rise steel frames made of cold-formed thin-walled steel sections. Ph.D. dissertation, Engineering Sciences Department, Middle East Technical University, Ankara, Turkey (2013)
15. Saka, M.P., Carbas, S., Aydogdu, I., Akin, A., Geem, Z.W.: Comparative study on recent metaheuristic algorithms in design optimization of cold-formed steel structures. In: Lagaros, N.D., Papadrakakis, M. (eds.) Engineering and Applied Sciences Optimization, pp. 145–173. Springer International Publishing (2015). Chapter: 9
16. ASCE 7-05: Minimum design loads for buildings and other structures. American Society of Civil Engineers (2005)
17. AISI (American Iron and Steel Institute) D100-08: Excerpts-Gross Section Property Tables, Cold-Formed Steel Design Manual, Part I; Dimensions and Properties (2008)

Exploring the Efficiency of Harmony Search Algorithm for Hydropower Operation of Multi-reservoir Systems: A Hybrid Cellular Automat-Harmony Search Approach

M.H. Afshar[1], M. Azizipour[1], B. Oghbaeea[1], and Joong Hoon Kim[2(✉)]

[1] School of Civil Engineering and Enviro-Hydroinformatic COE, Iran University of Science and Technology, Tehran, Iran

[2] School of Civil, Environmental and Architectural Engineering, Korea University, Seoul, South Korea

jaykim@korea.ac.kr

Abstract. Multi-reservoir systems are one the important infrastructures due to their role in the energy supply for human beings. Construction of these systems are very costly making their operation a delicate task due to their significant economic impacts. Optimal operation of hydropower reservoir systems is a complex task due to the non-convexity and nonlinearity of the problem involved. Conventional methods often fail to tackle the complexity of the problem while modern heuristic algorithms lack efficiency when solving this problem. This is further amplified when population based heuristic methods are to be used for large scale multi-reservoir real-time operation problems, where efficiency of the solution method is vital. This paper explores the hybridization of a newly proposed method namely Cellular Automata with the well-known Harmony Search algorithm for efficient solution of multi-reservoir hydropower operation problems. The HS method is embedded into a CA framework in which the CA is used to breakdown the large scale reservoir system operation into a series of small scale sub-problem with a size equal to the number of reservoirs in the system. HS method is then used to solve each sub-problem and the results are passed to the CA method. The proposed method is used to solve a nonlinear version of the well-known four reservoir problem and the results are presented and compared with the existing results.

Keywords: Multi-reservoir systems · Cellular automata · Harmony search · Hydropower operation

1 Introduction

Hydropower reservoir operation is a complex large-scale non-linear non-convex optimization problem which is known as a NP-hard (non-deterministic polynomial time hard) problem. Due to the complexity of the problem, most of the existing algorithms are restricted to consider some simplified form of the problem.

J. Del Ser (ed.), *Harmony Search Algorithm*, Advances in Intelligent Systems and Computing 514, DOI 10.1007/978-981-10-3728-3_25

Various optimization techniques employed by researchers for optimal operation of reservoir systems. Conventional optimization methods such as linear programming (LP) [1,2], nonlinear programming (NLP) [3–5] and dynamic programming (DP) [6–11] have been used to solve different forms of reservoir operation problems.

Recently, modern evolutionary algorithms (EAs) are being used more and more for solving reservoir operation optimization problems due to their ability to handle nonlinear and non-convex characteristics of the problems. Among EAs, genetic algorithms (GAs) are first applied to various forms of reservoir operation problems [12–16]. Ant colony optimization (ACO) [17–19], particle swarm optimization (PSO) [20–23], simulated annealing algorithms [24–26], honey bee algorithm [27,28], differential evolution algorithm [10,29] and invasive weed optimization [30,31] have also been used to solve optimal reservoir operation problems.

More recently, an efficient and effective optimization method namely cellular automata (CA)was introduced to solve water resources management problems. Afshar and Shahidi [32] proposed a CA for optimal water supply and hydropower operation of a single-reservoir system and compared the results with those obtained by GA, PSO and ACO, showing the superiority of the proposed CA to the existing evolutionary algorithms. Afshar [21] extended the cellular automata of Afshar and Shahidi [32] for solving multi-reservoir hydropower reservoir operation problems and concluded that CA was both more efficient and effective than GA and PSO for the case examples considered.

In this paper, a hybrid cellular automata-harmony search algorithm is proposed for efficient and effective solution of hydropower operation of multi reservoir systems. In the proposed method, CA is used to breakdown the large scale reservoir system operation into a series of small scale sub-problem with a size equal to the number of reservoirs in the system. HS method is then used to solve each sub-problem and the results are passed to the CA method. The iteration between CA and HS is continued until convergence is achieved. The proposed method is used to optimally solve a well-known four-reservoir system and the results are presented and compared with those obtained by CA-NLP, GA and PSO. The results indicate that the proposed method is much more efficient than existing methods in locating near-optimal solutions for the multi reservoir systems.

2 Proposed Hybrid Cellular Automata - Harmony Search (CA-HS) Algorithm

The single objective multi reservoir operation can be considered either as benefit based or reliability based operation. In benefit based case, the system is operated such that the net benefit of the system is maximized while in the reliability based case the system is operated to maximize the reliability of meeting a set of predefined demands over the operation period.

In this study, a benefit based hydropower reservoir operation is considered with the objective of maximizing the total benefit of system's energy production over the operation period. The optimization problem is mathematically defined as

$$Maximize(F) = \sum_{k=1}^{K}\sum_{t=1}^{T} b_{k,t}E_{k,t} \quad (1)$$

$$\mathbf{S_{t+1}} = \mathbf{S}_{\ bft} + \mathbf{I}_{\ bft} - \mathbf{M} \times \mathbf{R_t} \quad t = 1, 2, \ldots, T \quad (2)$$

$$S_{k,t}^{min} \leq S_{k,t} \leq S_{k,t}^{max} \quad t = 1, 2, \ldots, T+1; k = 1, 2, \ldots, K \quad (3)$$

$$R_{k,t}^{min} \leq R_{k,t} \leq R_{k,t}^{max} \quad t = 1, 2, \ldots, T+1; k = 1, 2, \ldots, K \quad (4)$$

$$E_{k,t} = R_{k,t}h_{k,t} \quad t = 1, 2, \ldots, T+1; k = 1, 2, \ldots, K \quad (5)$$

$$h_{k,t} = (\frac{H_{k,t} - H_{k,t+1}}{2}) - TW_{k,t} \quad t = 1, 2, \ldots, T+1; k = 1, 2, \ldots, K \quad (6)$$

$$H_{k,t} = a_k + b_k S_{k,t} + c_k S_{k,t}^2 + d_k S_{k,t}^3 \quad t = 1, 2, \ldots, T+1; k = 1, 2, \ldots, K \quad (7)$$

Where $b_{k,t}$ is the benefit function of reservoir k at period t, $E_{k,t}$ is the total energy potential of reservoir k at period t, K and T are the total number of reservoirs and operation periods, respectively. $\mathbf{S_t}$ is the vector of storage volumes at the beginning of period t, $\mathbf{R_t}$ and $\mathbf{I_t}$ are the vectors of releases from and inflow to the system over period t, respectively, and $\mathbf{M}$ is a $K \times K$ matrix describing the connectivity of reservoir network. $S_{k,t}^{min}$ and $S_{k,t}^{max}$ are the minimum and maximum allowable volume of reservoir k at period t, respectively; while $R_{k,t}^{min}$ and $R_{k,t}^{max}$ are the minimum and maximum allowable releases from reservoir k over period t, respectively. The effective head on the turbine, $h_{k,t}$, for reservoir k over period t is defined by Eq. 6 in which $TW_{k,t}$ is the tail-water elevation of the reservoir k at period t, $H_{k,t}$ and $H_{k,t+1}$ are water elevation of reservoir k at the beginning and the end of period t, respectively. The water elevation at each reservoir is obtained by an elevation-storage curve defined by Eq. 7 in which a_k, b_k, c_k and d_k are constants calculated via fitting Eq. 7 to the available data.

Application of CA to any optimization problem requires that four basic components of the CA method namely cells, cell state, cell neighborhood and the transition or updating rule are defined. Here, the beginning and the end of each period, represented by discrete points on the operation time span, is used as the CA cells and the corresponding storage volume of the reservoirs as the cell states [21]. The neighborhood of each cell is naturally defined as the previous and next period of operation. The local updating rule for an arbitrary cell t could be defined as the process of finding the updated value of the cell state $\mathbf{S}_{k,t}$ such that the produced energy of the system over the neighboring periods of $t-1$ and t is maximized:

$$Maximize(F) = \sum_{k=1}^{K}(b_{k,t-1}E_{k,t-1} + b_{k,t}E_{k,t}) \quad (8)$$

It should be mentioned that the constraints of original problems are only applied to considered cells and neighboring cells of $t-1$ and t, and mathematically defined as:

$$S_{k,t}^{min} \le S_{k,t} \le S_{k,t}^{max} \quad k = 1, 2, \ldots, K \tag{9}$$

$$S_{k,t} = S_{k,t-1} + I_{k,t-1} - M_{k,j}R_{j,t-1} \quad k = 1, 2, \ldots, K \tag{10}$$

$$S_{k,t+1} = S_{k,t} + I_{k,t} - M_{k,j}R_{j,t} \quad k = 1, 2, \ldots, K \tag{11}$$

$$R_{k,t-1}^{min} \le R_{k,t-1} \le R_{k,t-1}^{max} \quad k = 1, 2, \ldots, K \tag{12}$$

$$R_{k,t}^{min} \le R_{k,t} \le R_{k,t}^{max} \quad k = 1, 2, \ldots, K \tag{13}$$

This sub-problem can be solved with any of the conventional or modern optimization methods. Afshar [21] used first order NLP method to solve this sub-problem leading to a very efficient CA-NLP method. This sub-problem is, however, a multi-modal problem with the ability of trapping conventional methods in local optima while being a very small scale problem. It is, therefore, expected that solution of this sub-problem using modern methods can boost the performance of the CA method. Here, the resulting optimization sub-problem at each cell level is solved by one of the EAs namely harmony search algorithm proposed by Geem et al. [33]. HS is a population based algorithm and mimics the behavior of a music orchestra when aiming at composing the most harmonious melody, as measured by aesthetic standards.

The algorithm works through main four steps:

Step 1: Initiation the algorithm with random solutions.
Step 2: Improvisation of a new harmony.
Step 3: Inclusion the newly generated harmony provided that its fitness improves the worst fitness value of the previous population.
Step 4: Returning to step 2 until a termination criterion is satisfied.

There are two different probabilistic operators to control the improvisation procedure which are applied to each note to produce a new candidate solution, including Harmony Memory Considered Rate (HMCR) and Pitch Adjusting Rate (PAR). In order to find the proper values for HMCR and PAR, some preliminary runs carried out and the values of HMCR and PAR were set to 0.8 and 0.9, respectively. A harmony size equal to 5 and the number of HS iteration equal to 10 is used for all of the problems.

The proposed hybrid CA-HS algorithm starts with a randomly generated cell states, reservoir storages, for all periods of operation. The local sub-problems defined by Eqs. 8–13 is solved by HS to find the updated value of the cell states for all periods in turn using the current values of the neighboring cell states. Once all sub-problems, equal to the number of operation periods, are solved by HS, the cell states are updated simultaneously, and the results are checked for convergence and the algorithm is stopped if converged, and is continued if otherwise.

3 Model Application and Results

In this section, efficiency and effectiveness of the proposed CA-HS algorithm is illustrated for hydropower operation of a well-known four-reservoir system. Configuration of the reservoirs in the system is shown in Fig. 1. More details of the considered problem could be found in Afshar [21]. The effect of the problem scale on the algorithm's performance is assessed by applying the method for three different operation periods of 12, 60 and 240 months. Table 1 presents the maximum, minimum and average total benefit of the 10 runs, along with average computational time for 12, 60 and 240 operation periods, which are obtained in 200, 300, and 400 CA iterations, respectively.

Table 2 compares these results with those produced by CA-NLP of Afshar [21], GA and PSO. As shown in Table 2, CA-HS is more effective than other methods in locating near optimal solution while being more efficient than other metaheuristic search methods in particular for larger scale problem of 240 months of operation. Convergence characteristics of the proposed CA-HS is shown in Figs. 1 and 2 for the 12 and 240 monthly operation periods, respectively. It is seen that the number of CA iterations is not much affected by the scale of the

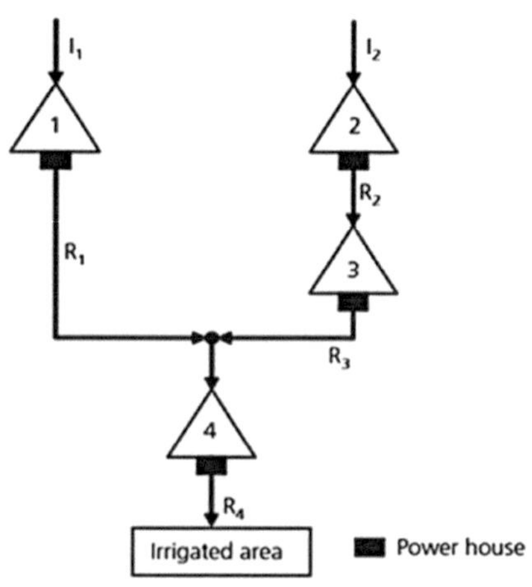

Fig. 1. Schematic representation of four-reservoir system

Table 1. Results of CA-HS algorithm for hydropower operation of four-reservoir system

Period	Total benefit			Average CPU time (sec)
	Max	Min	Average	
12	3.41E+04	3.35E+04	3.38E+04	7
60	1.71E+05	1.67E+05	1.70E+05	38
240	6.84E+05	6.75E+05	6.81E+05	210

Table 2. Comparison of CA-HS with existing results of CA-NLP, GA and PSO

Period	Method	Best solution	Average CPU time (sec)
12	CA-HS	3.42E+04	7.00
	CA-NLP	3.34E+04	0.34
	GA	3.34E+04	6.90
	PSO	3.34E+04	9.60
60	CA-HS	1.71E+05	38.00
	CA-NLP	1.62E+05	12.50
	GA	1.64E+05	126.70
	PSO 1	1.49E+05	181.80
240	CA-HS	6.84E+05	210.00
	CA-NLP	6.59E+05	73.50
	GA	6.50E+05	1487.00
	PSO	5.89E+05	2208.90

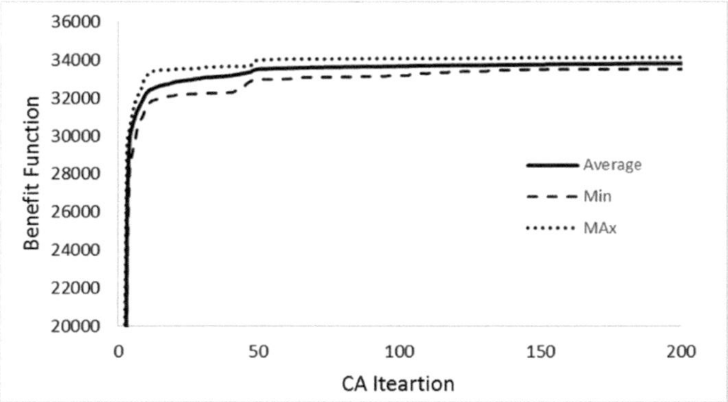

Fig. 2. Convergence curve for 12 periods of operation

problem being solved. This is the reason that the proposed method is much more efficient than the GA and PSO for larger scale problem of 240 months of operation. Furthermore, these Figs. Show that the method is not sensitive to the initial solutions created randomly making it a robust algorithm for the solution of reservoir operation problems (Fig. 3).

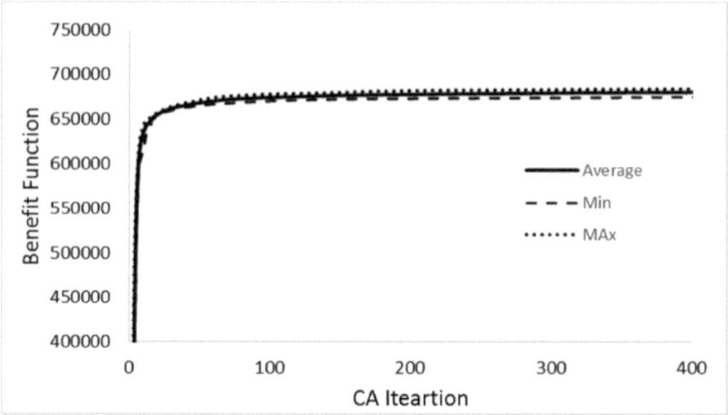

Fig. 3. Convergence curves for 240 periods of operation

4 Conclusion

A novel hybrid cellular automata-harmony search algorithm was developed for the efficient and effective solution of hydropower operation of multi reservoir system. The CA approach was used to break the main complicated problems to some optimization sub problem and each sub problems was, therefore, solved by employing harmony search algorithm. The proposed method were applied to optimal hydropower operation of a well-known four-reservoir system for different operation periods of 12, 60 and 240 and the results presented and compared with those of CA-NLP, GA and PSO, showing that the proposed algorithm is much more efficient than existing algorithm in locating near optimal solution for hydropower multi reservoir operation.

References

1. Ellis, J.H., ReVelle, C.S.: A separable linear algorithm for hydropower optimization. JAWRA – J. Am. Water Resour. Assoc. **24**, 435–447 (1988). doi:10.1111/j.1752-1688.1988.tb03003.x
2. Yoo, J.H.: Maximization of hydropower generation through the application of a linear programming model. J. Hydrol. **376**(1), 182–187 (2009)
3. Zambelli, M.S., Luna, I., Soares, S.: Long-term hydropower scheduling based on deterministic nonlinear optimization and annual inflow forecasting models. In: 2009 IEEE Bucharest PowerTech, pp. 1–8. IEEE (2009)
4. Arunkumar, R., Jothiprakash, V.: Optimal reservoir operation for hydropower generation using non-linear programming model. J. Inst. Eng. (India): Ser. A **93**(2), 111–120 (2012)
5. Shen, J., Cheng, C., Cheng, X., Wu, X.: A hybrid nonlinear optimization method for operation of large-scale cascaded hydropower plants. Scientia Sinica (Technologica) **3**, 010 (2014)
6. Allen, R.B., Bridgeman, S.G.: Dynamic programming in hydropower scheduling. J. Water Resour. Plan. Manag. **112**(3), 339–353 (1986)

7. Labadie, J.W.: Optimal operation of multireservoir systems: state-of-the-art review. J. Water Resour. Plan. Manag. **130**(2), 93–111 (2004)
8. Yurtal, R., Seckin, G., Ardiclioglu, G.: Hydropower optimization for the lower Seyhan system in Turkey using dynamic programming. Water Int. **30**(4), 522–529 (2005)
9. Zhao, T., Zhao, J., Yang, D.: Improved dynamic programming for hydropower reservoir operation. J. Water Resour. Plan. Manag. **140**(3), 365–374 (2012)
10. Li, Y., Zuo, J.: Optimal scheduling of cascade hydropower system using grouping differential evolution algorithm. In: 2012 International Computer Science and Electronics Engineering (ICCSEE) (2012)
11. Zhang, Z., Zhang, S., Wang, Y., Jiang, Y., Wang, H.: Use of parallel deterministic dynamic programming and hierarchical adaptive genetic algorithm for reservoir operation optimization. Comput. Ind. Eng. **65**(2), 310–321 (2013)
12. Oliveira, R., Loucks, D.P.: Operating rules for multireservoir systems. Water Resour. Res. **33**(4), 839–852 (1997)
13. Wardlaw, R., Sharif, M.: Evaluation of genetic algorithms for optimal reservoir system operation. J. Water Resour. Plan. Manag. **125**(1), 25–33 (1999)
14. Li, C.: Real coded genetic algorithm optimization of long term reservoir operation1. J. Am. Water Resour. Assoc. **39**(5), 1157–1165 (2003)
15. Cinar, D., Kayakutlu, G., Daim, T.: Development of future energy scenarios with intelligent algorithms: case of hydro in Turkey. Energy **35**(4), 1724–1729 (2010)
16. Tayebiyan, A., Ali, T.A.M., Ghazali, A.H., Malek, M.A.: Optimization of exclusive release policies for hydropower reservoir operation by using genetic algorithm. Water Resour. Manag. **30**, 1–14 (2016)
17. Kumar, D.N., Reddy, M.J.: Ant colony optimization for multi-purpose reservoir operation. Water Resour. Manag. **20**(6), 879–898 (2006)
18. Jalali, M.R., Afshar, A., Marino, M.A.: Reservoir operation by ant colony optimization algorithms. Iran. J. Sci. Technol. Trans. B: Eng. **30**(B1), 107–117 (2007)
19. Moeini, R., Afshar, M.H.: Extension of the constrained ant colony optimization algorithms for the optimal operation of multi-reservoir systems. J. Hydroinform. **15**(1), 155–173 (2013)
20. Cheng, C.T., Liao, S.L., Tang, Z.T., Zhao, M.Y.: Comparison of particle swarm optimization and dynamic programming for large scale hydro unit load dispatch. Energy Convers. Manag. **50**(12), 3007–3014 (2009)
21. Afshar, M.H.: A cellular automata approach for the hydro-power operation of multi-reservoir systems. In: Proceedings of Institution of Civil Engineers Water Management, vol. 166, no. 9, pp. 465–478 (2013)
22. Kiruthiga, D., Amudha, T.: Optimal reservoir release for hydropower generation maximization using particle swarm optimization. In: Snášel, V., Abraham, A., Krömer, P., Pant, M., Muda, A.K. (eds.) Innovations in Bio-Inspired Computing and Applications. AISC, vol. 424, pp. 577–585. Springer, Heidelberg (2016). doi:10.1007/978-3-319-28031-8_51
23. Zhang, X., Yu, X., Qin, H.: Optimal operation of multi-reservoir hydropower systems using enhanced comprehensive learning particle swarm optimization. J. Hydro-Environ. Res. **10**, 50–63 (2016)
24. Teegavarapu, R.S., Simonovic, S.P.: Optimal operation of reservoir systems using simulated annealing. Water Resour. Manag. **16**(5), 401–428 (2002)
25. Tospornsampan, J., Kita, I., Ishii, M., Kitamura, Y.: Optimization of a multiple reservoir system using a simulated annealing – a case study in the Mae Klong system Thailand. Paddy Water Environ. **3**(3), 137–147 (2005)

26. Kangrang, A., Compliew, S., Hormwichian, R.: Optimal reservoir rule curves using simulated annealing. In: Proceedings of ICE-Water Management, vol. 164, no. 1, pp. 27–34 (2010)
27. Afshar, A., Haddad, O.B., Mariño, M.A., Adams, B.J.: Honey-bee mating optimization (HBMO) algorithm for optimal reservoir operation. J. Franklin Inst. **344**(5), 452–462 (2007)
28. Niknam, T., Taheri, S.I., Aghaei, J., Tabatabaei, S., Nayeripour, M.: A modified honey bee mating optimization algorithm for multiobjective placement of renewable energy resources. Appl. Energy **88**(12), 4817–4830 (2011)
29. Lu, Y., Zhou, J., Qin, H., Wang, Y., Zhang, Y.: An adaptive chaotic differential evolution for the short-term hydrothermal generation scheduling problem. Energy Convers. Manag. **51**(7), 1481–1490 (2010)
30. Asgari, H.R., Bozorg Haddad, O., Pazoki, M., LoÃąiciga, H.A.: Weed optimization algorithm for optimal reservoir operation. J. Irrig. Drain. Eng. **142**, 04015055 (2015)
31. Azizipour, M., Ghalenoei, V., Afshar, M.H., Solis, S.S.: Optimal operation of hydropower reservoir systems using weed optimization algorithm. Water Resour. Manag. **30**(11), 3995–4009 (2016)
32. Afshar, M.H., Shahidi, M.: Optimal solution of large-scale reservoir-operation problems: Cellular-automata versus heuristic-search methods. Eng. Optim. **41**(3), 275–293 (2009)
33. Geem, Z.W., Kim, J.H., Loganathan, G.V.: A new heuristic optimization algorithm: harmony search. Simulation **76**(2), 60–68 (2001)
34. Cheng, C.T., Wang, W.C., Xu, D.M., Chau, K.W.: Optimizing hydropower reservoir operation using hybrid genetic algorithm and chaos. Water Resour. Manag. **22**(7), 895–909 (2007)

Optimization of Hydropower Storage Projects Using Harmony Search Algorithm

S.J. Mousavi[1(✉)], P. Nakhaei[1], Ali Sadollah[2], and Joong Hoon Kim[3]

[1] School of Civil Engineering,
Amirkabir University of Technology, Tehran, Iran
{jmosavi,p.nakhaee}@aut.ac.ir
[2] Nanyang Technological University, Singapore, Singapore
ali_sadollah@yahoo.com
[3] Department of Civil, Environmental and Architectural Engineering,
Korea University, Seoul, South Korea
jaykim@yahoo.com

Abstract. This paper proposes a harmony search algorithm-based optimization of design and operation of hydropower storage systems. The optimization formulation of the problem is a nonconvex nonlinear program difficult to solve by gradient-based nonlinear programming techniques. The search space of the problem is large due to number of operational variables. Harmony search optimization algorithm is applied in the problem of design and operation optimization of Bakhtiari Dam and its powerplant project in Iran. The problem is solved in two cases of optimizing only design variables and optimizing design and operational variables, simultaneously, and the promising results obtained are presented.

Keywords: Optimization · Harmony search algorithm · Reservoir operation · Hydropower

1 Hydropower Reservoir Operation Optimization

Optimization is widely used in engineering design and operation problems. Gradient-based optimization approaches, including successive linear programming, quadratic programming, dynamic programming, dual dynamic programming, etc., have been widely used in hydropower systems operation optimization (Kim and Palmer 1997; Barros et al. 2003; Yeh et al. 1979; Grygier and Stedinger 1985; Diaz and Fontane 1989; Mousavi et al. 2004). To estimate design parameters of a hydropower reservoir system, i.e. reservoir capacities, powerplant production capacity and the reservoir's minimum option level, for any combinations of these design parameters, the reservoir system operation needs to be simulated over a representative hydrologic period using sequential the streamflow routing (SSR) method (Afzali et al. 2007). To do so, the release volume from the reservoir must be determined in each time step of the simulation model through either an operating policy (design optimization problem) or an optimization scheme (design-operation, DO, optimization problem). The end-of-month storage is then calculated using mass balance equation, and the process

J. Del Ser (ed.), *Harmony Search Algorithm*, Advances in Intelligent Systems and Computing 514, DOI 10.1007/978-981-10-3728-3_26

continues until the last time step of the simulation. This paper shows how to solve design and DO optimization of hydropower storage systems by using harmony search algorithm as a metaheuristic optimization scheme.

2 Harmony Search Algorithm

Harmony search algorithm (HSA), developed by Geem et al. (2001), is derived from the concepts of musical improvisations and harmony knowledge, and is a widely used metaheuristic algorithm. The HSA and its improved variants have proved its advantages over other optimizers (Kim et al. 2001; Vasebi et al. 2007; Mahdavi et al. 2007; Chakraborty et al. 2009; Gao et al. 2009; Geem and Sim 2010). Recently, HS has been applied to various research areas and obtained considerable attention in different disciplines (Yoo et al. 2014). The HS intellectualizes the musical process of searching for a perfect state of harmony (Geem et al. 2001). Musicians seek a fantastic harmony determined by aesthetic estimation. Similarly, optimization techniques seek a best state (global optimum) determined by an objective function 47 value. Aesthetic estimation depends on the set of the sounds played by a musical ensemble, whereas the objective function is evaluated by a set of adjustable variables. More aesthetic sounds can be produced by constant practice, and the optimization of the objective function can generally be improved by repeated iterations. Further details of HS can be found in the work of Geem et al. (2001) and Kim et al. (2001). We use HSA in the real-world problem of optimal operation of Bakhtiari Dam hydropower project in Iran, and show the promising results obtained are presented.

3 Optimizing Hydropower Systems Design and Operation

Optimal selection of design variables such as normal and minimum reservoir operating levels as well as the powerplant production capacity is important in hydropower storage projects. The energy potential of the project for any possible combinations of the design variables is under hydrologic variability of inflows to the reservoir. This requires the analysis of the system performance over a long representative hydrologic period in order to assess the expected costs and benefits resulted from the project's construction and operations for those combinations as the candidate solutions. Since there are many alternatives for the values of the design variables to choose from, their best values resulting in the highest expected net benefit, can be determined by using an optimization model. On the other hand, as the model solution is under hydrologic uncertainty, the energy yield of the system may be assumed as a random variable with a certain probability to be realized. In this case, designer may like to have the model's solution at different levels of reliability of meeting the system's energy yield In other words, they like to formulate an optimization model in which the reliability of the meeting the energy yield can be specified by the designer. The general formulation of the optimization model for this purpose may be written as follows (Mousavi and Shourian 2010):

$$Min\ \mathrm{Cost}^{1}_{total} = CRF \times (DC + PC + PeC) - fvalue \times nyear \times \sum_{t=1}^{12} FE - svalue \times \sum_{t=1}^{T} SE(t) + \sum_{t=1}^{T} Spill(t) \tag{1}$$

S.t:

$$DC = d_1 \times S^2_{\max} + d_2 \times S_{\max} + d_3 \times z_1 \tag{2}$$

$$DC \le BigM \times z_1 \tag{3}$$

$$PC = p_1 \times Pcap^2 + p_2 \times Pcap + p_3 \times z_2 \tag{4}$$

$$PC \le BigM \times z_2 \tag{5}$$

$$S(t+1) = S(t) + Q(t) - R(t) - Spill(t) - L(t) \quad \forall t \tag{6}$$

$$E(t) = 2.73 \times R(t) \times (0.5 \times (h(t) + h(t+1)) - h_{tail}(t) - h_f(t)) \times e_p(t) \quad \forall t \tag{7}$$

$$h(t) = k_1 \times S(t)^3 + k_2 \times S(t)^2 + k_3 \times S(t) + k_4 \quad \forall t \tag{8}$$

$$h_{tail}(t) = k_5 \times (R(t) + Spill(t))^2 + k_6 \times (R(t) + Spill(t)) + k_7 \quad \forall t \tag{9}$$

$$FE = Pcap \times pf \times nhours \quad \forall t \tag{10}$$

$$E(t) \ge FE \times z(t) \quad \forall t \tag{11}$$

$$\frac{\sum_{t=1}^{T} z(t)}{T} \ge TarREL \quad \forall t \tag{12}$$

$$SE(t) = (E(t) - FE) \times z(t) \quad \forall t \tag{13}$$

$$E(t) \le \mathrm{Emax} = Pcap \times nhours \quad \forall t \tag{14}$$

$$S_{\min} \le S(t) \le S_{\max} \quad \forall t \tag{15}$$

$$R_{\min} \le R(t) \le R_{\max} \quad \forall t \tag{16}$$

The objective function of the above program is minimizing the total cost, $Cost^{1}_{total}$, (or to maximize the total net benefit) which includes the capital and variable costs of dam, *DC*, powerplant, *PC*, and tunnel, *PeC,* constructions and operations deducted by the benefits resulting from firm and secondary energies sale. $CRF = \frac{(1+r)^{nyear}-1}{(1+r)^{nyear}-1}$ is the capital recovery factor converting the present value costs to their equivalent uniform annual costs and *r* is the annual discount rate of money. *FE is* the monthly firm energy yield of the system that can be produced in adverse hydrologic conditions at a certain level of reliability. Although the firm energy can be a seasonal variable depending on each calendar month m in general, we define it as an amount of monthly energy that is produced reliably at least $TarREL\%$ of total months $T = m \times nyear$ over a planning horizon with $nyear$ years. $SE(t)$ is the secondary energy in period (month) t which is the energy produced in excess of the firm energy and *fvalue* and *svalue* are respectively the firm and secondary energy unit sales. *PeC* is the fixed cost of the penstock which does not affect the solution as it is assumed to be a constant value herein.

In above formulation, z_1 and z_2 are binary variables accounting for fixed cost of dam and powerplant constructions, respectively; $z(t)$ are binary variables which are equal to zero if the energy generated is less than the target energy yield and to one, otherwise. The constraint on meeting the energy yield at target reliability level ($TarREL$) is satisfied through defining these binary variables. d_1 to d_3 and p_1 to p_3 are constants determined by fitting the best quadratic curves to the cost functions of dam and powerplant constructions, respectively; k_1 to k_4 and k_5 to k_7 are respectively the constants of elevation-storage and tailwater-discharge equations; *BigM* is a positive big number; $S(t)$ is the beginning-of-month reservoir storage, $Q(t)$ is the inflow to reservoir, $R(t)$ is the turbine release, $Spill(t)$ is the spilled release, and $L(t)$ is the net evaporation and seepage losses all in period t; *Pcap* is the powerplant production capacity and *pf* is the specified plant factor that defines the number of hours per day in which the powerplant is operating at its production capacity; $nhours = 720$ is the number of hours per month; $E(t)$ is the energy generated, $e_p(t)$ is the powerplant efficiency, $h(t)$ is the beginning-of-month reservoir level, $h_{tail}(t)$ is the tailwater level and $h_f(t)$ is the total minor and frictional losses in conveyance structures all in month t; $S_{\min}$ and $S_{\max}$ are respectively the minimum and maximum storage volumes of the reservoir; and $R_{\min}$ and $R_{\max}$ are the minimum and maximum turbine release volumes, respectively. Note that the last term in the objective function is defined to ensure that spillage may occur only if necessary and when $S(t) = S_{max}$.

In practical hydropower storage systems, a reliability-based reservoir simulation (RBS) model is commonly employed for a limited number of design variables combinations rather than using an optimization scheme. In the following, the single-reservoir RBS model is presented first and then it is explained how it will be linked with the HS algorithm for solving the problem of optimal design and operation of hydropower systems as defined by Eqs. (1)–(16).

3.1 Reservoir Operation Simulation Model

Assume a hydropower single-reservoir system with the given normal and minimum reservoir operating levels and a specified plant factor. Then, a reliability-based reservoir simulation model may be used to determine the maximum system's firm energy yield as follows (Afzali et al. 2007):

An initial production capacity may be estimated by the following equation:

$$Pcap = \frac{2.73 \times Q_{ave} \times h_{\max} \times \alpha}{pf \times nhours} \tag{17}$$

where *Pcap* is the estimated production capacity, Q_{ave} is the mean monthly inflow to the reservoir, h_{max} is an initial estimation for maximum net head on turbines as the difference between the normal and tailwater levels, α is a decreasing factor considering the effect of dry periods ($0.5 < \alpha < 1$). *pf* and *nhours(t)* are defined previously.

Given the estimated production capacity (*Pcap*) and the specified plant factor (*pf*), the system's firm energy yield as $FE = Pcap \times nhours \times pf$ is estimated as defined in Eq. (10).

The reservoir system operation is simulated over a representative hydrologic period using sequential streamflow routing (SSR) method to estimate the energy-yield reliability. In each time period *t* of the simulation model, the energy generated, *E(t),* is set to be equal to the estimated firm energy yield, *FE:*

$$2.73 * R(t) \times (0.5 \times (ht) + h(t+1)) - h_{tail}(t) - h_f(t)) \times e_p(t) = FE$$
$$= Pcap \times nhours \times pf$$

Then, the turbine release in that period can be determined as:

$$R(t) = \frac{Pcap \times nhours \times pf}{2.73 * (0.5 * (h(t) + h(t+1)) - h_{tail}(t) - h_f(t)) * e_p(t)} \tag{18}$$

In Eq. (18), $h(t+1)$, $h_{tail}(t)$ and $h_f(t)$ depend on the turbine release making the equation implicit with respect to *R(t)*. Therefore, Eq. (18) is solved in each time period *t* iteratively (see Afzali et al. (2007) for details). Taking on the final end-of-month storage as the beginning-of-month storage of the next time period, the procedure continues up to the final period of the planning horizon. Subsequently, the energy-yield reliability is determined by counting the number of failure months in which firm energy yield has not been met. Subsequently, the production capacity is adjusted according to the three following cases. If the determined reliability is within the desired range specified for target reliability ($TarREL - \delta \leq REL \leq TarREL + \delta$), the estimated production capacity and energy yield are acceptable; otherwise, they increase ($TarREL + \delta \leq REL$) or decrease ($REL \leq TarREL - \delta$). All of the steps explained are repeated until the production capacity and energy yield values converge, and the reliability of meeting the energy yield arrives at its specified target value.

The RBS model may be repeated for different normal and minimum operating levels. Subsequently, an economic analysis can be used to choose the final normal and minimum operating levels of the reservoir, powerplant production capacity and other design and operational variables. However, such trial and error-based approach may not arrive at the best possible solution. It is, therefore, desired to employ the optimization model defined by Eqs. (1)–(16) to determine the optimal design configuration of the system. In order to solve this NP-hard optimization model, we will combine the HSA, as the optimizer module, and the reliability-based reservoir simulation (RBS) model as the simulator.

The problem is solved in two cases. In the first one, only the design variables are optimized, whereas the second case is about optimizing both design and operational variables.

3.2 Design Optimization Problem (Model A)

The aim in this problem is to find the best values of design parameters of the system maximizing the net benefit resulting from construction and operations of a hydropower system where the system operates based on a specified operating policy, which was explained in reservoir operation simulation module. The decision variables are the reservoir normal water level, corresponding to a specific height of the dam), the minimum reservoir operating level and the production capacity. The production capacity is found so that the maximum system's firm energy yield results at the specified level of reliability, *TarREL*. For any pairs of reservoir's normal and minimum water levels, the maximum achievable system's firm energy yield (powerplant production capacity) is determined as follow:

In this model the normal and minimum operating levels are searched for by HSA, and the reliability constraint on the energy yield is met using a penalty term in the objective function. In other words, the production capacity in model A is considered as a random-based search variable varying by HSA. The reservoir simulation model with a known operating policy, in which releases from the reservoir are determined by Eq. (18), is employed any candidate sets of design (decision) variables generated by HSA. The outputs of reservoir simulation model are the reliability level, annual firm and secondary energies associated with any candidate sets of these decision variables. A penalty term penalizing the deviation of the resulted reliability from the target one is added to the objective function as follows:

$$Cost^{2}_{total} = Cost^{1}_{total} + (|REL - TarREL|) * P \tag{19}$$

where $Cost^{2}_{total}$ is the total cost in model A, $Cost^{1}_{total}$ is as defined previously, *REL* is the level of reliability resulted from simulating the reservoir operation, *TarREL* is the target or desired reliability level, and *P* is a penalty factor that should be fine-tuned by a trial and error procedure. The penalty factor should not be very large to avoid premature convergence of the algorithm, and it should not be too small, resulting in a solution not satisfying the reliability constraint. In model A, the reservoir simulation model is the RBS model. However, adjusting (increasing or decreasing) the production capacity is not performed by the search procedure explained before; it is rather conducted by the HAS using the penalty approach.

3.3 Design-Operation Optimization Problem (Model B)

For the case of design-operation (DO) optimization, the operating policy is not specified a priori, but it is optimized by defining release rules whose parameters are considered as decision variables of the HS algorithm, in addition to the design variables considered before. In this model releases from the reservoir, *R(t)*, are also considered as the HSA decision variables. Therefore in our case study with a 41-years simulation horizon and 3 design variables, the optimization problem will include 495 decision variables, of which 492 variables are operational variables. This requires that special attention be given to generation of feasible solutions at early stages of the stochastic search algorithm.

4 Results

The case study of the paper is Bakhtiari Dam to be built on the Bakhtiari River in west of Iran. Tables 1 and 2 present input values to the models, i.e. basic characteristics of the system and the topographic data of Bakhtiari dam. Cost values for different dam and powerplant capacities are presented in Table 3. Firm and secondary energy unit sales, i.e. *fvalue* and *svalue*, are set to $160 * 10^{-4}$ monetary units per Mwh (Mousavi and Shourian 2010).

Table 1. Basic characteristics of Bakhtiari Dam and its powerplant

Maximum normal water level	830 Masl
Minimum water level	660 Masl
Tailwater elevation	533.5 Masl
Generator efficiency	92.12%
Head loss	3 M
Plant factor	0.25
Target reliability	90%

Table 2. Elevation-capacity relation at dam site

Elevation (Masl)	Capacity (MCM)
532	0
550	0.01
575	0.02
592	0.03
593	1.014
600	4.32
625	23.62
650	97
675	241.28
700	481.52
725	847.73
750	1371.57
775	2084.12
800	3031.3
825	4269
830	4582.37

Optimum values of objective function, design variables, average annual firm and secondary energies and the reliability levels obtained are presented in Table 4. Figure 1 shows the variation of the models' objective function against different iterations of the HSA. Model B outperforms model A in terms of objective function value. The difference, however, between the best objective function value of model B and that of model A is not significant. Although operational variables are optimized in model B, and its search space (495-dimensional) is much larger than that of model A (3-dimensional), model A has arrived at an objective value close to that of model B. Results presented are in good agreement with those obtained by particle swarm optimization (PSO) for the same problem reported in Mousavi and Shourian (2010).

Table 3. Cost of dam and powerplant construction for various capacities

Capacity (MCM)	Cost (monetary unit)	Installed capacity (MW)	Cost (monetary unit)
1941.61	781.44	913	152.45
2652.43	897.52	1014	162.40
3526.38	1016.07	1110	173.25
4587.37	1139.20	1220	189.30
4590.00	5500.00	1500	225.00

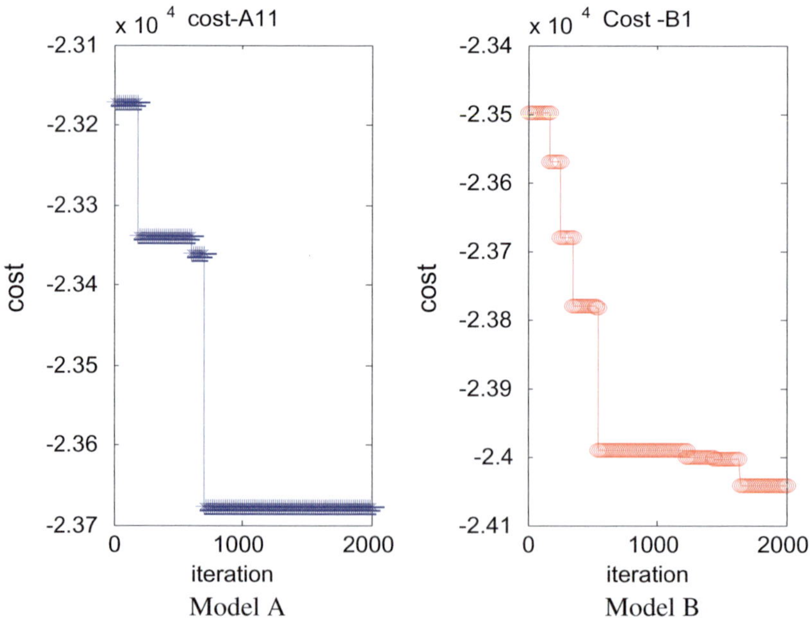

Fig. 1. Convergence curve of the objective function in HSA

Table 4. Optimal values of objective function, reliability and design variables resulted by the models

Item	Model A	Model B
Total cost	−23500.74	−24041.67
Reliability of meeting the energy demand (%)	90.85	91.26
Normal water level (Masl)	825.3	829.86
Min. operation level (Masl)	798.06	811.08
Powerplant production capacity (MW)	1194.3	1175.4
Annual firm energy generated (MWh)	2.5796×10^6	2.5389×10^6
Annual secondary energy generated (MWh)	3.112×10^6	3.183×10^6

5 Summary and Conclusions

The formulation of the problem of optimum design and operation of a hydropower reservoir system with a reliability constraint on meeting the energy yield is a non-convex MINLP. Therefore, we used harmony search algorithm (HSA) to solve this NP-hard problem. The objective function is to maximize the total net benefit resulted from design (constructing) and operation of the system over its life period. The problem was solved in two cases. In the first one, optimum design of Bakhtiyari hydropower system was considered, in which the decision variables were the normal and minimum operating levels of the reservoir as well as production capacity of the powerplant (model A). In the second case (model B), simultaneous optimization of design and operation variables of the system was carried out using HSA. In model B, the release decision variables were generated randomly in an incremental way at the first iteration of HSA so that a feasible solution can be obtained. HSA performed satisfactorily for both design and design-operation problems although model B was slightly better than model A.

References

Afzali, R., Mousavi, S.J., Ghaheri, A.: A reliability-based simulation optimization model for multi-reservoir hydropower systems operation: Khersan experience. J. Water Resour. Plan. Manag. **134**(1), 24–33 (2008). ASCE

Barros, M.T.L., Tsai, F.T.-C., Yang, S.-L., Lopes, J.E.G., Yeh, W.W.-G.: Optimization of large-scale hydropower system operations. Water Resour. Plan. Manag. **129**(3), 178–188 (2003). ASCE

Chakraborty, P., Roy, G.G., Das, S., Jain, D., Abraham, A.: An improved harmony search algorithm with differential mutation operator. Fundam. Inform. **95**, 1–26 (2009)

Diaz, G.E., Fontane, D.G.: Hydropower optimization via sequential quadratic programming. Water Resour. Plan. Manag. **115**(6), 715–733 (1989). ASCE

Gao, X.Z., Wang, X., Ovaska, S.J.: Uni-modal and multi-modal optimization using modified harmony search methods. Int. J. Innov. Comput. Inf. Control **5**(10A), 2985–2996 (2009)

Geem, G.W., Kim, J.H., Loganathan, G.V.: A new heuristic optimization algorithm: harmony search. Simulation **76**(2), 60–68 (2001)

Geem, Z.W., Sim, K.B.: Parameter-setting-free harmony search algorithm. Appl. Math. Comput. **217**(8), 3881–3889 (2010)

Grygier, J.C., Stedinger, J.R.: Algorithms for optimizing hydropower system operation. Water Resour. Res. **21**(1), 1–10 (1985)

Kim, J.H.Z., Geem, W., Kim, E.: Parameter estimation of the nonlinear Muskingum model using harmony search. J. Am. Water Resour. Assoc. **37**(5), 1131–1138 (2001)

Kim, Y.O., Palmer, R.N.: Value of seasonal flow forecasts in Bayesian stochastic programming. Water Resour. Plan. Manag. **123**(6), 327–335 (1997). ASCE

Mahdavi, V., Fesanghary, M., Damangir, E.: An improved harmony search algorithm for solving optimization problems. Appl. Math. Comput. **188**(2), 1567–1579 (2007)

Mousavi, S.J., Shokrvand, K., Seifi, A.: Application of an interior-points algorithm for optimization of a large79 scale reservoir system. Water Resour. Manag. **18**, 519–540 (2004)

Mousavi, S.J., Shourian, M.: Capacity optimization of hydropower storage projects using particle swarm optimization algorithm. J. Hydroinform. **12**(3), 275–291 (2010)

Yeh, W.G., Becker, G.L., Chu, W.S.: Real-time hourly reservoir operation. J. Water Resour. Plan. Manag. **105**(2), 187–203 (1979). ASCE

Yoo, D.G., Kim, J.H., Geem, Z.W.: Overview of harmony search algorithm and its applications in civil engineering. Evol. Intell. **7**, 3–16 (2014)

Vasebi, A., Fesanghary, M., Bathaee, S.M.T.: Combined heat and power economic dispatch by harmony search (2007)

Metaheuristic Based Optimization for Tuned Mass Dampers Using Frequency Domain Responses

Gebrail Bekdaş[1(✉)], Sinan Melih Nigdeli[1], and Xin-She Yang[2]

[1] Department of Civil Engineering, Istanbul University, 34320 AvcıLar, Istanbul, Turkey
{bekdas,melihnig}@istanbul.edu.tr

[2] Design Engineering and Mathematics, Middlesex University London, The Burroughs, London NW4 4BT, UK
x.yang@mdx.ac.uk

Abstract. The usage of tuned mass dampers is a practical technique for civil structures under undesired vibrations. Since earthquake or other excitations have random vibration with various frequencies, frequency domain responses of structures can be used in tuning of mass dampers. By the minimization of transfer function of the structure, the amplitude corresponding to the response frequency of the structure is minimized. Thus, the passive control of the structure is provided. The optimization problem is non-linear since the existence of inherent damping of the structure, possible solution range of tuned mass damper (TMD) and multiple modes of multiple degree of freedom structures. For that reason, three metaheuristic algorithms such as harmony search, flower pollination algorithm and teaching learning based optimization are compared in mean of performance and computation effort. All algorithms are feasible with significant possible advantages.

Keywords: Tuned mass damper · Optimization · Frequency domain responses · Earthquake · Harmony search · Flower pollination algorithm · Teaching learning based optimization

1 Introduction

Tuned mass damper (TMD), which are initially invented as a vibration absorber device without damping by Frahm [1], and modified for random vibrations [2] have been used in all types of systems using the principle of mechanics. Vibration mitigation of civil structures is also an area of tuned mass dampers (TMDs). In the existing structures, TMDs have been installed in order to maintain an effective reduction of structural vibration resulting from earthquake, winds and traffic. Soto and Adeli presented a list of structure including TMDs with short information [3]. In Fig. 1, the 361 m long television tower of Berlin is shown. The tower type structure contains a 1.5 ton TMD for preventing wind indicated vibrations.

J. Del Ser (ed.), *Harmony Search Algorithm*, Advances in Intelligent Systems and Computing 514, DOI 10.1007/978-981-10-3728-3_27

Fig. 1. The television tower in Berlin

Generally, TMDs are tuned by using frequency domain responses of the structure. In the documented methods, several formulations have been developed for frequency and damping ratio of TMD [4–8]. These formulations were derived for single degree of freedom structure and these formulation can be used for multiple degree of freedom structures by considering a single vibration mode. In order to consider the damping of the main structure and multiple modes, numerical algorithms are needed. Metaheuristic algorithms are very effective in tuning of mass dampers [8–23].

In this study, three different algorithms; Harmony search (HS), Flower pollination algorithm (FPA) and Teaching learning based optimization (TLBO) were employed for the optimum tuning of mass dampers. The optimization objective is related to the frequency domain responses of the structure. The aim is to minimize the top story acceleration transfer function. The proposed methods were applied for a 10 story structure and the comparisons of the algorithms were done.

2 The Optimization Problem

The objective of the optimization is to minimize the top story acceleration transfer function of the structure by optimizing the parameters of TMD positioned on the top of structures. The transfer function (TF) is the ratio of Laplace transformations of response (top story acceleration for the current problem) and an external excitation (ground acceleration for the current problem).

TF is a unitless value and the peak value of the transfer function is minimized. The maximum value is observed at the first natural frequency of the structure. The unitless TF is modified by taking base 10 logarithm and multiplying by 20. In that case, the unit of TF is dB. The TF formulation with respect to frequency (ω) is shown in Eq. (1) for the structure with TMD. This function contains the values of all stories and TMD. The objective function (f) is shown as Eq. (2). As the objective, the maximum results of the Nth story (top story) is used.

As seen in Fig. 1, the N-story structure is modeled as a shear building. The structure has N degree of freedom for all lateral motion of the stories and the number of degrees of freedom of the TMD controlled structure is $N+1$. The equations of motion of a shear building is written as follows:

$$TF(w) = \begin{bmatrix} TF_1(\omega) \\ TF_2(\omega) \\ \vdots \\ TF_N(\omega) \\ TF_d(\omega) \end{bmatrix} = [-M\omega^2 + C\omega j + K]^{-1} M\omega^2 \mathbf{1} \tag{1}$$

$$f = 20Log_{10}|max(TF_N(\omega))| \tag{2}$$

The value of TF contains imaginary (j) and real parts. For that reason, the absolute value of the function is taken and the unit of the function is modified as dB. M, C, and K are the mass, damping and stiffness matrices of a N-story shear structure with a TMD on the top. These matrices are shown in Eqs. (3)–(5). In these matrices, m_i, c_i and k_i are mass, damping coefficient, stiffness coefficients of ith story, respectively. The parameters of TMD such as mass, stiffness and damping coefficients are shown with m_d, k_d and c_d, respectively. The design variables are mass (m_d), period (T_d) and damping ratio (ξ_d) of TMD, respectively. The value of T_d and ξ_d are formulated in Eqs. (6) and (7), respectively. A vector ones is shown as **1**.

$$M = diag[m_1 \quad m_2 \ldots \quad m_N \quad m_d] \tag{3}$$

The mass, damping coefficient, stiffness coefficient and displacements are symbolized with m, c, k and x. The subscripts show the story number between 1 and N. The TMD parameters are mass (m_d), damping coefficient (c_d) and stiffness coefficient (k_d) while the displacement of the TMD is x_d.

The damping and stiffness coefficients can be written as period (T_d) and damping ratio (ξ_d) of TMD as shown in Eqs. (6) and (7). In the numerical example, these two values are taken as design variables together with mass of TMD (m_d) and TMD positioned story (i).

$$\begin{bmatrix} (c_1+c_2) & -c_2 & & & \\ -c_2 & (c_2+c_3) & -c_3 & & \\ & \vdots & \vdots & & \\ & \vdots & \vdots & & \\ & & c_N & (c_N+c_d) & -c_d \\ & & & -c_d & c_d \end{bmatrix} \tag{4}$$

$$\begin{bmatrix} (k_1+k_2) & -k_2 & & & \\ -k_2 & (k_2+k_3) & -k_3 & & \\ & \vdots & \vdots & & \\ & \vdots & \vdots & & \\ & & k_N & (k_N+k_d) & -k_d \\ & & & -k_d & k_d \end{bmatrix} \tag{5}$$

$$T_d = 2\pi\sqrt{\frac{m_d}{k_d}} \tag{6}$$

$$\xi_d = \frac{c_d}{2m_d\sqrt{\frac{k_d}{m_d}}} \tag{7}$$

3 Metaheuristic Methods

In this study, three different metaheuristic algorithms were separately employed. These algorithms are Harmony search (HS), Flower pollination algorithm (FPA) and teaching learning based optimization (TLBO). These methods are summarized in this section.

3.1 Harmony Search

The harmony search algorithm is the oldest one comparing to the other two algorithm. The process of a musician is a good match for optimization algorithms and HS employed in several TMD studies as stated in the introduction of the study. The algorithm developed by Geem et al. [24] has three parameters and the name of these parameters are harmony memory size (HMS: the number of harmony vectors in harmony memory), harmony memory consideration rate (HMCR: the rate of consideration of global and local searches) and pitch adjusting rate (PAR: the rate for adjusting the range of generated solutions).

In the global search of the HS algorithm, a new design variable (X_{new}) limited with upper (X_{max}) and lower (X_{min}) bounds can be randomly generated as shown in Eq. (8).

$$X_{new} = X_{min} + ran(0,1)(X_{max} - X_{min}) \tag{8}$$

In the equation, rand(0, 1) is a randomly generated number between 0 and 1. The local search of the algorithm is formulated as follows:

$$X_{new} = X_k + ran(1/2, -1/2)PAR(X_{max} - X_{min}) \tag{9}$$

X_k is k^{th} existing solution and k is randomly chosen. rand(−1/2, 1/2) is a random number between −1/2 and 1/2. In the study, the parameters: HMS, HMCR and PAR are taken as 5, 0.5 and 0.2, respectively.

3.2 Flower Pollination Algorithm

The inspiration of FPA is the reproduction process of flowering plants via pollination. Yang [25] formulated two types of pollination of flowering plants. These types are biotic (or cross) pollination done by pollinators and abiotic (or self) pollination done by the fertilization of flowers. Biotic pollination is used as a global optimization as shown in Eq. (10). Since the pollinators obey the rules of Lévy flight, a Lévy distribution (LD) is used. A ith new solution ($X_{new,i}$) is obtained by using i^{th} existing solution ($X_{old,i}$) and the best existing solution (X_{best}).

$$X_{new,i} = X_{old,i} + LD(X_{best} - X_{old,i}) \tag{10}$$

Abiotic pollination is used as a local optimization and it is formulated as follows:

$$X_{new,i} = X_{old,i} + rand(0, 1)(X_m - X_k) \tag{11}$$

In the local optimization, two existing solution (m^{th} and k^{th}) are randomly chosen and a linear distribution is used. The type generation is chosen by using a switch probability (p) and this value is taken as 0.5 in current study. The number of set of design variables in a solution matrices (population) is taken as 25.

3.3 Teaching Learning Based Optimization

TLBO was inspired from the education process of a class. The algorithm was developed by Rao et al. [26] and differently from the other algorithm, TLBO consequently use the two types of optimization phases which are teacher and student phases. For that reason, it is not needed to use a rate for choosing the generation type. This feature makes the algorithm easier than the others since the only user defined parameter is population (taken as 25 in current study).

The teacher phase is the process inspired by the education of a teacher. This phase is formulated as follows and the value of Teacher factor (T_F) is randomly assigned as 1 or 2. $X_{teacher}$ and X_{mean} are the best existing solution and average of all existing solutions, respectively.

$$X_{new,i} = X_{old,i} + rand(0, 1)(X_{teacher} - X_{mean}) \tag{12}$$

The student phase inspired from the self-study of student. It is formulated as seen in Eqs. (13) and (14). If m^{th} solution is better than k^{th} solution,

$$X_{new,i} = X_{old,i} + rand(0,1)(X_m - X_k) \tag{13}$$

if k^{th} solution is better than m^{th} solution

$$X_{new,i} = X_{old,i} + rand(0,1)(X_k - X_m). \tag{14}$$

4 Numerical Examples

The proposed methods were tested on a 10-story structure. The properties of stories of the structure are different. The mass, stiffness and damping coefficient parameters of the shear building are presented in Table 1.

The proposed method was applied to the optimization of a TMD on the 40-story structure [26]. The properties of the structure as shown in Table 1. The example structure is a shear building with different stiffness and damping coefficient for stories. These values are linearly decreased by increase of the number of stories. For limitation of the stroke capacity, st_max is taken as 1.5 and the optimization is done for 200 maximum iterations. The value of PAR is taken as 0.5 while the population number is 5 for the numerical example.

Table 1. The properties of the example building [6]

Story	m_i(t)	k_i(MN/m)	c_i(MNs/m)
1	179	62.47	0.81
2	170	52.26	0.67
3	161	56.14	0.72
4	152	53.02	0.68
5	143	49.91	0.64
6	134	46.79	0.60
7	125	43.67	0.56
8	116	40.55	0.52
9	107	37.43	0.48
10	98	34.31	0.44

The mass of TMD (m_d) was searched between 1% and 10% of the total mass of the structure. For the period of the TMD (T_d), it is searched between 0.8 and 1.2 times of the first natural frequency of the structure without TMD. The damping ratio of TMD (ξ_d) is randomly assigned between 0.05 and 0.3. The employed algorithms were tested by conducting 20 independent runs. In Table 2, the optimum design variables are presented with the best TF and average TF values of 20 runs. Additionally, standard derivative values and number of analyses for the optimum values are shown in Table 2.

Table 2. Optimum values for the proposed methods

	HS	FPA	TLBO
m_d (t)	138.5	138.5	138.5
T_d (s)	2.2926	2.2917	2.2917
ξ_d	0.2782	0.2763	0.2763
Best TF	11.7316	11.7303	11.7303
Average TF	11.7322	11.7303	11.7303
Standard derivative	$3.15 * 10\mathrm{m}^{-4}$	$1.05 * 10^{-6}$	$5.55 * 10^{-10}$
Number of analyses	16529	9125	49500

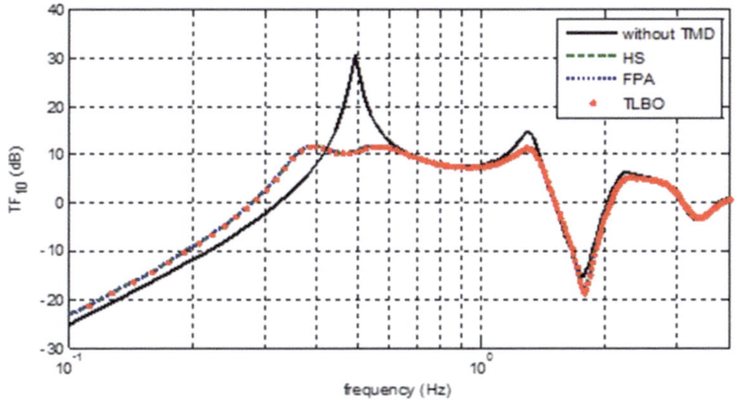

Fig. 2. TF plot of the top story of the structure

The optimum masses of TMD are at the upper bound of the solution range for all methods. Also, periods and damping ratios of TMD for different methods are very close to each other. In that case, all algorithms are effective on finding global optimum values for minimization of the top story acceleration transfer function. The plot of the transfer function is shown in Fig. 2 with the comparison of the structure without TMD and with TMD (for all methods). As seen in the figure, the optimum TMD is effective on the reduction of the peak amplitude of the first natural frequency of the structure. The plots of all methods are nearly the same.

Also, the average values of the TF values are nearly the same with the best values. For that reason, the standard derivative of 20 independent runs are very low. Especially, TLBO has very low standard derivative value, but the values of all methods are nearly zero. The analyses were iteratively done for 50000 function evaluation. All methods are effective to find the optimum values before the maximum number of iterations. The number of analyses to reach the optimum values are also presented and the values are 16529, 9125 and 49500, respectively.

5 Conclusions

The proposed algorithms, HS, FPA and TLBO are an effective source for the TMD optimization problem considering frequency domain responses. The effectiveness of the methods were proved by the results of the numerical example. The algorithms are also robust since the same optimum results were obtained for 20 independent runs.

Another important factor is the computation effort of the algorithms. According to the results, FPA is very quick in finding the optimum values. The worst one in computation effort is TLBO, but TLBO is an easy method to apply since it has no user defined parameters other than the population number. As a conclusion, all proposed methods are feasible for the optimization problem but all methods have different advantages.

References

1. Frahm, H.: Device for damping of bodies. U.S. Patent No: 989,958 (1911)
2. Ormondroyd, J., Den Hartog, J.P.: The theory of dynamic vibration absorber. T. ASME **50**, 9–22 (1928)
3. Soto, M.G., Adeli, H.: Tuned mass dampers. Arch. Comput. Methods Eng. **20**(4), 419–431 (2013)
4. Den Hartog, J.P.: Mechanical Vibrations, 3rd edn. Mc Graw-Hill, New York (1947)
5. Warburton, G.B.: Optimum absorber parameters for various combinations of response and excitation parameters. Earthq. Eng. Struct. Dyn. **10**, 381–401 (1982)
6. Sadek, F., Mohraz, B., Taylor, A.W., Chung, R.M.: A method of estimating the parameters of tuned mass dampers for seismic applications. Earthq. Eng. Struct. D. **26**, 617–635 (1997)
7. Chang, C.C.: Mass dampers and their optimal designs for building vibration control. Eng. Struct. **21**, 454–463 (1999)
8. Leung, A.Y.T., Zhang, H.: Particle swarm optimization of tuned mass dampers. Eng. Struct. **31**, 715–728 (2009)
9. Leung, A.Y.T., Zhang, H., Cheng, C.C., Lee, Y.Y.: Particle swarm optimization of TMD by non-stationary base excitation during earthquake. Earthq. Eng. Struct. Dyn. **37**, 1223–1246 (2008)
10. Hadi, M.N.S., Arfiadi, Y.: Optimum design of absorber for MDOF structures. J. Struct. Eng.-ASCE **124**, 1272–1280 (1998)
11. Marano, G.C., Greco, R., Chiaia, B.: A comparison between different optimization criteria for tuned mass dampers design. J. Sound Vib. **329**, 4880–4890 (2010)
12. Singh, M.P., Singh, S., Moreschi, L.M.: Tuned mass dampers for response control of torsional buildings. Earthq. Eng. Struct. Dyn. **31**, 749–769 (2002)
13. Desu, N.B., Deb, S.K., Dutta, A.: Coupled tuned mass dampers for control of coupled vibrations in asymmetric buildings. Struct. Control Health monit. **13**, 897–916 (2006)
14. Pourzeynali, S., Lavasani, H.H., Modarayi, A.H.: Active control of high rise building structures using fuzzy logic and genetic algorithms. Eng. Struct. **29**, 346–357 (2007)
15. Steinbuch, R.: Bionic optimisation of the earthquake resistance of high buildings by tuned mass dampers. J. Bionic Eng. **8**, 335–344 (2011)

16. Bekdaş, G., Nigdeli, S.M.: Estimating optimum parameters of tuned mass dampers using harmony search. Eng. Struct. **33**, 2716–2723 (2011)
17. Bekdaş, G., Nigdeli, S.M.: Optimization of tuned mass damper with harmony search. In: Gandomi, A.H., Yang, X.-S., Alavi, A.H., Talatahari, S. (eds.) Metaheuristic Applications in Structures and Infrastructures. Elsevier, Amsterdam (2013)
18. Bekdaş, G., Nigdeli, S.M.: Mass ratio factor for optimum tuned mass damper strategies. Int. J. Mech. Sci. **71**, 68–84 (2013)
19. Nigdeli, S.M., Bekdaş, G.: Optimum tuned mass damper design for preventing brittle fracture of RC buildings. Smart Struct. Syst. **12**(2), 137–155 (2013)
20. Nigdeli, S.M., Bekdaş, G.: Optimization of TMDs for different objectives. In: An International Conference on Engineering and Applied Sciences Optimization, Kos Island, Greece, 4–6 June 2014
21. Farshidianfar, A., Soheili, S.: Ant colony optimization of tuned mass dampers for earthquake oscillations of high-rise structures including soil-structure interaction. Soil Dyn. Earthq. Eng. **51**, 14–22 (2013)
22. Farshidianfar, A., Soheili, S.: ABC optimization of TMD parameters for tall buildings with soil structure interaction. Interact. Multiscale Mech. **6**, 339–356 (2013)
23. Farshidianfar, A., Soheili, S.: Optimization of TMD parameters for earthquake vibrations of tall buildings including soil structure interaction. Int. J. Optim. Civ. Eng. **3**, 409–429 (2013)
24. Geem, Z.W., Kim, J.H., Loganathan, G.V.: A new heuristic optimization algorithm: harmony search. Simulation **76**(2), 60–68 (2001)
25. Yang, X.-S.: Flower pollination algorithm for global optimization. In: Durand-Lose, J., Jonoska, N. (eds.) UCNC 2012. LNCS, vol. 7445, pp. 240–249. Springer, Heidelberg (2012). doi:10.1007/978-3-642-32894-7_27
26. Rao, R.V., Savsani, V.J., Vakharia, D.P.: Teaching-learning-based optimization: a novel method for constrained mechanical design optimization problems. Comput. Aided Des. **43**(3), 303–315 (2011)

Multidisciplinary Applications of Harmony Search

Optimal Phase Swapping in Low Voltage Distribution Networks Based on Smart Meter Data and Optimization Heuristics

Izaskun Mendia[1(✉)], Sergio Gil-López[1], Javier Del Ser[1,2,3], Ana González Bordagaray[4], Jesús García Prado[4], and Manuel Vélez[2]

[1] TECNALIA, 48160 Derio, Spain
{izaskun.mendia,sergio.gil,javier.delser}@tecnalia.com
[2] University of the Basque Country UPV/EHU, 48013 Bilbao, Spain
{javier.delser,manuel.velez}@ehu.eus
[3] Basque Center for Applied Mathematics (BCAM), 48009 Bilbao, Spain
[4] IBERDROLA Distribución Eléctrica, S. A., 48003 Bilbao, Spain
{ana.gb,jgarciapr}@iberdrola.es

Abstract. In this paper a modified version of the Harmony Search algorithm is proposed as a novel tool for phase swapping in Low Voltage Distribution Networks where the objective is to determine to which phase each load should be connected in order to reduce the unbalance when all phases are added into the neutral conductor. Unbalanced loads deteriorate power quality and increase costs of investment and operation. A correct assignment is a direct, effective alternative to prevent voltage peaks and network outages. The main contribution of this paper is the proposal of an optimization model for allocating phases consumers according to their individual consumption in the network of low-voltage distribution considering mono and bi-phase connections using real hourly load patterns, which implies that the computational complexity of the defined combinatorial optimization problem is heavily increased. For this purpose a novel metric function is defined in the proposed scheme. The performance of the HS algorithm has been compared with classical Genetic Algorithm. Presented results show that HS outperforms GA not only on terms of quality but on the convergence rate, reducing the computational complexity of the proposed scheme while provide mono and bi phase connections.

Keywords: Smart meter · Load curve · Harmony search

1 Introduction

In the electrical distributions systems of most utilities over the world, three phases of alternate current are utilized at each feeder aimed at increasing the energy efficiency in low-voltage distribution networks. During the past decades a huge increment in housing construction has occurred in many countries (especially in Spain), which has given rise to highly-populated low-voltage distribution networks. This noted fact, combined with the ever-growing use of home

J. Del Ser (ed.), *Harmony Search Algorithm*, Advances in Intelligent Systems and Computing 514, DOI 10.1007/978-981-10-3728-3_28

electronic devices, has heretofore unchained a dramatically sharp load growth. When held together, these two key points lay an operational challenge to distribution companies because the so-called feeder tripping problem [1] is even more involved due to in-excess neutral current caused by the unbalance loads among the three phases. Unbalanced loads between phases may lead to undesirable situations, including [2]: current increase in the most heavily loaded phase limits the amount of power transferred on a feeder; current increment in the neutral conductor; and problems with voltage drops in phase with higher loads, which ultimately results in a low quality of service. As a result, power losses in distribution networks may vary significantly depending on the load imbalance.

Accordingly, if loads on each phase are properly balanced technical losses will be notably reduced [3–5]. Consequently, a proper balance between the three phases will contribute to: (1) an optimized network infrastructure by increasing the capacity of distribution feeders, hence avoiding the deployment of unnecessary extra feeders and consequently reducing distribution costs; (2) a reduced monitoring complexity of the low-voltage distribution network due to the reduction of instabilities generated by the presence of high currents in the neutral conductor; and (3) an improved voltage profile due to the homogenization of the voltage drops at each stage of the distribution line.

There are two major techniques for phase balancing in the related literature [2]: feeder reconfiguration at the system level, and phase swapping at the feeder level. The former is a process of changing the topological structure of the distribution systems by altering the open/closed status of single phase sections and tie switches [6]. In phase swapping, however, the objective is to determine to which phase each load should be connected in order to reduce the unbalance when all phases are added into the neutral conductor. This objective can be formulated as a combinatorial optimization problem with exponential complexity growth with the number of possible loads: for the case of N loads to be connected to 6 phases (3 single phases and 3 complex phases), there is a total of 6^N combinations (possible solutions).

Many research contributions in the last years have dealt with the phase swapping problem [7], each resorting to different approaches and diverse methodologies such as Simulated Annealing [8–10], Neural Networks [5,11], Genetic Algorithms [12], Tabu Search [13], Greedy approaches and Dynamic Programming [2], among others. In [14] phase swapping is addressed as a load-to-line assignment problem and tackled under a mixed-integer programming formulation. In [15] the optimal load phase balance is obtained by solving the load redistribution problem by using a decaying self-feedback continuous Hopfield neural network (ADSCHNN). Likewise the work in [16] proposes a new approach for phase balancing planning using a specialized Genetic Algorithm which considers discretized load duration curve.

To the knowledge of the authors none of the above references accounts for two practical situations of real low-voltage distribution networks. The first one relates to the use of the information provided by Advanced Metering Infrastructures (AMIs) or Smart Meters, which provides a deeper, fine-grained knowledge of load patterns, far beyond the coarse-grained monitoring performed until their appearance. Indeed, by virtue of the hourly load patterns provided by the AMIs not only operational costs are reduced and the quality of service is improved, but also an evidence of paramount importance for our work has been unveiled: even if phase balancing can be met over a certain time span (i.e. yearly or monthly), the characteristics of connected loads vary continuously, which causes punctual, undesired situations of unbalanced phases. The second aspect which has not been taken into consideration in the literature as mentioned in [2] is that loads can be connected to two phases, as opposed to related contributions so far which consider only single phase connections. This point increases the computational complexity of the phase swapping problem by increasing the number of possibilities by which the load could be connected, i.e. from 3^N to 6^N for N loads.

The above challenges motivate the development of new heuristic procedures that efficiently tackle the phase swapping problem taking into consideration the time variance of load patterns and mono- and bi-phase connections. For this purpose in this paper we propose to apply the Harmony Search (HS [17]) algorithm as an heuristic procedure for solving the aforementioned problem. HS has been used for NP-hard problems providing a good balance between computational complexity and quality of the provided solutions [18, 19]. This manuscript delves into the adaptation of the HS characteristics to phase swapping introducing a novel metric definition and solution encoding for bi-phase connections. This analysis will provide a realistic procedure for optimizing the topology of low-voltage distribution networks in real Smart Grids minimizing, statistically, the load unbalance between the three phases. The performance of the proposed approach is assessed over a real use case comprising an entirely remotely-managed distribution substation, with hourly readings of 102 customers (82 residential, 5 industrial and 15 commercial) captured over a historical depth of one year, with more than 21 million watts managed during this period. As concluded from these simulations, the performance of the solver is confirmed to be promising and superior to other genetically inspired heuristics, hence paving the way towards its practical implementation in real energy distribution systems.

2 Problem Formulation

As mentioned in the introduction the overall goal of this research work is to develop an heuristic method for finding the mapping from loads to phases leading to a minimum imbalance between phases taking into account the hourly consumption traces provided by smart meters. A proper balance in the circuit is achieved by minimizing the Euclidean distance between residuals and phases, so that the lower the residuals are, the smaller the total imbalance of the electrical system will be. The application of these methods will focus on real hourly customers' load patterns. The periodicity at which the algorithm is executed is a strategic decision of the utility, which should consider both the operational cost of the phase reassignment and the seasonal treatment of the historical series. The optimization objective is to minimize the sum of Euclidean distances between the three aggregated loads of the users assigned to each phase, pair by pair. Such a difference is given, for 3 phases, N loads and time t (i.e. an index enumerating the 24 values for each of the 365 days in a year), by

$$Q(\mathbf{x},t) = \sum_{(\phi,\theta)\in\mathcal{P}} \sqrt{\left(\sum_{n=1}^{N} \mathbb{I}(x_n,\phi)E_n(t)\right)^2 - \left(\sum_{n'=1}^{N} \mathbb{I}(x_{n'},\theta)E_{n'}(t)\right)^2}, \quad (1)$$

where $\mathcal{P} \triangleq \{(R,S),(S,T),(R,T)\}$, $E_n(t)$ is the energy consumption of load n at time t, $\mathbf{x} \triangleq \{x_n\}_{n=1}^{N}$ is the mapping from loads to phases such that, by assuming a biphasic electric network, $x_n \in \{RR, RS, RT, SS, ST, TT\}$ $\forall n \in \{1,\ldots,N\}$, and $\mathbb{I}(x_n,\phi)$ is an indicator function taking value 1 if $\phi \in x_n$ and 0 otherwise. In words, each phase is represented by the sum of the hourly energy consumption of the loads assigned to that phase. The case when $Q(\mathbf{x},t) = 0$ means that the energy between phases is perfectly balanced at time t, hence lower values of $Q(\mathbf{x},t)$ represent a better load balance.

Before proceeding further, it is worth to delve into the rationale why a balanced distribution network is desirable for the distributor. A perfect balance implies an electric system with minimal energy losses. Indeed, load imbalance may yield up to three times more losses through an imbalanced distribution line when compared to a balanced one. This can be argued, on the one hand, by the application of Joule's Law, which rules the conversion of energy into heat with a consequent increase in the temperature of the conductor,

$$P = RI^2 = \frac{E}{\Delta t} \quad (2)$$

which, by applying Ohm's Law (i.e. the potential V arising between the extremes of a conductor is proportional to the electric current I going through it), yields

$$V = ZI\cos\alpha \rightarrow I = \frac{E}{\Delta t\, V \cos\alpha} \quad (3)$$

where Z denotes the impedance of the conductor (line) and α is the angle between the current phase vector and the voltage V, also referred to as power factor. If we note that the current I is given by the sum of the individual currents over each phase, i.e. $I = I_R + I_S + I_T$, a perfectly balanced line will satisfy $I_R = I_S = I_T = I/3$. By contrast, an imbalanced line undergoing a strong phase imbalance with all the current circulating through a single phase fulfills that e.g. $I = I_R$ and $I_S = I_T = 0$. Bearing this in mind, the power losses for a perfectly balanced system are given by

$$P_{bal} = Z\left(I_R^2 + I_S^2 + I_T^2\right) = Z\left(\frac{I^2}{9} + \frac{I^2}{9} + \frac{I^2}{9}\right) = \frac{ZI^2}{3}, \quad (4)$$

Since the load curve reports the measured energy consumption, the technical power losses P_{bal} for three phases and perfect balance are given by

$$P_{bal} = \frac{RE^2}{3(\Delta t\, V\cos\alpha)^2}. \quad (5)$$

On the other hand, if we deal with an imbalanced system where all energy is conducted over a single phase, technical power losses P_{unbal} will increase up to

$$P_{imbal} = RI^2 = \frac{RE^2}{(\Delta t\, V \cos\alpha)^2} = 3P_{bal}, \qquad (6)$$

i.e. they can potentially as high as three times the losses for the perfectly balanced case. There lies the interest of the energy distributor in balancing the consumption among phases, and the rationale for the application of the heuristic solver explained in the next section.

3 Description of Harmony Search

The Harmony Search is a metaheuristic algorithm based on the emulation of the music improvisation process observed in jazz bands, whose members use to combine different musical notes based on the historical record of notes played by each musician followed by an occasional, random yet slightly pitch tuning. HS maintains a pool of K candidate solutions or harmonies $\{\mathbf{x}^k\}_{k=1}^K = \{\{x_n^k\}_{n=1}^N\}_{k=1}^K$, each comprising N optimization variables or notes (i.e. the number of loads associated to the feeder at hand whose optimal phase assignment is to be discovered by the algorithm). The main steps of the standard HS solver [19] are as follows:

- Step 1 (initialization): this first step is only considered at the first iteration. The pool of K harmonies (also referred to as Harmony Memory, HM) is initially filled, which is done uniformly at random if no a priori knowledge about the solution space is assumed. This represents the starting point for the set of candidate harmonies. For the problem at hand the alphabet of notes has 6 possible values, 3 for single-phase connections and 3 for bi-phase connections, namely $x_n^k \in \{1, 2, 3, 4, 5, 6\}$ corresponding with each possible connection assignment $\{RT, RR, RS, SS, ST, TT\}$. It should be noted that the mapping from the note encoding to the sequence of phases is designed so that when tones need to be adjusted to any other tone in their vicinity, changes in the phase mapping are not drastic, as neighboring notes in the integer alphabet correspond to phases with at least one phase in common with that of the original note. In this case it is assumed that in bi-phase connections the current is equally distributed over each individual phase.
- Step 2 (Improvisation): for each of the iterations a new harmony is generated. To this end two are the operators defined in the naive HS solver:
  1. Harmony Search Considering Rate (HMCR $\in \mathbb{R}[0, 1]$), which sets the probability that the value of a new proposed note is drawn from the set of values that such a note has in the rest of harmonies, i.e. HMCR $= 0.9$ involves that 90% of the new notes are drawn from the whole set of harmonies at each iteration. As will be later explained, based on previous studies [20] a linear variation of the HMCR parameter along iterations has been adopted so as to enhance the convergence of the search process. Any component not selected for memory consideration will be randomly set to a value between the lower and upper bounds of its possible range in the defined alphabet.

2. Pitch Adjustment Rate (PAR $\in \mathbb{R}[0,1]$), which establishes the probability that a given note is set to one of its neighboring values within the note alphabet. i.e., PAR = 0.2 involves that 20% of the new notes are drawn from the neighboring (lower or higher with equal probability) set of notes defined on the integer alphabet on which they are encoded.

- Step 3 (HM update): the new improvised harmony is now evaluated according to its value of the objective function which, for the case study tackled in this paper, is given by Expression (1) aggregated over a certain time horizon,

$$f(\mathbf{x}^k) = \sum_{t=1}^{T} \sum_{(\phi,\theta)\in\mathcal{P}} \sqrt{\left(\sum_{n=1}^{N} \mathbb{I}(x_n^k,\phi)E_n(t)\right)^2 - \left(\sum_{n'=1}^{N} \mathbb{I}(x_{n'}^k,\theta)E_{n'}(t)\right)^2}, \quad (7)$$

where T denotes the time span over which the phase balance provided by $\mathbf{x}^k$ is evaluated. If the objective function value for the new harmony is better (lower) than the objective function value for the worst harmony in the HM, then such a worst harmony is replaced with the newly improvised harmony.
- Step 4 (Stop criteria): if a maximum number of improvisation is reached, then stop, otherwise step 3 and step 4 are repeated.

The HMCR and PAR operators aid the algorithm in the search for better solutions, and even affect the speed of convergence. The values for these two parameters had to be optimized to find the best set of parameters in terms of balancing explorative versus exploitative character of the exploratory finding, but this study is omitted for lack of space. The work in [21] proposes to improve the performance of the HS algorithm hinges on imposing a certain progression along iterations on the values of its operational parameters of HMCR and PAR. For instance, the value of the PAR is dynamically updated according to:

$$\text{PAR}(iter) = \text{PAR}_{\min} + (\text{PAR}_{\max} - \text{PAR}_{\min}) \cdot \xi(iter) \quad (8)$$

where $\text{PAR}_{\min}$ and $\text{PAR}_{\max}$ are minimum and maximum values of the adjusting rate, and $\text{PAR}(iter)$ is the pitch adjusting rate for iteration $iter \in \{1, 2, \ldots, niter\}$. The coefficient $\xi(iter)$ is calculated based on the iteration $iter$ and the maximum number of iterations $numMaxIter$ as

$$\xi(iter) = \frac{iter}{numMaxIter}. \quad (9)$$

This modification of the HS algorithm, by an iterative process, establishes dynamic values for HMCR and PAR with each new iteration, seeks to avoid premature convergences in suboptimal regions. Thus, this approach prioritizes the explorative capability of the search process rather than its exploitative behavior.

4 Experiments and Results

In order to evaluate the performance of the proposed scheme, a set of experiments based on real data has been designed for dealing with the phase swapping

Table 1. Refined parameters of the HS and GA solvers.

No.	HS parameter	Value	GA parameter	Value
1	K	50	Population size	100
2	$[\text{HMCR}_{\text{min}}, \text{HMCR}_{\text{max}}]$	[0.7–0.9]	Crossover rate	0.9
3	$[\text{PAR}_{\text{min}}, \text{PAR}_{\text{max}}]$	[0.01–0.1]	Mutation rate	0.2
4	$numMaxIter$	200	Max. generation	100

problem discussed above taking into consideration the time variance of load patterns and mono- and bi-phase connections. The data is provided by an remotely managed distribution substation. Only one feeder with three phases is considered accounting up to 102 customers, from which 82 are residential, 5 industrial and 15 commercial. One-year-long load patterns with one-hour granularity are considered in the experiments, which yields 365×24 data points per customer. The complexity of the proposed experiments is 6^{102} considering three monophase connections and other three bi-phase connections. The paper compares the proposed HS algorithm described in Sect. 3 with the Genetic Algorithm (GA) proposed in [12]. GA is controlled by mutation and crossover operators which drive the behavior of the iterative search procedure. A tournament selection is selected in an attempt at reinforcing the capacity of the GA to escape from local optima during its search. One of the most evident differences between both schemes is that GA creates new chromosomes by using only one (mutation) or two (crossover) individuals, as opposed to the HS solver which exploits the knowledge embedded in the entire set of harmonies stored in the HM.

The comparative study is discussed in statistical terms motivated by the heuristic nature of the algorithms in the benchmark and the stochasticity imposed by their probabilistic operators. A computationally fair comparison is guaranteed by setting equal the number of metric evaluations for both schemes. The values of the improvisation operators for both optimization approaches are refined based on a previous grid search not shown due to lack of space. The values for which the best balance between explorative versus exploitative behaviour of the proposed schemes was found are listed in Table 1. In order to compute performance statistics 50 Monte Carlo simulations are performed. It is important to observe in Table 1 that GA prioritizes a memory that doubles in size that of the HS approach. Therefore, for a fair comparison the maximum number of generations is set to half the one for its HS counterpart.

Figure 1 shows the best metric obtained for each algorithm in the iterative process in the statistics of the 50 Monte Carlo simulations. It can be noted that HS outperforms GA, i.e. for the 50 independent runs of the algorithms the fitness values attained by HS are smaller than GA, not only in terms of its mean value but also in what regards to their dispersion (a standard deviation of 11000 W for HS and 12508 W for GA). Indeed only in 10 of the 50 experiments HS falls within the value range bounded as mean $\pm$ standard deviation computed for GA. In the rest of experiments results of HS are outside this GA confidence region.

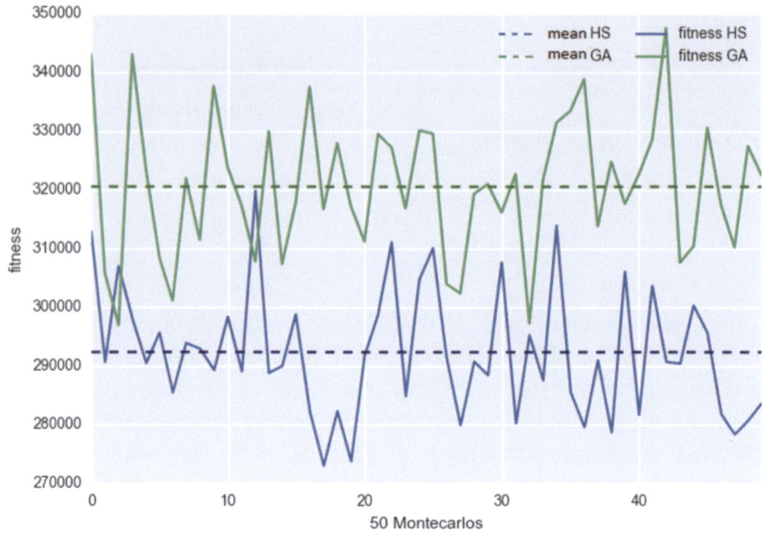

Fig. 1. Fitness value obtained over the 50 Monte Carlo experiments performed.

The discussion follows in Fig. 2, where it is shown that the convergence of the proposed HS scheme is faster than that of GA. The plot depicts the average metric evolution (solid lines) throughout the 50 Monte Carlos during the iterative process of both schemes, whereas the standard deviation of the mean at each iteration is marked showing the independence of the obtained results.

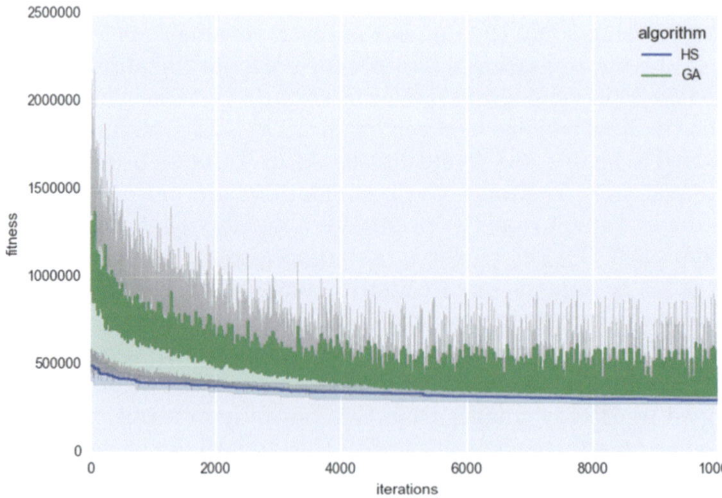

Fig. 2. Fitness convergence along iterations for both HS and GA algorithms.

The ideal case – a perfect balance between phases – is obtained when the metric as per Expression (1) is $Q(\mathbf{x}, t) = 0$ at each time stamp (i.e. each hour during the whole considered year). Integrating the energy in time along the whole considered year $t \in \{1, \ldots, 365 \times 24\}$, a perfect balance means 33.3% for each phase. Figure 3 shows the integrated energy distribution per phase obtained with HS and GA; while GA renders a maximum difference of 6.8% between two phases, in the case of HS this score is limited to 4.7%, which corresponds to an absolute value of 441000 W over the whole year. This noted fact means that there is no clear compensation of seasonality effects along the year and through users, because the hourly study shown in Figs. 1 and 2 is corroborated with the yearly one in Fig. 3. This does not have to be extrapolated to all situations.

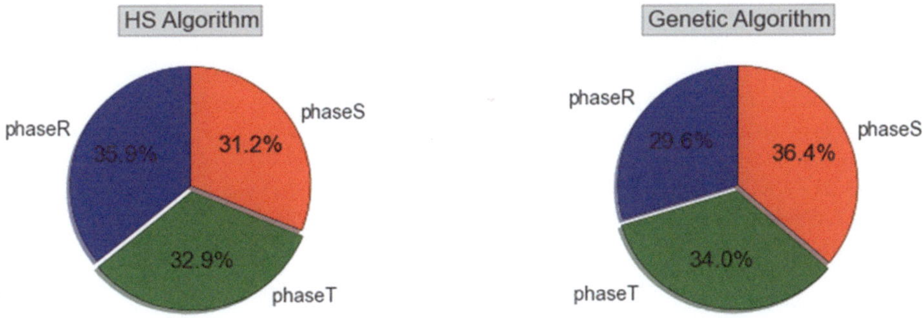

Fig. 3. Percentage distribution of energies per phase for the two algorithms.

5 Conclusions and Future Research Lines

This paper has proposed a meta-heuristic scheme specially tailored for efficiently finding the mapping from loads to phases that leads to a minimal energy imbalance between phases. The study builds upon real consumption traces hourly provided by smart meters. On this purpose a novel fitness function is proposed, and modifications of the naive Harmony Search algorithm scheme are introduced. The proposed methodology is compared with a standard Genetic Algorithm (GA) from the literature, showing that the HS outperforms GA with statistical significance over 50 independent Monte Carlo experiments, not only on its hourly component but on the yearly one. Future research will focus on multi-objective formulations of this problem in order to improve further the practical benefits of the method proposed in this manuscript. The considered metrics will span beyond the reduction of the technical losses tackled in this paper towards the minimization of the operational costs derived from the reassignment of different users. In addition, the obtained conclusion – i.e. hourly results are buttressed by the yearly phase balances – will be generalized using different, more diverse substations demonstrating that hourly knowledge is necessary to avoid the seasonality compensation effect when optimizing over the whole year.

Acknowledgments. This paper includes partial results of the UPGRID project. This project has received funding from the European Union's Horizon 2020 research and innovation programme under grant agreement No 646.531), as well as by the Basque Government through the ELKARTEK programme (BID3A and BID3ABI projects). For further information check the website: http://upgrid.eu/.

References

1. Lin, C.H., Chen, C.S., Chuang, H.J., Ho, C.Y.: Heuristic rule based phase balancing of distribution systems by considering load patterns. IEEE Trans. Power Syst. **20**(2), 709–716 (2005)
2. Wang, K., Skiena, S., Robertazzi, T.G.: Phase balancing algorithms. Electr. Power Syst. Res. **96**, 218–224 (2013)
3. Lee, C.: Feeder reconfiguration and capacitor setting for loss reduction of distribution systems. Electr. Power Syst. Res. **58**(2), 97–102 (2001)
4. Chen, T.H., Cherng, J.T.: Optimal phase arrangement of distribution transformers connected to a primary feeder for system unbalanced improvement and loss reduction using a genetic algorithm. IEEE Trans. Power Syst. **15**(3), 994–1000 (2000)
5. Kim, H., Ko, Y., Jung, K.: Artificial neural network based feeder reconfiguration for loss reduction in distribution systems. IEEE Trans. Power Deliv. **8**(3), 1356–1366 (1993)
6. Civanlar, S., Grainger, J., Yin, H., Lee, S.: Distribution feeder reconfiguration for loss reduction. IEEE Trans. Power Deliv. **3**(3), 1217–1223 (1988)
7. Dolatdar, E., Soleymani, S., Mozafari, B.: A new distribution network reconfiguration approach using a tree model. World Acad. Sci. Eng. Technol. **58**(34), 1186 (2009)
8. Jeon, Y.J., Kim, J.C., Kim, J.O., Lee, K.Y.: An efficient simulated annealing algorithm for network reconfiguration in large-scale distribution systems. IEEE Trans. Power Deliv. **17**(4), 1070–1078 (2002)
9. Chiang, H., Jean-Jameau, R.: Optimal network reconfiguration in distribution systems. IEEE Trans. Power Deliv. **5**(4), 1902–1909 (1990)
10. Zhu, J., Bilbro, G., Chow, M.Y.: Phase balancing using simulated annealing. IEEE Trans. Power Syst. **14**(4), 1508–1513 (1999)
11. Bouchard, D., Chikhani, V., John, V., Salama, M.: Applications of hopfield neural networks to distribution feeder reconfiguration. In: International Forum on Applications of Neural Networks to Power Systems, pp. 311–316 (1993)
12. Vulasala, G., Sirigiri, S., Thriruveedula, R.: Feeder reconfiguration for loss reduction in unbalanced distribution system using genetic algorithm. Int. J. Electr. Electron. Eng. **3**(12), 754–762 (2009)
13. Lafortune, M., Bouchard, D., Morelli, J.: Phase swapping for distribution system using Tabu search. In: WSEAS International Conference on Energy Planning, Energy Saving, Environmental Education, pp. 67–71 (2007)
14. Zhu, J., Chow, M.Y., Zhang, F.: Phase balancing using mixed-integer programming. IEEE Trans. Power Syst. **13**(4), 1487–1492 (1998)
15. Fei, C.G., Wang, R.: Using phase swapping to solve load phase balancing by ADSCHNN in LV distribution network. Int. J. Control Autom. **7**(7), 1–14 (2014)
16. Granada, M., Gallego, R., Lopez Lezama, J.M.: Optimal phase balancing planning for loss reduction in distribution systems using a specialized genetic algorithm. Ingenieria y Ciencia **8**(15), 121–140 (2012)

17. Geem, Z.W., Kim, J.-H., Loganathan, G.V.: A new heuristic optimization algorithm: harmony search. Simulation **76**(2), 60–68 (2001)
18. Manjarres, D., Landa-Torres, I., Gil-Lopez, S., Del Ser, J., Bilbao, M.N., Salcedo-Sanz, S., Geem, Z.W.: A survey on applications of the harmony search algorithm. Eng. Appl. Artif. Intell. **26**(8), 1818–1831 (2013)
19. Geem, Z.W.: Music-Inspired Harmony Search Algorithm: Theory and Applications. Springer, Heidelberg (2009)
20. Del Ser, J., Matinmikko, M., Gil-Lopez, S., Mustonen, M.: Centralized and distributed spectrum channel assignment in cognitive wireless networks: a harmony search approach. Appl. Soft Comput. **12**(2), 921–930 (2012)
21. Gil-Lopez, S., Del Ser, J., Landa, I., Garcia-Padrones, L., Salcedo-Sanz, S., Portilla-Figueras, J.A.: On the application of a novel grouping harmony search algorithm to the switch location problem. In: International Conference on Mobile Lightweight Wireless Systems, pp. 62–672 (2010)

Novel Light Coupling Systems Devised Using a Harmony Search Algorithm Approach

Imanol Andonegui[1(✉)], Itziar Landa-Torres[2], Diana Manjarres[2], and Angel J. Garcia-Adeva[1]

[1] Department of Applied Physics I., University of the Basque Country UPV/EHU, 01006 Vitoria, Spain
imanol.andonegui@ehu.eus
[2] TECNALIA, 48160 Derio, Spain
itziar.landa@tecnalia.com

Abstract. We report a critical assessment of the use of an Inverse Design (ID) approach steamed by an improved Harmony Search (IHS) algorithm for enhancing light coupling to densely integrated photonic integratic circuits (PICs) using novel grating structures. Grating couplers, performing as a very attractive vertical coupling scheme for standard silicon nano waveguides are nowadays a custom component in almost every PIC. Nevertheless, their efficiency can be highly enhanced by using our ID methodology that can deal simultaneously with many physical and geometrical parameters. Moreover, this method paves the way for designing more sophisticated non-uniform gratings, which not only match the coupling efficiency of conventional periodic corrugated waveguides, but also allow to devise more complex components such as wavelength or polarization splitters, just to cite some.

Keywords: Grating structure · Photonic integrated circuit · Improved harmony search

1 Introduction

Electronics made possible to achieve a massive shrinking of the core electrical elements into chip scale building blocks. Likewise, silicon on insulator (SOI) technology has the capacity to overcome the integration scaling of electronics by enabling the mass production of large scale and dense integration of PICs. These high index contrast materials are an outstanding platform for designing compact devices consisting on low loss waveguides, splitters/combiners, wavelength multiplexers and i.e. a wide variety of integrated components. In the same line, the ongoing trend is to integrate these components on a single platform. However, the high integration density of such devices and i.e. their small feature size complicates the light coupling interfacing between standard single mode optical fibres (SMF) and SOI circuits, causing high insertion losses and high packaging

J. Del Ser (ed.), *Harmony Search Algorithm*, Advances in Intelligent Systems and Computing 514, DOI 10.1007/978-981-10-3728-3_29

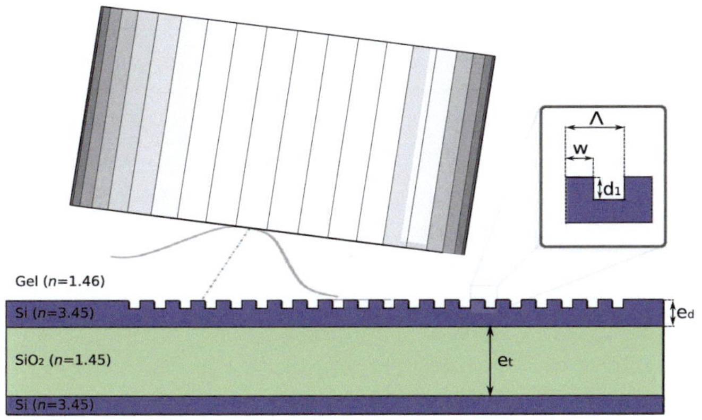

Fig. 1. Basic configuration for a custom fiber-to-chip periodic grating coupler. A standard SMF interfaces near vertically with a diffractive grating structure defined on the surface of the silicon waveguide. The grating coupler structure is determined by the duty cicle (W), the filling factor (*ff*), the period (Λ) and the depth of the grooves (d_1). The rib waveguide consists of a silicon layer of thickness, etching depth (e_d) on top of a buried oxide layer (BOX) of thickness e_t. To clearly illustrate the design, the diagram is not to scale.

costs. This decoupling is mainly due to the huge mode mismatch between the cross-sections of fibres and SOI guides [1]. To accomplish the fibre-chip coupling issue, two main strategies have been proposed so far: butt-coupling schemes and out-of-plane coupling solutions. Butt-coupling methods provide wide bandwidth and low insertion loss operation [2], but these kind of solutions require a lensed, high numerical aperture fibre or an inverted taper with a very sensitive fiber-guide alignment [3,4] to match the optical mode size between two light guiding materials. In contrast, out-of-plane coupling presents some major advantages over the former method: using this approach eludes the need of cleaved facet fibres and there is no limitation for extracting/coupling light everywhere on the chip, which is a critical advantage for large scale wafers [2]. Regular solutions provide guided-mode resonances using periodic gratings which have been widely demonstrated [5,6]. Nevertheless, periodic grating designs present a small coupling strength, are rather long, working on not wide enough bandwidth.

The principle of periodic diffraction gratings for coupling light into a SMF is shown in Fig. 1. Throughout this work we present a broad comparative study of non uniform grating structures for boosting the light coupling under different criteria beyond the limitations of standard periodic designs.

2 A Deterministic Search: Finding the Optimum

Figure 1 shows the schematic structure of a naive grating coupler defined on a regular SOI wafer with a 250 nm thick crystalline silicon layer which refractive index n is, $n = 3.467$, over a 2 µm thick buried oxide $n = 1.46$ layer. A single mode fibre is positioned with a tilt angle to the normal direction above the grating to avoid backreflection. In order to eliminate the fiber's facet reflection, an index-matching gel $n = 1.46$ is utilized to fill the space between the fibre and the grating. Aside from the consideration of these parameter, it is of practical relevance to choose the etching depth and the period of the grating. Aside from this toy-model case, when the amount of degrees of freedom involving the design process is large, it is impractical to determine the optimum, neither it is practical to do a thorough exploratory search among the whole set of feasible parameters. Thus, in order to show a simplified stage, wherein a brute-force hard search is still feasible, we delimited the search to the exploration of two coupled trivial parameters, i.e. to the duty cycle of the grating coupler W that defines the width of the grating tooth and the fill factor *ff* defined as the ratio of the grating period and the duty cycle ($ff = \Lambda/W$). Nonetheless, simplifying the designing parameters and assuming only a two parameter space search means that one should intuitively fix any other parameters. In this case we fixed the value of some parameters according to pure intuition and experimental lore, such as the position of the SMF fibre to 1.8 µm from the silicon top layer and to 5 µm in the horizontal axis from the grating edge, with a the tilt angle of 8°. The grating is devised by performing a 70 nm deep etching. Thus, we constrained the search to only these two parameters and we mapped the maximum coupling efficiency between a transverse electricaly (TE) polarized Gaussian beam centred at $\lambda = 1.55$ µm, where λ is the light's wavelength, spanning 0.1 µm, *intoa* 10.4 µm cross-section SMF. This configuration was modelled by means of two-dimensional Finite-Differential Time-Domain (FDTD) simulations with a uniform grid featuring elements with sizes bellow $\lambda/10$ and we used Perfectly Matched Layer (PML) boundary conditions surrounding the whole grating and fibre domains. The coupling efficiency (η) can be calculated by means of the following integral [7]:

$$\eta = |\int\int E_x E_{(y=y_0,z)} A e^{\frac{(x-x_0)^2+(z-z_0)^2}{\omega_0^2}} e^{jn\frac{2\pi}{\lambda}z\sin\Theta} dxdz|, \quad (1)$$

where Θ is the fibre tilt angle, (x_0, y_0, z_0) is the position of the fibre with respect to the grating coupler, n is the refractive index on top of the grating and the constant A represents the normalization of the Gaussian beam.

Figure 2a depicts the coupling proficiency of 10.000 combinations of *ff* and W computed in a regular *i*3 computer where it took around 72 h to complete it. As expected in this trivial case, an almost linear relation between the filling factor and the duty cycle can be appreciated along with some global maxima remarked in black. In addition, we assumed that the oxide thickness, fibres tilt angle, etching deep and some other aforementioned parameters were almost optimal for a minimal downward leakage, but that is not necessarily true since all these parameters are coupled to each other. Therefore, it is infeasible to do these kind of intensive calculations when the parameter space is increased using a deterministic search.

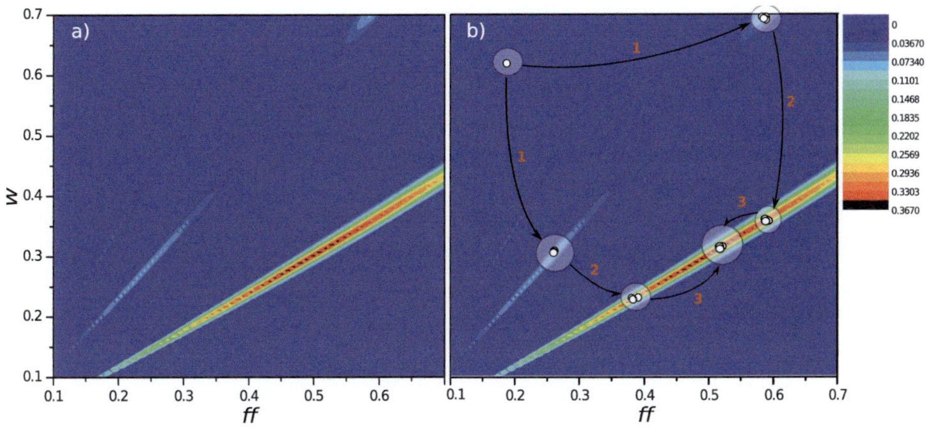

Fig. 2. (a) The coloured chart depicts the coupling efficiency at $\lambda = 1550$ nm for the entire set of W and ff combinations. (b) The solution dataset calculated by the IHS algorithm explained in Sect. 3 in (a) is over-imposed by circled set of points surrounded by a white cloud, showing the solutions given by this ID method at each iteration. The ID process clearly converges to the maximum coupling parameter pairs with low number of iterations in this toy model case, as detailed in Sect. 4.

3 The IHS Algorithm for Non-deterministic Search of Optima

Except of the trivial case comprised by only 2 parameter data sets discussed in Sect. 2, when more parameters are taken into account, a deterministic search for optimum coupling becomes futile. To counterbalance this issue, we have recently proposed a more efficient inverse design approach for designing photonic components that facilitates the prospect of optimum topologies in a rather variety of fields. This time, the ID method uses an IHS [8] algorithm as a core engine.

The HS algorithm [9] is a meta-heuristic optimization algorithm inspired by the observation of musical improvisation process, in which the aim is to search for a perfect state of harmony. HS was proposed by Geem et al. in 2001 and thenceforth it has been widely applied to a variety of combinatorial optimization problems [10–12]. During the process of musical improvisation, suggesting a new melody is subject to three different circumstances, a musician can either play a tune exactly in the way he has memorized it or he can play something similar to a stored melody (eg. slightly adjusting the pitch), or compose new notes randomly. Based on these three options, Geem et al. made a new quantitative optimization process. The three components analogous to the ones of musical improvisation became: memory usage of harmonies (HMCR), pitch adjustment (PAR) determined by a pitch bandwidth bw_{range} and randomization (RSR) [13].

In the original HS algorithm proposed by Geem in [9], the PAR and bw parameters are fixed and do not evolve during the optimization process. Therefore, the election of appropriate values for these operators becomes critical and the algorithm demands a high number of iterations for locating the global optimum state. In reference [8], however, the *PAR* and *bw* parameters are cyclically tuned in order to achieve a smoother search, in a manner analogous to what the Cauchy-Lorentz probability distribution does in a Fast Simulated Annealing (FSA) algorithm [14]. In other words, the *bw* parameter must be set to high values in initial iterations in order to increase the diversity of solutions, but in final generations it must be set to low values to enhance the intensification and to sacrifice the diversification, that is, to search locally instead of globally. Throughout this work, we set a PAR_{min} and PAR_{max} parameters to 0.4 and 0.9. Then the bandwidth of each iteration *bw(n)* is given non-linearly as:

$$bw(n) = bw_{max} \exp(c.n), \tag{2}$$

and the coefficient c is recalculated in each iteration as

$$c(n) = \frac{\ln \frac{bw_{min}}{bw_{max}}}{N}. \tag{3}$$

In order to cover a large number of solutions and to expand the search space in which diversification has a significant relevance, we set the minimum bandwidth, BW_{min}, to 0.0001; and the maximum bandwidth, BW_{max}, to 1. The *RSR* $\in$ [0,1], sets the probability of picking the value for the new note randomly from the domain of the input variables. It performs a role similar to the uphill jump probability function in the Simulated Annealing algorithm, but in this case the alteration in the harmony set stored in the memory of harmonies (*HM*) is performed regardless of its fitness with probability *RSR*. In this work we set the *RSR* parameter to 0.1. The *HM* is comprised of 20 harmonies, with 22 notes dilling each harmony. In the general case, where the etching deep of the groves is fixed, the first 20 notes belong to the period of each trench, bounded between a value of 0.15 µm and 0.5 µm. The two additional parameters are the SMF separation distance and the fibre tilt angle, bounded betwwen 4 and 7 µm, and 8 and 15 degrees, respectively.

4 Speeding Up: An Heuristic Engine for Non-assisted Exploration

It seems of practical relevance to check whether an ID process like the one described in [15] and using the IHS method of 3, gets to improved grating couplers or, conversely, if conventional periodic grating couplers fare better. In Fig. 2b the ID method results are over-imposed to the exhaustive lookup–map computed in Sect. 2. Noticeably, when using the ID methodology to this problem, the solution is quickly achieved taking into account only those combinations of parameters that yield to promising results. The ID method, boosted by IHS jumps quickly from average fitness solutions to local optima points and subsequently splits the search through local optima points to global optimum in just 3 iterations. The whole process takes only few minutes to converge to the solution remarked by white markers. Hence, even in this simplified model, introducing the ID method drastically reduces the computation time and prevents the calculation of non-profitable solutions by determining the combination of parameters that enhances light overlapping between guide and fibre.

Some other optimization techniques could deal with this issues as well but either are based on gradient methods and tend to stack at local maxima or they provide lattices that are often not practical with current fabrication tolerances. Examples include exotic shaped designs that can barely be turned into operative designs using most advanced experimental prototyping techniques.

Due to this fact, it is convenient to constraint the optimization of theoretical designs to some parameters that render to a topology comprised by naive elements, that are closer to current fabrication facilities. In spite of this, there is no certainty of achieving a resulting device that operates as good as theoretical designs predictions. This is due to the fact that devising process always introduce some random errors that are independent of the design. In nanophotonics these errors can not be neglected as they may lead to produce totally different structures. In this regards, some authors have proposed to give design these devices by imposing in their designs the worst case analysis [16, 17]. The worst case analysis is a widely used approach in nowadays industry, but is still highly ineffective. On the one hand, it is not always easy to determine which of the configuration parameters yields to the worst case and in the other hand, a high restrictive proceeding forces to disregard many potentially good solutions.

5 Periodicity and Aperiodicity, a Comparison

The key features of an effective light coupling system are compactness, low insertion loss, large alignment tolerance, and broadband operation [2]. In a first attempt we took a 20 periods length grating profile with the parameters specified in Sect. 2, except that this time we led the ID algorithm decide the horizontal position of the SMF fibre, its tilt angle as well as the grating period, Λ. With this regards, we run the simulation and the grating was optimized for enhancing the coupling of light from the integrated waveguide to the SMF fibre. The final simulation set-up was found to be conformed by $\Lambda = 640\,\text{nm}$, with a fibre displaced $5.23\,\mu\text{m}$ from the grating edge and tilted 15°. By means of these values a maximum theoretical coupling efficiency of 66% is achieved with an estimated 1 dB bandwidth of 40 nm. According to the principle of interference, the Bragg conditions is fulfilled when

$$k_{in} \sin \Theta + m\frac{2\pi}{\Lambda} = \beta, \tag{4}$$

where $k_{in} = 2\pi/\lambda$ and $\beta = (2\pi/\lambda)\, n_{eff}$, being n_{eff} the effective refractive index of the structure. All in all, according to Eq. 4 the grating period should be, for this particular case (for diffraction order m = +1, an angle of 15° and n_{eff} around 2.8):

$$\Lambda = \frac{\lambda}{n_{eff} - \sin \Theta}. \tag{5}$$

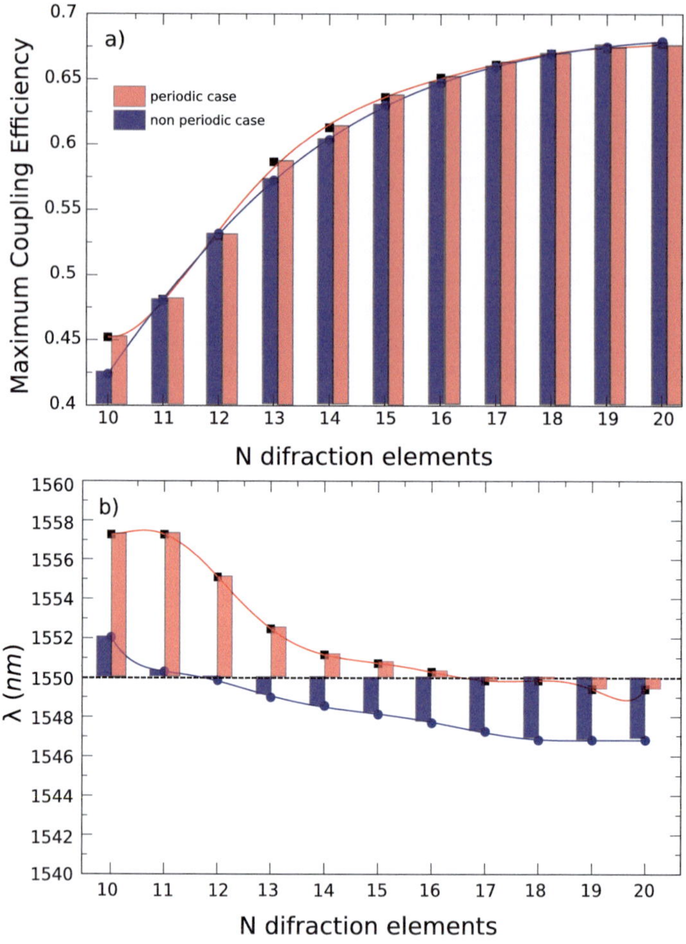

Fig. 3. (a) Maximum coupling efficiency with regards to the number of grooves in the periodic and non-periodic gratings, sketched in red and blue respectively and (b) deviation of the peak coupling efficiency in each case.

Equation 5 states that the optimum Λ for a λ of 1.55 μm should be 610 nm. This result is very close to the solution given by the ID method (640 nm). In fact, it should be pointed out that the Bragg condition is only exact for infinite structures, thus for finite structures, as log as in finite gratings there is not exactly one discrete wave-vector for which diffraction occurs, but for a range of wave-vectors. However, in periodic gratings the tuning of the wave vectors is limited by the period of the grating. The decay of the coupling performance for this structure is shown in Fig. 3a. The performance of the periodic grating converges when 20 grooves are etched in the silicon layer and the removal of grooves presents a polynomial decay of the coupling efficiency (see Fig. 3a). However, as one disposes of just 4 periods, the peak coupling efficiency starts to deviate from

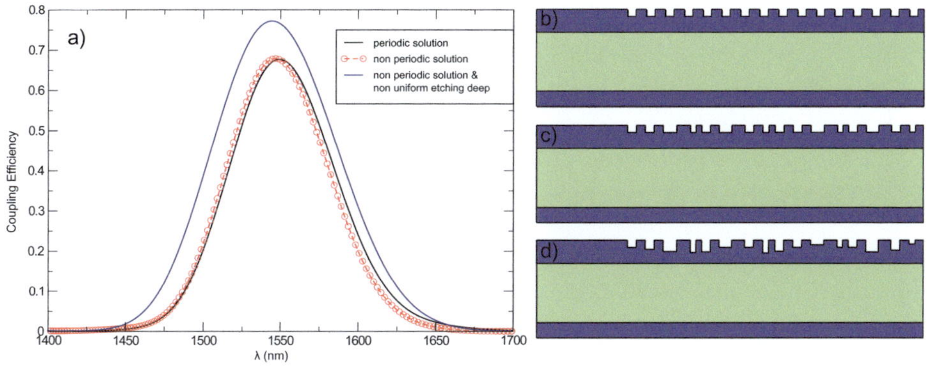

Fig. 4. (a) Simulations results for coupling efficiency spectrum for each case of study. (b)-(d) The schematic representation of the optimized periodic grating, the aperiodic grating and the non-uniform grating made by also tuning the etch depth, respectively. The dimensions of the grooves have been exaggerated for clarity purposes.

the target λ, as it is shown in Fig. 3b. This deviation amounts to 8 nm when only 10 periods are considered in the periodic grating.

Although the etching nonuniformity is usually detrimental for most applications, it can offer more freedom to realize a nonuniform grating coupler. However, designing such non periodic clusters entails dealing with several correlated parameters at the same time and cannot be designed using current analytic methods, or by tuning a small number of parameters by hand [18]. Nonetheless the ID method provides a prospect way for dealing with N-parameter look-up issues. In Fig. 4c a non-uniform grating coupler scheme is represented. In this particular case, every groove is detuned at the same time, constrained to fabrication limitations. The efficiency of the final corrugated waveguide yields to slightly better coupling results when compared to the former periodic case, depicted in Fig. 4b.

The solution obtained with the IHS algoritm provides a peak coupling efficiency of 69 % with a 1 dB bandwidth close to 38 nm. Noticeably the grating structure depicted in Fig. 4c shows an smooth profile that ensures a good stability when the grating is shorten (see Fig. 3a). In this case, the efficiency of the coupling almost corresponds to the periodic case. On the other hand, this grating profile enhances the coupling of light close to the 1.55 µm with a maximum deviation of 3 nm.

To further increase the efficiency of periodic gratings we should include the etching deep of the periodic grooves into the design process. However, according to Eq. 5, these kind of gratings should be those with shallow deep etched profiles. With this regards, we headed on a different approach, i.e. we constrained the etching deep and the duty cycle of a non-uniform grating and we utilized the ID method for achieving a grating coupler that further enhances the light overlapping preventing full etching of the silicon layer or narrow grooves. The resulting grating profile is schematically shown in Fig. 4d. The simulated coupling spectrum for the TE polarization is shown in Fig. 4a for comparison. One can see that a maximum coupling efficiency of 78% is achieved at 1.55 μm and the 1 dB bandwidth is about 40 nm. The coupling efficiency is much larger than that for a standard uniform grating coupler and the wideband makes it very promising for C-band applications.

6 Concluding Remarks

In this manuscript we demonstrate the great usefulness of using an inverse design approach to the process of exploring new grating coupler designs. By means of this method the designer can handle more parameters even when there is no theoretical background leading to any hints in the process of designing better couplers. We found that this methods are easily applicable to grating coupler design problems since they do not require great detailed knowledge of the structure of the problem and they can rely on the solver of the system as a blackbox. By this insight we found that the coupling efficiency of non-uniform grating couplers can outmatch custom periodic designs. Moreover, the structures proposed in this work are fully compatible with conventional manufacturing processes as they prevent the formation of exotic geometries or shallow edges. The length of the non-uniform gratings is slightly shorter than that of canonical gratings. Besides, if these gratings are shortened, then they still hold for a reliable coupling ratio with a minimal deviation in the C-band.

References

1. Galan, J.V.: Addressing fiber-to-chip coupling issues in silicon photonics. Ph.D. thesis, Universidad Politecnica de Valencia (2010)
2. Laere, F.V., Bogaerts, W., Taillaert, D., Dumon, P., Thourhout, D.V., Baets, R.: Grating couplers for coupling between optical fibers and nanophotonic waveguides. J. Lightwave Tech. **25**, 151–156 (2007)
3. Shoji, T., Tsuchizawa, T., Wanatabe, T., Tamada, K., Morita, H.: Low loss mode size converter from 0.3 μm square Si wire waveguides to singlemode fibers. Electron. Lett. **38**, 1669–1670 (2002)
4. Taillaert, D., Laere, F.V., Ayre, M., Bogaerts, W., Van Thourhout, D., Bienstman, P., Baets, R.: Grating couplers for coupling between optical fibers and nanophotonic waveguides. Jpn. J. Appl. Phys. **45**, 6071–6077 (2006)
5. Wang, Y., Wang, X., Flueckiger, J., Yun, H., Shi, W., Bojko, R., Jaeger, N.A.F., Chrostowski, L.: Focusing sub-wavelength grating couplers with low back reflections for rapid prototyping of silicon photonic circuits. Opt. Express **22**, 20652–20662 (2014)
6. Topley, R., O'Faolain, L., Thomson, D.J., Gardes, F.Y., Mashanovich, G.Z., Reed, G.T.: Planar surface implanted diffractive grating couplers in SOI. Opt. Express **22**, 1077–1084 (2014)
7. Taillaert, D.: Grating couplers as interface between optical fibres and nanophotonic waveguides. Ph.D. thesis, Universiteit Gent (2005)
8. Mahdavi, M., Fesanghary, M., Damangir, E.: An improved harmony search algorithm for solving optimization problems. Appl. Math. Comput. **188**, 1567–1579 (2007)

9. Geem, Z.W., Kim, J.H., Loganathan, G.V.: A new heuristic optimization algorithm: harmony search. Simulation **76**, 60–68 (2001)
10. Forsati, R., Haghighat, A.T., Mahdavi, M.: Harmony search based algorithms for bandwidth-delay-constrained least-cost multicast routing. Comput. Commun. **31**, 2505–2519 (2008)
11. Geem, Z.W.: Harmony search in water pump switching problem. Adv. Nat. Comput. **3612**, 751–760 (2005)
12. Del Ser, J., Bilbao, M.N., Gil-López, S., Matinmikko, M., Salcedo-Sanz, S.: Iterative power and subcarrier allocation in rate-constrained orthogonal multicarrier downlink systems based on hybrid harmony search heuristics. Eng. Appl. Arti. Intell. **24**(5), 748–756 (2011)
13. Landa-Torres, I., Gil-Lopez, S., Salcedo-Sanz, S., Del Ser, J., Portilla-Figueras, J.A.: A novel grouping harmony search algorithm for the multiple-type access node location problem. Expert Syst. Appl. **39**(5), 5262–5270 (2012)
14. Szu, H., Hartley, R.: Fast simulated annealing. Phys. Lett. A **122**, 157–162 (1987)
15. Andonegui, I., Calvo, I., Garcia-Adeva, A.J.: Inverse design and topology optimization of novel photonic crystal broadband passive devices for photonic integrated circuits. Appl. Phys. A **115**(2), 433–438 (2014)
16. Sigmund, O.: Manufacturing tolerant topology optimization. Acta. Mech. Sin. **25**(2), 227–239 (2009)
17. Momeni, B., Adibi, A.: Optimization of photonic crystal demultiplexers based on the superprism effect. Appl. Phys. B **77**(6–7), 555–560 (2003)
18. Piggott, A.Y., Lu, J., Babinec, T.M., Lagoudakis, K., Petykiewicz, J., Vuckovic, J.: Inverse design and implementation of a wavelength demultiplexing grating coupler. In: CLEO: Science and Innovations, pp. SM3I-2 (2015)

An Improved Harmony Search Algorithm for Protein Structure Prediction Using 3D Off-Lattice Model

Nanda Dulal Jana[1(✉)], Jaya Sil[2], and Swagatam Das[3]

[1] Department of IT, National Institute of Technology, Durgapur 713209, India
nanda.jana@gmail.com
[2] Department of CST, Indian Institute of Engineering Science and Technology, Shibpur 711103, India
js@cs.iiests.ac.in
[3] ECS Unit, Indian Statistical Institute, Kolkata 700108, India
swagatam.das@isical.ac.in

Abstract. Protein structure prediction (PSP) is an important research area in bio-informatics for its immense scope of application in drug design, disease prediction, name a few. Structure prediction of protein based on sequence of amino acids is a *NP*-hard and multi-modal optimization problem. This paper presents an improved harmony search (ImHS) algorithm to solve the PSP problem based on 3D off lattice model. In the proposed method, the basic harmony search (HS) algorithm combined with dimensional mean based perturbation strategy to avoid premature convergence and enhance the capability of jumping out from the local optima. The experiments are carried out on a set of real protein sequences with different length collected from the Protein Data Bank (PDB) to validate the efficiency of the proposed method. Numerical results show that the ImHS algorithm significantly outperforms compared to other algorithms on protein energy minimization.

Keywords: Protein structure prediction · Harmony search · Off-lattice model · Premature convergence · Difference mean based perturbation

1 Introduction

Proteins are the main building blocks for all living beings and composed of 20 amino acids by a ploy-peptide bond [1]. Protein folded into a structure that provide the essential functional features of the protein which are plying important role in biological science, medicine, disease prediction and many more [2]. Therefore, structure prediction of a protein is an important research area in bio-informatics. Most of the protein structures deposited in PDB were determined by the experimental methods like X-ray crystallography and Nuclear Magnetic Resonance (NMR) [3]. The experimental techniques are not always feasible due

J. Del Ser (ed.), *Harmony Search Algorithm*, Advances in Intelligent Systems and Computing 514, DOI 10.1007/978-981-10-3728-3_30

to very expensive, time consume, strict laboratory requirements and heavy operational burdens [4]. Thus, many researchers focused their interest on PSP from primary sequence of amino acids using computational approach [5].

Anfinesen theory of thermodynamic [6] states that conformation of a protein corresponds to the global minimum free energy surface and therefore, structure prediction of a protein can be transformed into a global optimization problem. The PSP using computational methods reveals two important key parts. The first part is to state the simplified mathematical model that corresponds to a protein energy function. The second part is to develop a computationally efficient optimization algorithm which can find the global minimum of the potential energy function. Currently, a coarse-grained model like AB off-lattice model [7] has been widely used for PSP. The model consists two types of amino acid residues such as hydrophobic (A) and hydrophilic (B) and provide main driving force of the protein energy function. In the past, large number of heuristic or meta-heuristic algorithms have been developed based on 3D AB off-lattice model for PSP [8–14]. The harmony search (HS) algorithm is an optimization algorithm that based on the music improvisation processes where musicians improvise the pitches of the instruments to search for a perfect state of harmony [15]. A new harmony improvisation and harmony memory updation are the main components of the HS algorithm. Various successful studies have been confirmed that the HS algorithm paid more and more attention to solve a wide range of engineering optimization problems [16]. However, less diversification can be found in the population when solving multi-modal optimization problems.

In this paper, we proposed an improved HS (ImHS) algorithm for solving PSP problem based on 3D AB off-lattice model. In ImHS algorithm, a new harmony vector is generated based on the best harmony and modified with normal distribution without using pitch adjustment rate (PAR) and the harmony memory is updated using greedy selection mechanism. In addition, a simple dimensional mean based perturbation scheme is integrated to all members of the harmony memory for better exploration to ensure faster convergence and efficiency. Experiments are carried on a set of real protein sequences with various lengths collected from PDB. The results show that the proposed algorithm significantly outperforms to the CPSO [10], BEABC [13] and simple HS [17] for all the real protein instances.

The paper is organized as follows: In Sects. 2 and 3, basic fundamentals of 3D AB off-lattice model and HS algorithm are describe, respectively. The details of proposed algorithm is presented in Sect. 4, followed by experimental results and analysis in Sect. 5. Finally, the conclusion is drawn and future works are highlighted in the Sect. 6.

2 AB Off-Lattice Model

The 3D AB off-lattice model proposed by Stilinger in [7] that represents the intra-molecular interactions between amino acids of a sequence. All amino acids are classified into hydrophobic and hydrophilic residues, labeled as ‘A’ and ‘B’ respectively and linked up by rigid unit-length bonds. The three dimensional structure of an n length protein is specified by the $(n-1)$ bond length, $(n-2)$ bond angles θ_1, θ_2,..., θ_{n-2} and $(n-3)$ torsion angles β_1, β_2, ..., β_{n-3}. Therefore, the n length protein conformation is determined by $(2n-5)$ angle parameters $(\theta_1, \theta_2, ..., \theta_{n-2}, \beta_1, \beta_2, ..., \beta_{n-3})$ where $\theta_i, \beta_i \in [-180°, 0)$ and $\theta_i, \beta_i \in (0, 180°]$.

The protein energy function for an n length sequence have two parts: one is the bending energy (V_1) of protein backbone and the other is non-bonded potential energy (V_2). Amino acids along the backbone encoded as bipolar variables ξ_i. If ξ_i =1, the ith amino acid is A and $\xi_i = -1$ for B. The potential energy function E for an n length protein sequence is expressed as follows:

$$E = \sum_{i=2}^{n-1} V_1(\theta_i) + \sum_{i=1}^{n-2} \sum_{j=i+2}^{n} V_2(r_{ij}, \xi_i, \xi_j). \tag{1}$$

Where $V_1(\cdot)$ is the bending potential energy as defined in Eq. (2).

$$V_1(\theta_i) = \frac{1}{4}(1 - cos\theta_i). \tag{2}$$

The non-bonded interactions $V_2(\cdot)$ have a species - dependent Lennard - Jones 12, 6 form, described in Eqs. (3) and (4), respectively.

$$V_2(r_{ij}, \xi_i, \xi_j) = 4[r_{ij}^{-12} - C(\xi_i, \xi_j) r_{ij}^{-6}] \tag{3}$$

$$C(\xi_i, \xi_j) = \frac{1}{8}(1 + \xi_i + \xi_j + 5\xi_i\xi_j). \tag{4}$$

Where r_{ij} denotes the distance between i^{th} and j^{th} residue of the sequence. For an AA pair, $C(\xi_i, \xi_j) = 1$ regarded as strongly attracting, for an AB or BA pair, $C(\xi_i, \xi_j) = -0.5$, regarded as weakly repelling and for a BB pair, $C(\xi_i, \xi_j) = 0.5$, regarded as weakly attracting. The basic objective of the PSP for n length protein sequence is to finding the optimal $(n-2)$ bond angles and $(n-3)$ torsional angles in order to minimize the energy function E defined in Eq.(1), representing lowest free energy of the structure of a protein.

3 Harmony Search (HS) Algorithm

Harmony search (HS) algorithm is a new optimization algorithm that based on the improvisation processes of musicians proposed by Greem et al. [15] in 2001. During the improvisation processes, musicians search for a perfect state of a harmony by slightly adjusting the pitch of their instruments. The HS is a population based meta-heuristic algorithm where population represent as harmony memory (HM) and there are few control parameters such as harmony memory size (HMS), harmony memory considering rate (MHCR) used for determining whether a decision variable is to be selected from HM or not, pitch adjusting rate (PAR) used for deciding whether the selected decision variable is to be modified or not. The basic HS algorithm consists of five phases which are initialization of HM, improvisation of a new harmony, updating of HM and stopping criterion.

Initialization of Harmony Memory: The HM is a solution matrix of size HMS where each harmony represents a D-dimensional solution vector in the search space. In general, the ith harmony is represented by $X_i = (x_{i1}, x_{x_i2}, ..., x_{iD})$ and generated randomly as follows:

$$x_{ij} = x^j_{min} + (x^j_{max} - x^j_{min}) \times \ rand(0,1). \tag{5}$$

where $i = 1, 2, ..., HMS$, $j = 1, 2, ..., D$; x^j_{min} and x^j_{max} are the lower and upper bound of jth component, respectively.

New harmony improvisation: A new harmony vector, $Y' = (y'_1, y'_2, ..., y'_D)$ is generated from the HM on the basis of memory considerations, pitch adjustments and randomization. In the memory consideration, j^{th} component of Y' is selected randomly from $\{x_{1j}, x_{2j}, ..., x_{HMSj}\}$ with a probability of $HMCR \in [0,1]$ which ensure good harmony is considered in the new harmony vector. The memory consideration can be expressed as

$$y'_j = \begin{cases} y'_j \in \{x_{1j}, x_{2j}, ..., x_{HMSj}\} & if \quad rand \leq HMCR \\ x^j_{min} + (x^j_{max} - x^j_{min}) \times \ rand(0,1) & otherwise \end{cases} . \tag{6}$$

Furthermore, the component in the new harmony vector obtained from the HM is modified using PAR for finding good solution in the search space. These components are examined and tuned with the probability PAR as follows:

$$y'_j = \begin{cases} y'_j \pm bw \times rand(0,1) & if \quad rand \leq PAR \\ y'_j \quad otherwise, \end{cases} , \tag{7}$$

where bw is a bandwidth distance used for variation in the decision variables and $rand(0,1)$ is a uniformly distributed random number in $[0,1]$.

Updation of the harmony memory: In order to update the HM, the new harmony vector (Y') is evaluated using the given objective function. If the objective function value $f(Y')$ of Y' is better than $f(X_{worst})$ of the worst harmony (X_{worst}) stored in the HM, then X_{worst} is replaced by Y'. Otherwise, the generated new vector is ignored.

Stopping criterion: The HS algorithm terminates when the stopping condition, maximum number of improvisation is reached. In this work, the improvisation or generation term is refer to a single iteration of the search algorithm.

4 Proposed Methodology

4.1 Difference Mean Based Perturbation (DMP)

DMP strategy is used in particle swarm optimization (PSO) and shown the effective performance in global optimization [18]. The DMP is a learning strategy amounts to simply perturbing a newly generated population member with

Algorithm 1. DMP implementation

Calculate the dimensional mean of the best individual (X_{best}) with dimension, D of a population:

$best\_avg = \frac{1}{D}\sum_{j=1}^{D} X_{best,j}$

for each harmony $i = 1, 2, ..., HMS$ **do**

  Calculate the dimension mean of ith individual, X_i of the population:

$ind\_avg_i = \frac{1}{D}\sum_{j=1}^{D} x_{i,j}$

  Calculate the difference mean of ith individual, X_i:

$diff\_mean_i = best\_avg - ind\_avg_i$

  Generate a vector V with components chosen randomly unit normal distribution with zero mean and unit variance:

$V = \{v_1, v_2, ..., v_D\}$

  Calculate $\hat{V}$: $\hat{V} = \frac{V}{\|V\|}$

  Perturb amount of ith individual X_i:

$X_i = X_i + diff\_mean_i \times \hat{V}$

end for

a scaled unit vector along any random direction. The unit vector is scaled by the difference mean formed by the subtracting the dimensional mean of the individual to be perturbed from the current best individual of the population. Note that dimensional mean is a scalar quantity obtained by averaging the components of a vector individual from the population. After updating the HM, DMP is applied to all the harmony/individuals as shown in Algorithm 1.

4.2 The ImHS Algorithm

ImHS algorithm aims to overcome the limitations associated with the basic HS algorithm. In order to do so, some modifications have been made on the HS algorithm. In the proposed algorithm, pitch adjustment operation is not considered in such a way that the components chosen in new harmony vector with HMCR are directly modified. In the memory consideration process, j^{th} component of the new harmony vector is selected from the jth component of the best harmony with HMCR. Then the selected component of new harmony vector modified as follows:

$$X_{i,j}^{New} = X_{best}(j) + bw \times N(0,1). \tag{8}$$

Where bw is a distance bandwidth, the amount of changes or movement that may have occurred to the components of the new harmony vector. $N(0,1)$ is normal distribution with zero mean and unit variance.

On the other hand, new vector components generated randomly using Eq. (5) with probability $(1-HMCR)$. After improvisation of the new harmony vector, a greedy selection scheme is applied between X_i and X_i^{New} for updating the HM. DMP strategy is applied to all individuals after the HM updation to explore the

Algorithm 2. ImHS algorithm

```
Initialize the HM size (HMS), memory consideration rate (HMCR), Dimension (D)
Initialize the HM randomly using Eq. (5) and evaluate energy for each harmony
  using Eq. (1)
while stopping criterion is not satisfied do
    Calculate best harmony X_best from the HM
    for each harmony i = 1, 2, ..., HMS do
        for j = 1, 2, ..., D do
            if rand(0, 1) ≤ HMCR then
                New harmony vector, X_{i,j}^{New} is generated using Eq. (8)
            else
                X_{i,j}^{New} is generated randomly using Eq. (5)
            end if
        end for
        if f(X_i^{New}) ≤ f(X_i) then
            X_i in the HM is replaced by X_i^{New}
        end if
    end for
    for each harmony i = 1, 2, ..., HMS do
        Perform DMP strategy using Algorithm (1)
    end for
    Update the HM
    Record the best harmony so far
end while
```

search space for finding the global optimum solution. Finally, the pseudo code of the ImHS algorithm shown in Algorithm 2.

5 Experimental Results and Analysis

5.1 Protein Sequence

The proposed ImHS algorithm is applied on real protein sequences (RPs) which are collected from Protein Data Bank (PDB, http://www.rcsb.org/pdb/home/home.do). The detailed description of these sequences are given in Table 1. In the experiment, K-D method [19] is used to distinguish the hydrophobic (A) and hydrophilic (B) residues of 20 amino acids in real protein sequences. The amino acids I, V, L, P, C, M, A, G are belongs to class A and D, E, F, H, K, N, Q, R, S, T, W, Y are belongs to class B.

5.2 Parameter Settings

In the experimental study, all individuals are encoded in angels within the range $[-180°, 180°]$. The dimension, D of an individual will vary from sequence to sequence based on the respective PSL which set as $(n-2+n-3)$ (n is the length of the sequence). All the experiments are executed on the basis of stopping

Table 1. Real protein sequences

PS No	PSL	PDB ID	Sequence
RS1	13	1BXP	MRYYESSLKSYPD
RS2	16	1BXL	GQVGRQLAIIGDDINR
RS3	18	2ZNF	VKCFNCGKEGHIARNCRA
RS4	21	1EDN	CSCSSLMDKECVYFCHLDIIW
RS5	25	2H3S	PVEDLIRFYNDLQQYLNVVTRHRYX
RS6	29	1ARE	RSFVCEVCTRAFARQEALKRHYRSHTNEK
RS7	35	2KGU	GYCAEKGIRCDDIHCCTGLKCKCNASGYNCVCRKK
RS8	38	1AGT	GVPINVSCTGSPQCIKPCKDQGMRFGKCMNR KCHCTPK
RS9	46	1CRN	TTCCPSIVARSNFNVCRLPGTPEAICATYTGCIII PGATCPGDYAN
RS10	60	2KAP	KEACDWLRATGFPQYAQLYEDFLFPIDISLVKREHD FLDRDAIEALCRRLNTLNKCAVMK

criterion which is the maximum number of function evaluations (FEs). The FEs is set to `Max_FES` $= 10^4 \times D$ defined in [20] which is same for other algorithms used for comparisons. The ImHS algorithm compared with CPSO, BEABC and the basic HS algorithm. The parameters for CPSO, BEABC and the basic HS algorithms are the same as those recommended in [10,13,17]. The parameters for the ImHS algorithm are set as follows: $HMS = 50$, $HMCR = 0.9$ and $bw = 0.2$. Each algorithm is run 30 times independently for each RPS with the same initial population. All the algorithms are implemented in MATLAB R2010a and executed on an Intel Core (TM) I7-2670 QMCPU running at 2.20 GHz with 8 GB of RAM with Windows XP platform.

5.3 Results and Discussions

Table 2 shows the best, mean and standard deviation (std.) of energy values achieved by CPSO, BEABC, HS and ImHS algorithm on the RPS with 30 independent runs. Wilcoxon'srank sum test [21] is conducted at the 5% significance level in order to judge whether the mean result obtained with the best performing algorithm differ from the mean result of the other compared algorithms in a statistically significant way. '+' marks in Table 2 indicate that ImHS statistically outperformed with compared algorithms based on mean result. All marks are labeled with parenthesis in Table 2. The best results are shown in bold.

The comparisons in Table 2 show that ImHS achieved the highest accuracy performance in terms of best and mean results on the RPS's. In particular, the performance of ImHS significantly outperformed compared to all other algorithms for higher length protein sequences (RS7, RS8, RS9 and RS10) in terms of best and mean results. In addition, results obtained by ImHS algorithm is

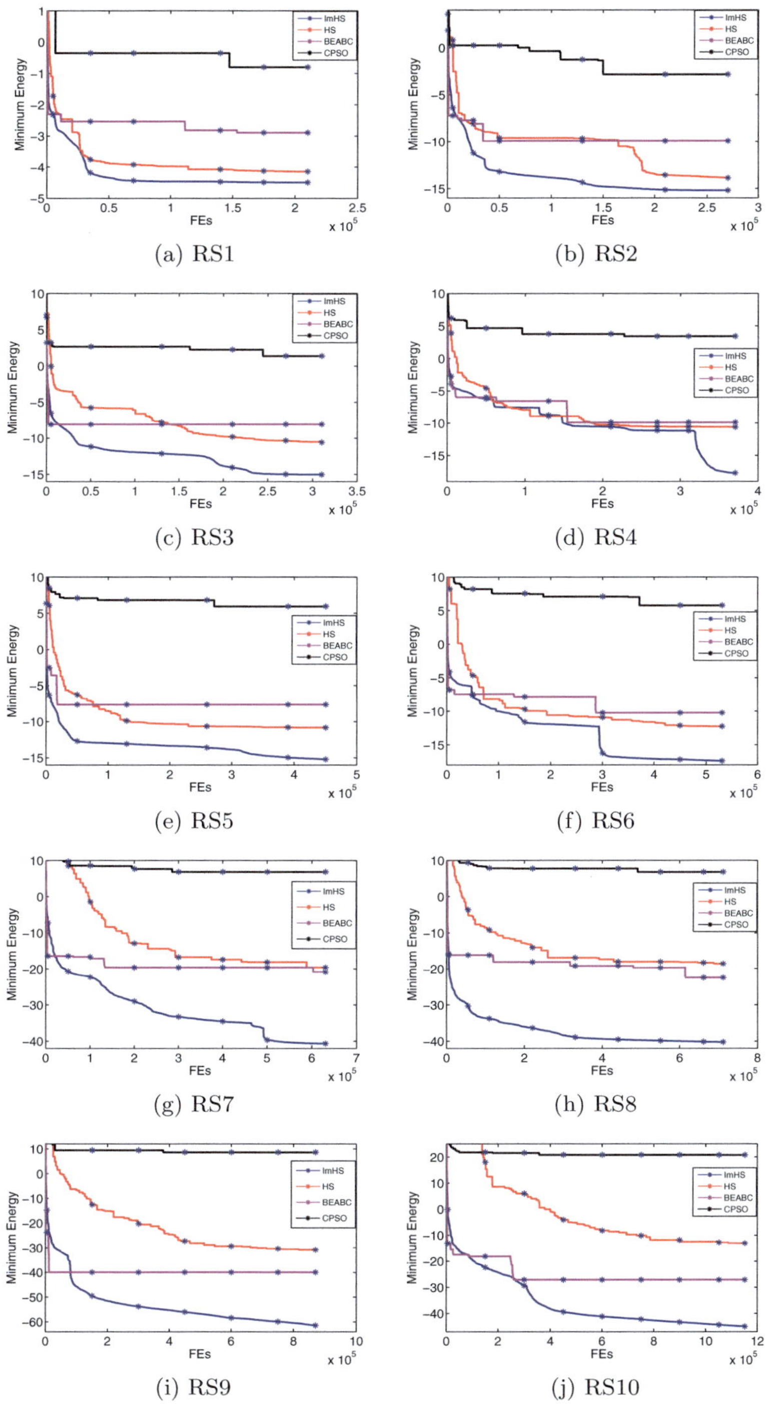

(a) RS1 (b) RS2

(c) RS3 (d) RS4

(e) RS5 (f) RS6

(g) RS7 (h) RS8

(i) RS9 (j) RS10

Fig. 1. Convergence curve of best solutions of different algorithms on RPS's.

Table 2. Comparisons of best, mean and standard deviation results for RPS's

PS No.	PSL	CPSO			BEABC			HS			ImHS		
		Best	Mean	Std.	Best	Mean	Std.	Best	Mean	Std.	Best	Mean	Std.
RS1	13	-0.802	-0.022 (+)	0.590	-2.893	-2.160 (+)	0.734	-4.150	-3.640 (+)	0.424	**-4.498**	**-4.193**	**0.306**
RS2	16	-2.845	-0.932 (+)	1.007	-9.949	-9.084 (+)	0.695	-13.886	-9.534 (+)	2.415	**-15.200**	**-14.880**	**0.373**
RS3	18	1.377	2.418 (+)	0.681	-8.103	-5.815 (+)	1.648	-10.607	-7.455 (+)	1.813	**-15.056**	**-14.108**	**1.096**
RS4	21	3.397	4.622 (+)	0.900	-9.914	-7.689 (+)	0.866	-10.660	-8.005 (+)	1.518	**-17.721**	**-16.519**	**0.760**
RS5	25	5.941	6.413 (+)	0.358	-7.670	-6.803 (+)	**0.834**	-10.857	-8.205 (+)	1.815	**-15.240**	**-13.905**	1.058
RS6	29	5.766	7.120 (+)	1.004	-10.258	-8.126 (+)	1.002	-12.277	-9.827 (+)	1.290	**-17.416**	**-15.965**	**1.058**
RS7	35	6.830	8.739 (+)	1.116	-20.891	-18.612 (+)	1.508	-19.614	-16.709 (+)	1.369	**-40.696**	**-37.945**	**0.901**
RS8	38	6.775	9.086 (+)	1.207	-22.442	-20.614 (+)	1.410	-18.616	-16.277 (+)	1.641	**-40.300**	**-39.637**	**0.426**
RS9	46	8.601	11.085 (+)	2.113	-39.998	-38.008 (+)	1.391	-30.876	-23.538 (+)	5.450	**-61.426**	**-59.142**	**1.083**
RS10	60	20.730	23.675 (+)	1.572	-27.140	-23.107 (+)	1.967	-13.142	-8.991 (+)	2.678	**-44.972**	**-42.634**	**0.772**

also outperformed on the remaining RPS. BEABC and HS algorithms achieved approximately same on mean result for the sequence RS2, RS3, RS4, RS5, RS6 and RS7 whereas BEABC outperformed HS on RS8, RS9 and RS10. The merits of CPSO are worst than the others algorithms on the all sequences. The strong significant improvement has been observed on ImHS algorithm than others as the protein sequence length increases. Moreover, obtained results are very robust based on std. by ImHS algorithm except for RS5. Therefore, the proposed approach provides better performance indicated that it is an effective algorithm for determining the structure of real proteins using 3D AB off-lattice model.

Best convergence characteristic obtained by ImHS and other algorithms on RPS's are shown in Fig. 1. In the y-axis, the minimum energy value is plotted and in x-axis, the function evaluations is plotted. The graph conclusively establish the faster convergence of ImHS over the other algorithms. It is evident from the figures that only the proposed method has been able to obtain high quality solutions.

6 Conclusion

In this paper, we have presented an ImHS algorithm which uses best harmony in improvisation of a new harmony vector and difference mean based perturbation to all the harmony to improve the exploration capability for preventing premature convergence and ability to jumping out from the local optima. The proposed algorithm evaluated on the real protein sequences with varying sequence length. The analysis and experiments show that the ImHS algorithm is significantly effective for PSP of the real protein sequences. It should be noted that our study concerns the few number of real protein sequences with smaller lengths.

Our future works will include investigation of the ImHS algorithm on more real-world protein sequences with higher sequence length in 3D AB off-lattice model.

References

1. Dehzangi, A., Paliwal, K., Lyons, J., Sharma, A., Sattar, A.: A segmentation-based method to extract structural and evolutionary features for protein fold recognition. IEEE/ACM Trans. Comput. Biol. Bioinform. (TCBB) **11**(3), 510–519 (2014)
2. Hendy, H., Khalifa, W., Roushdy, M., Salem, A.B.: A study of intelligent techniques for protein secondary structure prediction. Int. J. Inf. Mod. Anal. **4**(1), 3–12 (2015)
3. Bagaria, A., Jaravine, V., Güntert, P.: Estimating structure quality trends in the protein data bank by equivalent resolution. Comput. Biol. Chem. **46**, 8–15 (2013)
4. Sousa, S.F., Fernandes, P.A., Ramos, M.J.: Protein-ligand docking: current status and future challenges. Protein Struct. Funct. Bioinform. **65**(1), 15–26 (2006)
5. Dorn, M., e Silva, M.B., Buriol, L.S., Lamb, L.C.: Three-dimensional protein structure prediction: methods and computational strategies. Comput. Biol. Chem. **53**, 251–276 (2014)
6. Anfinsen, C.B.: Principles that govern the folding of protein chain. Science **181**, 223–230 (1973)
7. Stillinger, F.H., Head-Gordon, T., Hirshfeld, C.L.: Toy model for protein folding. Phys. Rev. E **48**(2), 1469 (1993)
8. Kalegari, D.H., Lopes, H.S.: An improved parallel differential evolution approach for protein structure prediction using both 2D and 3D off-lattice models. In: IEEE Symposium on Differential Evolution, pp. 143–150 (2013)
9. Wang, Y., Guo, G.D., Chen, L.F.: Chaotic artificial bee colony algorithm: a new approach to the problem of minimization of energy of the 3D protein structure. Mol. Biol. **47**(6), 894–900 (2013)
10. Parpinelli, R.S., Benitiez, C.M., Cordeiro, J., Lopes, H.S.: Performance analysis of swarm intelligence algorithms for the 3D-AB off-lattice protein folding problem. Mult. Valued Logic Soft Comput. **22**(3), 267–286 (2014)
11. Lin, X., Zhang, X.: Protein structure prediction with local adjust tabu search algorithm. BMC Bioinform. **15**(Suppl 15), S1 (2014)
12. Zhou, C., Hou, C., Wei, X., Zhang, Q.: Improved hybrid optimization algorithm for 3D protein structure prediction. J. Mol. Model. **20**(7), 1–12 (2014)
13. Li, B., Chiong, R., Lin, M.: A balance-evolution artificial bee colony algorithm for protein structure optimization based on a three-dimensional AB off-lattice model. Comput. Biol. Chem. **54**, 1–12 (2015)
14. Li, B., Lin, M., Liu, Q., Li, Y., Zhou, C.: Protein folding optimization based on 3D off-lattice model via an improved artificial bee colony algorithm. J. Mol. Model. **21**(10), 1–15 (2015)
15. Geem, Z.W., Kim, J.-H., Loganathan, G.V.: A new heuristic optimization algorithm: harmony search. Simulation **76**(2), 60–68 (2001)
16. Xiang, W.L., An, M.Q., Li, Y.Z., He, R.C., Zhang, J.F.: An improved global-best harmony search algorithm for faster optimization. Expert Syst. Appl. **41**(13), 5788–5803 (2014)
17. Ashrafi, S.M., Dariane, A.B.: Performance evaluation of an improved harmony search algorithm for numerical optimization: Melody Search (MS). Eng. Appl. Artif. Intell. **26**(4), 1301–1321 (2013)
18. Kundu, R., Das, S., Mukherjee, R., Debchoudhury, S.: An improved particle swarm optimizer with difference mean based perturbation. Neurocomputing **129**, 315–333 (2014)
19. Mount, D.W.: Bioinformatics: Sequence and Genome Analysis. Cold Spring Harbor Laboratory Press, Cold Spring Harbor, New York (2001)

20. Liang, J.J., Qu, B.Y., Suganthan, P.N., Hernández-Díaz, A.G.: Problem definitions and evaluation criteria for the CEC 2013 special session on real-parameter optimization. Computational Intelligence Laboratory, Zhengzhou University, Zhengzhou, China and Nanyang Technological University, Singapore, Technical Report 201212 (2013)
21. Derrac, J., García, S., Molina, D., Herrera, F.: A practical tutorial on the use of nonparametric statistical tests as a methodology for comparing evolutionary and swarm intelligence algorithms. Swarm Evol. Comput. **1**(1), 3–18 (2011)

Gender-Sensitive Disaster Vulnerability Using Analytic Hierarchy Process and Genetic Algorithm

Gunhui Chung(✉)

Department of Civil Engineering, Hoseo University,
Asan, Chungcheongnam-do, South Korea
gunhuic@gmail.com

Abstract. Disasters might cause different impacts for women, girls, or men be-cause of their capability to evacuate. These differences are initiated by cultural constraints on female mobility, lack of physical skills or strength. In South Korea, floods have been the most impacted natural disaster resulting in the loss of people's lives and economic damage. Therefore, the gender-sensitive disaster vulnerability factors were proposed in this study. The weights of the factors were calculated using Analytic Hierarchy Process (AHP) and combination of AHP and Genetic Algorithm (GA). As a result, AHP-GA method provided more consistent results and gender related factors had higher weights in AHP-GA method than conventional AHP. Therefore, more detailed analysis is required to gender sensitive vulnerability index for the natural disaster prepare and response.

Keywords: Gender-sensitive · Disaster vulnerability · Analytic hierarchy process · Genetic algorithm

1 Introduction

Disaster vulnerability has been estimated using multi-criteria decision making process to prioritize the importance of impacted factors. Analytic Hierarchy Process (AHP) is one of the methods popularly used for the estimation. However, the personnel selection is an important part in AHP and that is a major drawback because it is hard to make a consistent decision. Therefore, the development of vulnerability index is not easy and public consensus is hard to reach.

Aly and El-hameed [1] proposed a new approach to estimate the weights of AHP using Genetic Algorithm (GA). The proposed AHP-GA method was applied in a couple of examples and the better results than conventional AHP was shown. In the AHP-GA method, the personal selection was excluded and the optimal weights were found. Therefore, it this study, the AHP-GA method was applied to estimated weights of gender-sensitive disaster vulnerability index. Many disaster vulnerability indices considering different risk such as flood, drought, or social risk have been developed. However, not many indices have considered gendered factors. UNDP [2] proposed Flood Vulnerability Index (FVI)

J. Del Ser (ed.), *Harmony Search Algorithm*, Advances in Intelligent Systems and Computing 514, DOI 10.1007/978-981-10-3728-3_31

considered gendered sensitivity factors. The percent of female employees in urban areas and the number of female representatives in the National Parliament were considered as the gender factors.

In this study, factors for the gender-sensitive disaster vulnerability were proposed and the weights were calculated using the AHP-GA method and the results were compared with the conventional AHP method.

2 Analytic Hierarchy Process

AHP is a Multi-Attribute Decision-Making (MADM) model proposed by Saaty [3]. The prioritization is determined using pair-wise comparisons between alternatives. The final weights could be calculated in an efficient way when the personal selection is made by experts with consistency. However, when the number of alternatives is large, it is hard to be consistent. Therefore, the consistency is checked by the Consistency Ratio (CR). The rule of thumb is that the comparison is consistent when a CR is less than 0.1.

3 AHP-GA Model

To cope with the drawback of AHP, Genetic Algorithm (GA) is implemented to evaluate the alternatives in a consistent way. Aly and El-hameed [1] introduced the combination of AHP and GA. The AHP-GA procedure is shown in Fig. 1.

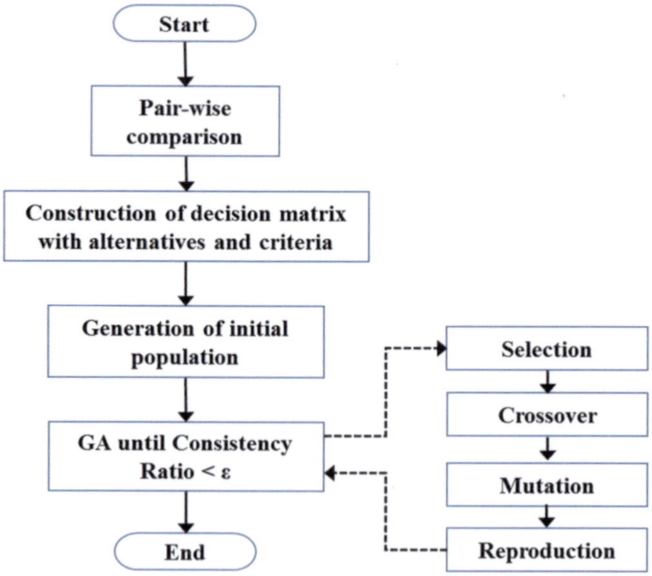

Fig. 1. HP-GA procedure for deriving priorities.

The objective is to minimize the Euclidean distance of Least Square Method as follows:

$$\min F(x) = \sum_{i=1}^{n} \left(\sum_{j=1}^{n} \left(a_{ij} - \frac{w_i}{w_j} \right)^2 \right)^{1/2}, \tag{1}$$

$$\text{subject to} \quad \sum_{i=1}^{n} w_i = 1 \tag{2}$$

Where a_{ij} is the normalized weights of j alternative and i criteria, and w_i is normalized Eigenvalue.

4 Disaster Vulnerability

4.1 Definition

UNESCO-IHE defined vulnerability using Flood Vulnerability Index (FVI) as the extent of harm, which can be expected under certain conditions of exposure, susceptibility and resilience.

Human being is vulnerable to disasters due to three main factors; exposure, susceptibility and resilience. Exposure is the values at the location where disasters can occur. These values can be goods, infrastructure, cultural heritage, agricultural fields or mostly people. Gendered sensitive data could be considered in this category. Susceptibility is system characteristics such as the awareness and preparedness of affected people regarding the risk they live with. Lastly, Resilience is the systematic effort to mitigate disasters. Vulnerability is defined as below using exposure, susceptibility, and resilience.

$$\text{Vulnerability} = \text{Exposure} + \text{Susceptibility} - \text{Resilience}. \tag{3}$$

4.2 Weight Results

Conventional AHP and AHP-GA methods were applied to calculate weights of the vulnerability factors. In GA, total number of decision variables is 36 pairwise comparison. The number of population was 200, crossover and mutation rate were 50% and 10% (Table 1).

Gendered sensitive disaster vulnerability is estimated using vulnerability indices.

The overall weights are shown in Table 2. And the category results are shown in Fig. 2. As shown in this plot the resilience category has larger weight in AHP-GA. Gender related factors have larger weights in AHP-GA. The consistency ratio of AHP and AHP-GA were 0.0658 and 0.0394, respectively. AHP-GA showed the much better consistent results (Fig. 3).

Table 1. Disaster vulnerability factors.

Exposure	Susceptibility	Resilience
Population density	Urban population under the national poverty curve	% of female employees in urban areas
% of female population	Gross domestic product in urban areas	Female representatives in the National Parliament
% of persons over 65		# of government workers in disaster management
% of persons over 19		

Table 2. Weights of disaster vulnerability factors using AHP and AHP-GA.

Category	Factors	AHP	AHP-GA
Exposure	Population density	0.0345	0.0395
	% of female population	0.1232	0.0831
	% of persons over 65	0.0932	0.0740
	% of persons over 19	0.0932	0.0740
Susceptibility	Urban population under the national poverty curve	0.1628	0.0971
	Gross domestic product in urban areas	0.0522	0.0655
Resilience	% of female employees in urban areas	0.0200	0.0655
	Female representatives in the National Parliament	0.0162	0.0655
	# of government workers in disaster management	0.4049	0.4359

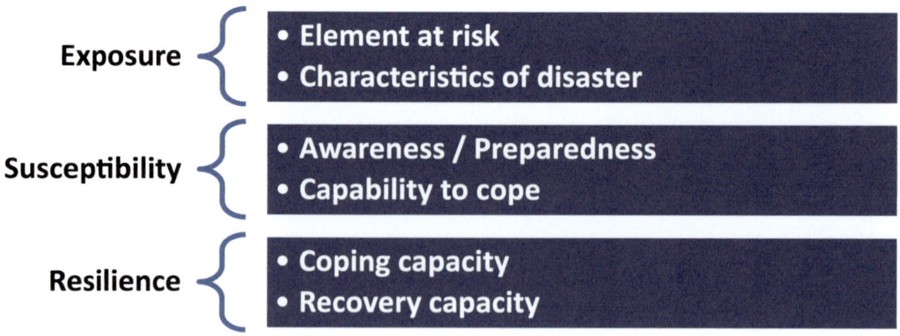

Fig. 2. Factors of vulnerability.

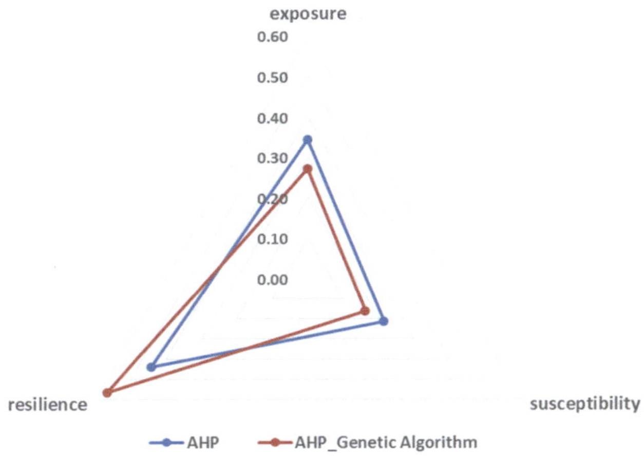

Fig. 3. Weights using entropy and AHP_Genetic Algorithm.

4.3 Summary and Conclusions

This study compared the weight results from AHP and AHP-GA methods for gender sensitive disaster vulnerability. The proposed vulnerability factors in this study considered gender related issues such as the percent of female employees in urban areas and the female representatives in the National Parliament. These factors were suggested by UNDP [2] as the gender related vulnerability factors.

The proposed vulnerability factors were evaluated using AHP and AHP-GA methods. The AHP-GA was applied to overcome the drawback of AHP when the personal selection is made. The results shows AHP-GA provided more consistent weights than conventional AHP. The gender related factors have lager weights in the AHP-GA results. Therefore, further research has to be implemented on the effect of gender sensitive vulnerability for the natural disaster.

Acknowledgement. This research was supported by Support Program for Women in Science, Engineering and Technology through the National Research Foundation of Korea (NRF) funded by the Ministry of Science, ICT and Future Planning (reference number 2016H1C3A1903202).

References

1. Aly, M.F., El-hameed, H.M.A.: Integrating AHP and genetic algorithm model adopted for personal selection. Int. J. Eng. Trends Technol. **6**(5), 247–256 (2013)
2. UNDP. Adaptation Policy Frameworks for Climate Change: Developing Strategies, Policies, and Measures, Cambridge University Press (2004)
3. Saaty, T.L.: How to make a decision: the analytic hierarchy process. Euro. J. Oper. Res. **48**(1), 9–26 (1990)

A Multi-objective Harmony Search Algorithm for Optimal Energy and Environmental Refurbishment at District Level Scale

Diana Manjarres(✉), Lara Mabe, Xabat Oregi, Itziar Landa-Torres, and Eneko Arrizabalaga

TECNALIA Research and Innovation, 48160 Derio, Spain
{diana.manjarres,lara.mabe,xabat.oregi,itziar.landa, eneko.arrizabalaga}@tecnalia.com

Abstract. Nowadays municipalities are facing an increasing commitment regarding the energy and environmental performance of cities and districts. The multiple factors that characterize a district scenario, such as: refurbishment strategies' selection, combination of passive, active and control measures, the surface to be refurbished and the generation systems to be substituted will highly influence the final impacts of the refurbishment solution. In order to answer this increasing demand and consider all above-mentioned district factors, municipalities need optimisation methods supporting the decision making process at district level scale when defining cost-effective refurbishment scenarios. Furthermore, the optimisation process should enable the evaluation of feasible solutions at district scale taking into account that each district and building has specific boundaries and barriers. Considering these needs, this paper presents a multi-objective approach allowing a simultaneous environmental and economic assessment of refurbishment scenarios at district scale. With the aim at demonstrating the effectiveness of the proposed approach, a real scenario of Gros district in the city of Donostia-San Sebastian (North of Spain) is presented. After analysing the baseline scenario in terms of energy performance, environmental and economic impacts, the multi-objective Harmony Search algorithm has been employed to assess the goal of reducing the environmental impacts in terms of Global Warming Potential (GWP) and minimizing the investment cost obtaining the best ranking of economic and environmental refurbishment scenarios for the Gros district.

Keywords: Energy · Environmental · Refurbishment · District · Multi-objective · Optimization

1 Introduction

Energy security and climate change are driving a future that implies important improvements in the energy performance of the building sector. The 28 Member States of the European Union (EU) have set a Global Warming Potential

J. Del Ser (ed.), *Harmony Search Algorithm*, Advances in Intelligent Systems and Computing 514, DOI 10.1007/978-981-10-3728-3_32

(GWP) reduction target of 20% by 2020, which has to be reached mainly through energy efficiency measures [1]. The building sector is one of the major sources of environmental impacts worldwide, as well as in the EU.

In order to support the energy transition of EU towards a low carbon economy, municipalities have a key role to play. Within the Covenant of Mayors initiative, thousands of local and regional authorities voluntarily committed to implement EU climate and energy objectives on their territory [2]. A Sustainable Energy Action Plan (SEAP) [3] is the key document in which the Covenant signatory outlines how it intends to reach its CO2 reduction target by 2020. Through the development of SEAPs, local and regional authorities have defined targets and developed plans and choose specific energy efficiency measures to attempt these targets.

However, there is a lack of connection between global objectives at city level and the implementation of energy strategies at district level. Specific solutions defined at global scale usually underestimate barriers at district and building level. Besides, the introduction into the decision making process of different factors, such as: CO2 emissions and budget, greatly complicates the problem.

In this framework, authors in [4] present an evolutionary multi-objective optimization algorithm (NSGA-III) that optimizes four objectives at a time for a public school retrofit planning. Among the diverse options and alternatives found in the literature for generating appropriate retrofit scenarios, the aspects considered herein are: (1) minimize energy consumption; (2) minimize CO2 emissions; (3) minimize retrofit costs; and (4) maximize thermal comfort. Additionally, population-based meta-heuristic algorithms including Non-dominated Sorting Genetic Algorithm II (NSGA-II), Pareto-Archived Evolution Strategy (PAES) and Particle Swarm Optimization (PSO), have been found in a number of building optimization studies [5,6]. Also related to multi-objective genetic based algorithms, the work in [7] proposes two improvement strategies for building system design optimization. With the aim of modifying the behavior of conventional evolutionary algorithms, adaptive operators and the meta-model approach have been modified in order to improve the optimization convergence and speed performance. Along the optimization process, a set of optimal solutions are generally generated; as this process usually takes a number of energy simulations at each generation, the optimization time of this algorithms in retrofit planning problems is increased. Improvement of this process includes tuning algorithm parameters and hybrid local search algorithms with meta-heuristic algorithms [8,9]. Tuning parameters is unfeasible in actual case studies since one optimization process can take days to complete. The other approach based on hybrid optimization algorithms, try to narrow down the search space and utilize fast and accurate gradient-based search algorithms to converge on the optimal region [10,11].

In this regard, this paper advances over the state of the art by proposing a novel multi-objective heuristic method based on the Harmony Search algorithm specially tailored for obtaining optimal refurbishment solutions at district level. Although the presented approach can be used for different municipalities and indicators according to specific policy goals, a case study of the district of Gros in Donostia-San Sebastian (Spain) is employed in which two different objectives are considered: (1) the maximization of the reduction of CO2 emissions (i.e. minimization of the GWP) and (2) the minimization of the initial economic investment.

2 Gros District Case Study Definition

The district of Gros (Fig. 1) is selected for the case study due to its representativeness for the city as the east enlargement of it that has progressively gained space to the river and to the sea. Moreover, Gros is one of the most chaotic examples of the urbanism of Donostia - San Sebastian with very irregular blocks and a maze of streets saturated by a mix of uses of its ground floors (commercial, workshops, garages and small industries that are being gradually replaced by new residential buildings). These characteristics combined with other aspects, such as: the different ages, thermal properties, energy generation systems and protection level of its buildings, offer an ideal context for the development of a broad variety of scenarios for the retrofit optimization process.

Fig. 1. Screenshot of the Gros district case study. The modelization and the environmental assessment has been carried out by NEST tool [12].

The information obtained in close collaboration with the city of Donostia-San Sebastian (see Table 1) enabled the definition of the majority of required inputs for modelling and assessing the baseline scenario of Gros district. Some of this data is based on assumptions from previous studies [12].

Table 2 shows the main characteristics of Gros district regarding the relationship between the energy certification of the buildings, their Energy Demand (ED), the quantity of buildings, the Heated Floor Area (HFA) and the amount of surface per building typology. The residential buildings analyzed for the district of Gros have been grouped depending on their efficiency level, their energy consumption per final use and other building envelope characteristics indicated in Table 2. The systems considered for the calculation of heating and Domestic Hot Water (DHW) consumptions are gas boilers for the buildings with a building typology between C and E and electricity for a building typology between F and G. Finally, due to the climatic zone of Donostia-San Sebastian no cooling system has been considered for residential buildings.

Table 1. Summary of aspects of the district of Gros.

General data	
Built surface	1,153,443 m2
Number of dwellings	9,581
Residential building surface	1,126,050 m2
Office buildings surface	8,299 m2
Other tertiary building surface	17,512 m2
Wall surface of the buildings	392,087 m2
Building characteristics	
Energy labelling	C-G [24]
Heating and DHW system	Natural gas and electricity
Architectural protection grade	Some buildings [0–4] (according urban rules)
Renewable energy potential: Useful surface for solar technologies	
Total roof surface	159,964 m2
Flat roof surface	71,215 m2
North oriented roof surface	47,696 m2
Maximum useful roof surface for solar technologies	50,726 m2
Total roof surface	159,964 m2
Flat roof surface	71,215 m2
North oriented roof surface	47,696 m2
Maximum useful roof surface for solar technologies	50,726 m2

Table 2. Main characteristics of the district of Gros per building typology. (ED - Energy Demand, HFA - Heated Floor Area, L - Lighting, A - Appliances, TS - Total Surface).

N of blocks - Building typologies	HFA (m^2)	ED (Kwh/m^2.year)				TS (m^2)		
		Heating	DHW	L	A	Opaque facade surface - %75 (m2)	Openings surface - %25 (m2)	Total usefull roof surface (m2)
3 - C	6.671	39.4	13.0	6.4	39.0	1.676	559	486
42 - D	89.350	49.2	13.0	6.4	39.0	23.391	7.797	3.567
396 - E	781.204	64.5	13.0	6.4	39.0	214.207	71.402	37.705
55 - F	93.684	84.7	13.0	6.4	39.0	27.394	9.131	4.376
51 - G	80.998	103.2	13.0	6.4	39.0	22.122	7.374	4.569

Table 2 shows that the energy labelling level of the 92% of the buildings of Gros in less than the D energy rating. That is, the energy and environmental performance of the 92% of the buildings is worse than the limit value defined by the Spanish legislation. Based on these baseline values of the district of Gros, it is necessary to evaluate, optimize and apply different refurbishment strategies to improve their energy and environmental performance.

3 District Energy Retrofitting Problem Formulation

As stated in previous section, the majority of buildings of Gros district has a poor energy labelling level (below D). Therefore, it is of great importance to obtain optimal district retrofitting solutions in terms of cost-effective and CO2 efficient designs. In order to accomplish this, this paper proposes a multi-objective heuristic approach that simultaneously minimize the investment cost and the GWP while considering a set of different constraints between the selected refurbishment strategies.

Despite the numerous existing technologies for energy refurbishment and the current trends towards "Net Zero Energy Buildings" [25], this work is limited to assessing only most common approaches.

In the case of the passive refurbishment strategies, two different efficiency levels per each refurbishment strategy are however proposed to take into account trends toward more energy efficient buildings: basic (b) and advanced (a) levels. The basic efficiency level is based on refurbishment strategies that enforce the minimum thermal requirements determined by the existing regulations and standards. The advanced efficiency level strategies improve the building thermal properties to very high values, such as those used in standards like the Passive House [26].

The first solutions (1B, 1A) corresponds to *ventilated facade system*, which is composed of an aluminum substructure, a layer of insulation and a ceramic outlayer. The second strategy (2B, 2A) is an *indoor thermal improvement solution* consisting of a layer of insulation and plasterboard. Different insulation thicknesses are proposed for basic (1B, 2B) and advanced (1A, 2A) efficiency levels. The projected insulation thicknesses for the basic efficiency energy level are 5 cm for the façade, 8 cm for the deck and 6 cm for the first floor slab. The thicknesses proposed for the advanced energy efficiency level are 25, 30 and 15 cm, respectively.

The third refurbishment strategy (3B, 3A) focuses on the *replacement windows* with a new frame and glazing. The windows for the basic energy efficiency level (3B) consist of a double glazing (2.7 W/(m2 K)) and aluminum frame (2.9 W/(m2 K)), meeting the minimum thermal requirements for refurbishments in Spain. The windows for advanced level (3A) consist of a low-emissivity coated glazing (1.4 W/(m2 K)) and wooden frames (1.2 W/(m2 K)).

Along with energy conservation refurbishment systems, this work has evaluated different strategies based on the use of energy from renewable sources. The first renewable strategy is the *installation of a solar thermal system* on the roof of the building (4), which uses solar energy to generate heat that is then used to produce hot water, reducing the electricity and natural gas use of current water heaters. The second renewable strategy (5) is the *installation of photovoltaic panels* on the roof of the building, generating and exporting electricity to the national grid.

Table 3 depicts a symmetric matrix representing the constraints for each refurbishment strategy. In this table, the set of strategies that can be jointly employed are represented by a value of *1*, whereas the strategies that cannot be

Table 3. Symmetric matrix for refurbishment strategies' constraints.

Refurbishment strategies	1B	1A	2B	2A	3B	3A	4	5
1B	0	0	0	0	1	1	1	1
1A	0	0	0	0	1	1	1	1
2B	0	0	0	0	1	1	1	1
2A	0	0	0	0	1	1	1	1
3B	1	1	1	1	0	0	1	1
3A	1	1	1	1	0	0	1	1
4	1	1	1	1	1	1	0	1
5	1	1	1	1	1	1	1	0

applied together are represented by *0*. Note that in solar and photovoltaic strategies (4, 5) the total useful surface must be less that the 100% of the non-north oriented roof surface.

4 Proposed Multi-objective Harmony Search Algorithm

Let us start by briefly sketching the fundamentals of Harmony Search, which was first coined by Zong et al. in [27] and thereafter applied to a wide number of applications and problems, such as: the Combined Heat and Power Economic Dispatch problem (CHPED) [28], variants of the Traveling Salesman Problem (TSP) [29,30], tour routing [31], Sudoku puzzle solving [32], distribution of 24 h energency units [33] and Grouping problems [34,35], among others [36].

This paper elaborates further on the multi-objective view of the problem and presents a two-objective Harmony Search algorithm that attempts at simultaneously minimizing two (possibly conflicting) fitness functions: Investment Cost (IC) and Global Warming Potential (GWP). By this way, instead of finding a single solution of the problem, it obtains a set of good compromises or trade-offs called the *Pareto optimal* set. Due to the population-based rationale of the Harmony Search algorithm, it relies on a set of candidates $\{\mathbf{H}(k)\}_{k=1}^{K}$ (Harmony Memory), which are iteratively refined by means of intelligent combinations and mutations applied note-wise. Assuming the classical notation related to HS, we will hereafter refer to a possible candidate set $\mathbf{H}(k)$ as *harmony* or *melody*, whereas *note* denotes any of its compounding entries $h(k)$, with $k \in \{1, \ldots, K\}$. In our optimization framework, each melody encodes a refurbishment strategy for a building typology in the district $\in \{C, D, E, F, G\}$ and each note represents the percentage of application $perc \in \{0, 10, \ldots, 100\}\%$ of each refurbishment strategy $\in \{1B, 1A, 2B, 2A, 3B, 3A, 4, 5\}$.

The refinement procedure is controlled by three different parameters: (1) the *Harmony Memory Considering Rate*, HMCR; (2) the *Pitch Adjusting Rate*, PAR and (3) the *Random Selection Rate*, RSR. After the improvisation procedure, the value of the two objective functions (IC and GWP) are separately computed for every improvised melody and the best (with respect to fitness values and spread) K melodies – out of the newly produced ones and those from the previous iteration – compose the Harmony Memory for the next iteration. Note that this procedure is repeated until a fixed number of iterations $\mathcal{I}$ is completed. In the following, the steps of the proposed multi-objective HS algorithm are described in detail:

A. The *initialization* process is only executed at the first iteration. At this step, the entries of the Harmony Memory $\mathbf{H}(k)$ are randomly generated within the range {0, 10,..., 100} while meeting the constraints of the District Energy Retrofitting problem.
B. In the *improvisation* procedure, three different probabilistic operators are sequentially applied to each note so as to produce a new set of K improvised harmonies, namely:
   - The Harmony Memory Considering Rate, HMCR $\in [0, 1]$, sets the probability that the new value for a certain note is drawn uniformly from the values of this same note in all the other $K - 1$ melodies.
   - The Pitch Adjusting Rate, PAR $\in [0, 1]$, refers to the probability that the new value for a given note is taken from its neighbouring values. A step of 10% is added or subtracted with probability $\frac{1}{2}$.
   - The Random Selection Rate, RSR $\in [0, 1]$, establishes the probability to pick a random value for the new note from the subset $[0, 10, \ldots 100]$.
C. The algorithm *checks* whether the newly improvised energy retrofitting solutions are valid in terms of active and passive refurbishment strategies. Regarding active strategies the sum of non-north oriented roof must be less or equal 100% and in all cases the district strategies' constraints (Table 3) must be fulfilled.
D. At each iteration the new generated candidate solutions are then evaluated in terms of both Investment Cost (IC) Eq. (1) and reduction of Global Warming Potential (GWP) Eq. (2).

$$IC = \sum_{k=0}^{K} h(k) \cdot c(k) \cdot A \tag{1}$$

$$GWP = \sum_{k=0}^{K} \frac{h(k) \cdot HD(k) \cdot EH(K) \cdot A \cdot \rho_{GWP}}{\rho}, \tag{2}$$

   where $h(k)$ represents a note in the harmony memory, $c(k)$ the cost value per refurbishment startegy, A the area of useful surface per strategy, i.e. opaque (1B, 1A, 2B, 2A) or openning (3B, 3A) surfaces or non-north oriented surfaces (4, 5), $HD(K)$ the heating demand reduction after the application of the strategies, $EH(K)$ the energy consumption associated to heating demand, ρ_{GWP} the GWP factor per thermal generation system and ρ the performance of the thermal generation system.

   Based on such metric values, a rank and a crowding distance value are assigned at each solution (as explained in [37]). Candidate solutions with less rank value and largest crowding distance value are preferred in order to fill the harmony memory for subsequent iterations. That is, between two solutions with different non-domination ranks, the point with the lower rank is selected. Otherwise, if both of them belong to the same front, then the point located in a region with lesser number of solutions (larger crowding distance) is preferred. If $nIter < \mathcal{I}$, the algorithm iterates by setting $nIter = nIter + 1$ and by returning to step B. Otherwise, the algorithm stops and the set of candidate solutions that conformed the dominant pareto front is given as possible outcomes of the energy retrofitting solution.

5 Simulation Results

In order to obtain the optimal refurbishment scenarios in terms of investment cost and reduction of GWP, the Multi-Objective HS algorithm has been applied at each building typology representing the Gros case study. Figure 2 depicts the Pareto front approximation per building typology after 20 Monte Carlo simulations in terms of initial economic investment (€/m2) for the avoided GWP (kg CO2 eq./m2 year). The Pareto front has been expressed in unit of surface in order to allow comparing the cost-effectiveness of the scenarios at each building typology.

It can be inferred that most efficient building typologies (building typologies C and D) have less cost-effective refurbishment potential than building typologies

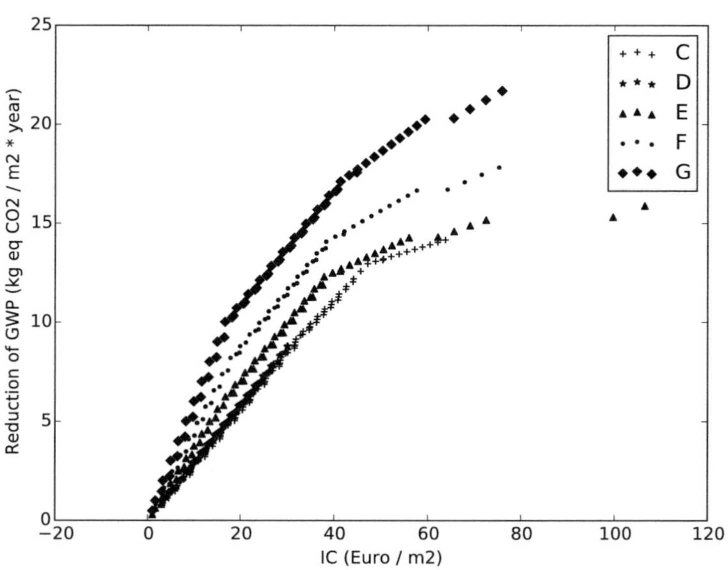

Fig. 2. Pareto front approximation of each building typology in terms of IC (€/m2) and reduction of GWP (kg CO2 eq./m2 year).

Table 4. Investment cost and scenario description for specific GWP reduction goals for each building typology.

GWP reduction: 5 kg CO2 eq./m2 year		
Building typology	IC (€/m2)	Refurbishment scenario
C	18.85	60% (2B) +30% (4)
D	17.49	70% (2B) +20% (4)
E	13.40	80% (2B)
F	11.96	60% (2B) +10% (5)
G	8.34	50% (2B)
GWP reduction: 13 kg CO2 eq./m2 year		
Building typology	IC (€/m2)	Refurbishment scenario
C	50.52	100% (2B) +100% (4) + 10% (5)
D	N.A	N.A
E	37.86	100% (2B) + 100% (4)
F	36.25	100% (2B) + 90% (4)
G	25.58	100% (2B) + 30% (4)
GWP reduction: 18 kg CO2 eq./m2 year		
Building typology	IC (€/m2)	Refurbishment scenario
C	N.A	N.A
D	N.A	N.A
E	N.A	N.A
F	104.24	100% (2B) + 100% (4) + 80% (3B)
G	46.76	100% (2B) + 100% (4) + 30% (3A)

with a worse energy labelling (E, F, G). The algorithm proposes the application of internal thermal insulation and solar thermal panels implementation as the less initial economic investment solutions. Once the maximum surface has been covered by both solutions, the algorithm adds the replacement of windows. Within the Pareto

Table 5. Final cost-effective energy refurbishment scenarios to achieve GWP reduction goals at district level.

	GWP reduction: 20%	GWP reduction: 30%	GWP reduction: 50%
	IC = 15.67M€ GWP red.= 6,504 (tn CO2 eq./year)	IC = 25.19M€ GWP red.= 10,150 (tn CO2 eq./year)	IC = 56.42M€ GWP red.= 14,883 (tn CO2 eq./year)
C	40% (2B) + 20% (4)	20% (2B) + 60% (4)	40% (2B) + 20% (4)
D	40% (2B) + 70% (4)	50% (2B) + 40% (4) + 10% (5)	100% (2B) + 20% (4)
E	80% (2B)	100% (2B) + 60% (4) + 10% (5)	100% (2B) + 100% (4) + 100% (3A)
F	100% (2B) + 20% (4)	100% (2B) + 30% (4)	100% (2B) + 100% (4) + 80% (3B)
G	100% (2B) + 70% (4)	100% (2B) + 10% (4) + 10% (3A)	100% (2B) + 100% (4) + 80% (3A)

Front, some scenarios for building typologies C and D accept the implementation of solar thermal panels without the application of passive strategies. It is known [38] that for buildings with a good thermal insulation, renewables can be a most cost-effective solution to achieve environmental goals than passive strategies.

For building typologies E, F and G the evolution is different than for the most efficient ones (C, D) described previously. In this case, the algorithm proposes a thermal internal insulation until covering the whole opaque surface area. With a higher investment cost, the thermal internal insulation is complemented with solar thermal panels implementation. Once the hundred percent of useful roof's surface and of the opaque surface are covered, the replacement of windows is proposed. Finally, it also infers scenarios replacing the thermal internal insulation strategy by the external one that implies more initial economic investment per surface area but higher insulation potential.

In order to explain in more detail the results obtained, Table 4 shows the scenarios obtained by the algorithm for each building typology considering a low, a medium and a high GWP reduction. It can be shown that for buildings with better energy labelling, a higher investment cost is required to achieve the same CO2 savings.

Finally, regarding the final cost-effective energy refurbishment scenarios at district level, Table 5 depicts the obtained solutions for each building typology in order to acquire a GWP reduction goal of 20%, 30% and 50%.

6 Concluding Remarks

This paper proposes a Multi-Objective HS algorithm for the optimal district's energy refurbishment design. The developed approach is applied to a real case study of Gros in order to define the best energy refurbishment scenarios to reach different GWP reduction goals. The achieved results elucidates the goodness of the proposed Multi-Objective HS algorithm during a district refurbishment's decision making process. It is capable of obtaining a wide range of feasible scenarios in terms of environmental aspects and economic investment allowing the selection of the optimal scenario considering the available budget and the architectural boundaries. Even if the robustness of this algorithm is rigurously assessed within this paper, the scope of the study can be extended by means of including more refurbishment strategies and boundaries. This approach will be developed within the OptEEmAL project [39] where the Multi-Objective HS algorithm will be integrated within a platform that will allow its interoperability with the Energy Conservation Measures database developed in the project in which the objective functions will be calculated with a simulation module based on Energyplus [18].

Acknowledgment. Part of this work was developed from results obtained during the "Optimised Energy Efficient Design Platform for Refurbishment at District Level" (OptEEmAL) project, Grant Agreement Number 680676.

References

1. Directive 2010/31/EU. European parliament and of the council of 19 on the energy performance of buildings, May 2010
2. Covenant of Mayors for Climate and Energy. http://www.covenantofmayors.eu/The-Covenant-of-Mayors-for-Climate.html
3. Minuartia Enea, RSM Gasso Auditores, CIMNE. Sustainable Energy Action Plan of Donostia (2011). http://www.donostia.eus/info/ciudadano/ma_areas.nsf/voWebContenidosId/80FE86B9FAB3DD7DC1257FF6002C5878/
4. Son, H., Kim, C.: Evolutionary multi-objective optimization in building retrofit planning problem. Procedia Eng. **145**, 565–570 (2016)
5. Machairas, V., Tsangrassoulis, A., Axarli, K.: Algorithms for optimization of building design: a review. Renew. Sustain. Energy Rev. **31**, 101–112 (2014)
6. Karaguzel, O.T., Zhang, R., Lam, K.P.: Coupling of whole-building energy simulation and multi-dimensional numerical optimization for minimizing the life cycle costs of office buildings. Build. Simul. **7**, 111–121 (2014)
7. Xu, W., Chong, A., Karaguzel, O.T., Lam, K.P.: Improving evolutionary algorithm performance for integer type multi-objective building system design optimization. Energy Build. **127**, 714–729 (2016)
8. Alajmi, A., Wright, J.: Selecting the most efficient genetic algorithm sets insolving unconstrained building optimization problem. Int. J. Sustain. Built Environ. **3**(1), 18–26 (2014)
9. Nguyen, A.T., Reiter, S., Rigo, P.: A review on simulation-based optimization methods applied to building performance analysis. Appl. Energy **113**, 1043–1058 (2014)
10. Greiner, D., Galván, B., Périaux, J., Gauger, N.: Advances in Evolutionary and Deterministic Methods for Design, Optimization and Control in Engineering and Sciences., vol. 36. Springer, Heidelberg (2016)
11. Juan, Y.K., Gao, P., Wang, J.: A hybrid decision support system for sustainable office building renovation and energy performance improvement. Energy Build. **42**(3), 290–297 (2010)
12. Oregi, X., Pousse, M., Mabe, L., Escudera, A., Mardaras, I.: Sustainability assessment of three districts in the city of Donostia through the NEST simulation tool. Natural Resources Forum (2016). doi:10.1111/1477-8947.12104
13. GeoEuskadi. http://www.geo.euskadi.net/s69-bisorea/es/x72aGeoeuskadiWAR/index.jsp

14. Cadaster of Gipuzkoa. http://www4.gipuzkoa.net/ogasuna/catastro
15. Basque Government. Energy Efficiency Certificate. http://www.industria.ejgv.euskadi.eus/r44-in0100/es/contenidos/informacion/
16. Ministry of Housing, Order FOM/1635/2013 of 10th September. CTE-DB-HE technical building code - energy saving document updated. BOE n 219, 12 September 2013a
17. Design Builder Simulation Tool. http://www.designbuilder.es
18. Energy Plus Energy simulation tool. U.S. Department of EnergyâĂŹs (DOE) Building Technologies Office (BTO). https://energyplus.net/
19. ASHRAE. International Weather Files for Energy Calculations 2.0 (IWEC2). https://www.ashrae.org/resources-publications/
20. CTE. Spanish Technical Building Code (2013). http://www.codigotecnico.org/
21. Metz, B., Davidson, O., Bosch, P., Dave, R., Meyer, L.: Contribution of Working Group III to the Fourth Assessment Report of the Intergovernmental Panel on Climate Change. Cambridge University Press, Cambridge (2007)
22. Weidema, B.P., Bauer, C., Hischier, R., Mutel, C., Nemecek, T., Reinhard, J., Vadenbo, C.O., Wernet, G.: Overview and methodology. Data quality guideline for the ecoinvent database version 3. Ecoinvent Report 1(v3). St. Gallen: The ecoinvent Centre (2013)
23. REE. Spain Electricity Network (2015). http://www.ree.es/en
24. Ministry of Housing. Royal Decree 235/2013, 5th of approving the Spanish Energy Certification of Buildings. BOE n89, 13th April 2013
25. Cellura, M., Guarino, F., Longo, S., Mistretta, M.: Different energy balances for the redesign of nearly net zero energy buildings: an Italian case study. Renew. Sustain. Energ. Rev. **45**, 100–112 (2015)
26. IPHA. International Passive House Association. http://www.passivehouse-international.org/
27. Geem, Z.W., Kim, J.-H., Loganathan, G.V.: A new heuristic optimization algorithm: harmony search. Simulation **76**(2), 60–68 (2001)
28. Vasebia, A., Fesangharyb, M., Bathaeea, S.M.T.: Combined heat and power economic dispatch by harmony search algorithm. Int. J. Electr. Power Energy Syst. **29**(10), 713–719 (2007)
29. Geem, Z.W., Hoon Kim, J., Loganathan, G.V.: A new heuristic optimization algorithm: harmony search. Simulation **76**(2), 60–68 (2001)
30. Ser, J., Bilbao, M.N., Perfecto, C., Salcedo-Sanz, S.: A harmony search approach for the selective pick-up and delivery problem with delayed drop-off. In: Kim, J.H., Geem, Z.W. (eds.) Harmony Search Algorithm. AISC, vol. 382, pp. 121–131. Springer, Heidelberg (2016). doi:10.1007/978-3-662-47926-1_13
31. Geem, Z.W., Tseng, C.-L., Park, Y.: Harmony search for generalized orienteering problem: best touring in China. In: Wang, L., Chen, K., Ong, Y.S. (eds.) ICNC 2005. LNCS, vol. 3612, pp. 741–750. Springer, Heidelberg (2005). doi:10.1007/11539902_91
32. Geem, Z.W.: Harmony search algorithm for solving sudoku. In: Apolloni, B., Howlett, R.J., Jain, L. (eds.) KES 2007. LNCS (LNAI), vol. 4692, pp. 371–378. Springer, Heidelberg (2007). doi:10.1007/978-3-540-74819-9_46
33. Landa-Torres, I., Manjarres, D., Salcedo-Sanz, S., Ser, J., Gil-Lopez, S.: A multiobjective grouping harmony search algorithm for the optimal distribution of 24-hour medical emergency units. Expert Syst. Appl. **40**(6), 2343–2349 (2013)

34. Landa-Torres, I., Ser, J., Salcedo-Sanz, S., Gil-Lopez, S., Portilla-Figueras, J.A., Alonso-Garrido, O.: A comparative study of two hybrid grouping evolutionary techniques for the capacitated P-median problem. Comput. Oper. Res. **39**(9), 2214–2222 (2012)
35. Landa-Torres, I., Gil-Lopez, S., Del Ser, J., Salcedo-Sanz, S., Manjarres, D., Portilla-Figueras, J.A.: Efficient citywide planning of open WiFi access networks using novel grouping harmony search heuristics. Eng. Appl. Artif. Intell. **26**(3), 1124–1130 (2013)
36. Manjarres, D., Landa-Torres, I., Gil-Lopez, S., Ser, J., Bilbao, M.N., Salcedo-Sanz, S., Geem, Z.W.: A survey on applications of the harmony search algorithm. Eng. Appl. Artif. Intell. **26**(8), 1818–1831 (2013)
37. Kalyanmoy, D., Agrawal, S., Pratap, A., Meyarivan, T.: A fast elitist non-dominated sorting genetic algorithm for multi-objective optimization: NSGA-II. IEEE Trans. Evol. Comput. **6**(2), 182–197 (2000)
38. Oregi, X., Hernandez, P., Hernandez, R.: Analysis of life-cycle boundaries for environmental and economic assessment of building energy refurbishment projects. Energy Build. **136**, 12–25 (2017). doi:10.1016/j.enbuild.2016.11.057
39. Garcia-Fuentes, M.A., Hernandez, G., Serna, V.I., Vicente, J.M.: OptEEmAL: una herramienta de diseño para proyectos de rehabilitación en áreas urbanas hacia distritos de Energía Casi Nula. Congreso EECN III (Edificios Energía Casi Nulo) (2016)

Computing Self-Similar Contractive Functions for the IFS Inverse Problem Through the Cuckoo Search Algorithm

Javier Quirce[1], Akemi Gálvez[1,2], and Andrés Iglesias[1,2(✉)]

[1] Department of Applied Mathematics and Computational Sciences, University of Cantabria, Avenida de los Castros s/n, 39005 Santander, Spain
iglesias@unican.es

[2] Department of Information Science, Faculty of Sciences, Narashino Campus, Toho University, 2-2-1 Miyama, Funabashi 274-8510, Japan
http://personales.unican.es/iglesias

Abstract. One of the most powerful and popular methods to generate fractal images is the so-called iterated function systems (IFS). Given a finite system of contractive maps $\{w_i\}_{i=1,\ldots,n}$ on the compact metric space $\mathbb{R}^2$, this system has a unique non-empty compact fixed set $\mathcal{A}$, called the attractor of the IFS. The graphical representation of this attractor is a self-similar fractal image. The opposite is also true: each self-similar fractal image in $\mathbb{R}^2$ can be mathematically represented as the only attractor of an IFS. Obtaining the parameters of the IFS system (called the IFS inverse problem) is a very difficult issue. A good strategy to address it consists of solving firstly the sub-problem of computing a suitable set of self-similar contractive functions to be further applied to obtain the optimal IFS for the inverse problem. In this paper we address this sub-problem by using a powerful metaheuristic technique called cuckoo search algorithm. Our experimental results show that the method performs quite well for several self-similar fractal images.

Keywords: Swarm computation · Cuckoo search algorithm · Iterated function systems · Fractal images · Self-similar contractive functions

1 Introduction

Fractals are intriguing mathematical objects exhibiting a repeating pattern that displays at every scale. They differ from other mathematical shapes in that the scale of this replicating pattern is not necessarily an integer number, but a real one. This value is called the *fractal dimension* and very often it is larger than the topological dimension of the fractal [3,10]. Amazingly, these properties can be found in many natural objects; natural structures such as the branches of the trees, the river networks, the coastlines, the snowflakes, or the mountain ranges, to mention just a few, can be properly represented by fractals. Owing to this reason, fractal images have been widely used in fields such as computer graphics, dynamical systems, bioinformatics, and many others [1,3,10,12,13].

J. Del Ser (ed.), *Harmony Search Algorithm*, Advances in Intelligent Systems and Computing 514, DOI 10.1007/978-981-10-3728-3_33

Several methods can be found in the literature to obtain fractal images, such as Brownian motion, escape-time fractals, L-systems, and others [4,6–8]. Among them, the *Iterated Function Systems* (IFS) have been widely recognized as a popular, efficient, and simple way to obtain self-similar fractal images [1,14]. IFS consists of a finite system of contractive maps $\{w_i\}_{i=1,\dots,n}$ on the compact metric space $\mathbb{R}^2$. This system has a unique non-empty compact fixed set $\mathcal{A}$ (called the attractor of the IFS), whose graphical representation is a self-similar fractal image. The opposite is also true: each self-similar fractal image in 2D can be represented by an IFS. Obtaining the parameters of such IFS (including the total number of contractive functions) is called the *IFS inverse problem.*

This IFS inverse problem is very difficult and only partial solutions have been reported in the literature so far. One of the most promising strategies to tackle this issue consists of solving firstly the sub-problem of computing a suitable collection of self-similar contractive functions for the IFS (the SSCF problem). This set can then be further applied to obtain the optimal IFS for the general IFS inverse problem. This paper is focused on this SSCF problem. In this work we solve it by using a powerful metaheuristic technique called *cuckoo search algorithm.* The method has been applied to several instances of 2D self-similar fractal images from a benchmark with satisfactory results.

The structure of this paper is as follows: in Sect. 2 introduces the main concepts and definitions about the iterated function systems along with a brief description of the IFS inverse and the SSCF problems. Then, Sect. 3 describes the cuckoo search algorithm, the metaheuristics used in this paper. Our proposed method is described in detail in Sect. 4, while the experimental results are briefly discussed in Sect. 5. The paper closes with the main conclusions and some ideas about future work in the field.

2 Mathematical Background

2.1 Iterated Function Systems

An *Iterated Function System* (IFS) is a finite set $\{w_i\}_{i=1,\dots,n}$ of contractive maps $w_i : X \longrightarrow X$ defined on a complete metric space (X, d). We refer to the IFS as $\mathcal{W} = \{X; w_1, \dots, w_n\}$. In the two-dimensional case, the metric space (X, d) is typically $\mathbb{R}^2$ along with the Euclidean distance d_2, which is a complete metric space, so the affine transformations w_i are of the form:

$$\begin{bmatrix} x^* \\ y^* \end{bmatrix} = w_i \begin{bmatrix} x \\ y \end{bmatrix} = \begin{bmatrix} a_i & b_i \\ c_i & d_i \end{bmatrix} \cdot \begin{bmatrix} x \\ y \end{bmatrix} + \begin{bmatrix} e_i \\ f_i \end{bmatrix} \tag{1}$$

or equivalently: $\mathbf{w}_i(\mathbf{x}) = \mathbf{A}_i.\mathbf{x} + \mathbf{b}_i$ where $\mathbf{b}_i$ is a translation vector and $\mathbf{A}_i$ is a 2×2 matrix with eigenvalues λ_1, λ_2 such that $|\lambda_i| < 1$. In fact, $s_i = |det(\mathbf{A}_i)| < 1$ meaning that w_i shrinks distances between points. Let us now define a transformation, T, in the compact subsets of X, $\mathcal{H}(X)$, by

$$T(A) = \bigcup_{i=1}^{n} w_i(A). \tag{2}$$

If all the w_i are contractions, T is also a contraction in $\mathcal{H}(X)$ with the induced Hausdorff metric [1,14]. Then, T has a unique fixed point, $\mathcal{A}$, called the *attractor of the IFS*.

Let us now consider a set of probabilities $\mathcal{P} = \{p_1, \dots, p_n\}$, with $\sum_{i=1}^{n} p_i = 1$. We refer to $\{\mathcal{W}, \mathcal{P}\} = \{X; w_1, \dots, w_N; p_1, \dots, p_n\}$ as an *IFS with probabilities* (IFSP). There exists an efficient method, known as *probabilistic algorithm*, for the generation of the attractor of an IFS. This algorithm follows from the result $\overline{\{x_k\}_{k>0}} = \mathcal{A}$ provided that $x_0 \in \mathcal{A}$, where (see, for instance, [2]):

$$x_k = w_i(x_{k-1}) \text{ with probability } p_i > 0. \tag{3}$$

Picking an initial point, one of the mappings in the set $\{w_1, \dots, w_n\}$ is chosen at random using the weigths $\{p_1, \dots, p_n\}$ according to Eq. (3). The selected map is then applied to generate a new point, and the same process is repeated again with the new point. As a result of this stochastic iterative process, we obtain a sequence of points. This resulting sequence converges to the fractal as the number of points increases. This algorithm, called the *chaos game* [1], generates a sequence of points randomly distributed over the fractal, according to the chosen set of probabilities. Thus, the larger the number of iterations, the better the resolution of the resulting fractal image.

The fractal image is determined only by the set of contractive mappings; the set of probabilities gives the efficiency of the rendering process. Thus, one of the main problems of the chaos game algorithm is that of finding the optimal set of probabilities to render the fractal attractor associated with an IFS [3,9]. The most standard method was suggested by Barnsley [1]: for each of the mappings, this method (called *Barnsley's algorithm*) selects a probability value that is proportional to the area of the figure associated with the mapping. Since the area filled by a linear mapping w_i is proportional to its contractive factor, s_i, this algorithm proposes to take:

$$p_i = \frac{s_i}{\sum\limits_{j=1}^{n} s_j} \quad ; \quad i = 1, \dots, n. \tag{4}$$

2.2 The IFS Inverse Problem

The *IFS inverse problem* can be stated as follows: suppose that we are given a self-similar fractal image $\mathcal{F}$. The goal is to obtain an IFS whose attractor has a graphical representation $\mathcal{F}'$ that approximates $\mathcal{F}$ accurately according to a metrics function φ, which measures the graphical distance between $\mathcal{F}$ and $\mathcal{F}'$. Mathematically, it means that:

$$\varphi(\mathcal{F}, \mathcal{F}') < \delta \tag{5}$$

for a prescribed value of δ. Equation (5) can be transformed into the optimization problem:

$$\underset{\{\mathbf{A}_i, \mathbf{b}_i\}_{i=1,\dots,n}}{minimize} \quad [\varphi(\mathcal{F}, \mathcal{F}')] \tag{6}$$

The problem (6) is a continuous optimization problem, because all free variables in $\{\mathbf{A}_i, \mathbf{b}_i\}_i$ are real-valued. It is also a constrained problem, since all those free variables must satisfy the condition that the corresponding functions w_i have to be contractive. It is also a multimodal problem, since there can be several global or local minima of the fitness function. Therefore, we have to solve a difficult multimodal and multivariate constrained continuous optimization problem. The problem is so difficult that only partial solutions have been reported so far, but the general problem still remains unsolved to a large extent.

2.3 The SSCF Problem

One of the most promising approaches to overcome this limitation consists of solving firstly the sub-problem of computing a suitable collection of self-similar contractive functions for the IFS (this is called the *SSCF problem*). This set can then be further applied in a second step to obtain the optimal IFS for the IFS inverse problem. However, even this SSCF problem becomes very challenging because we do not have any information a priori about the number of contractive functions and their parametric values. A way to overcome this limitation is to apply a given number of contractive maps w_i onto the original fractal image $\mathcal{F}$ and compare the resulting images with $\mathcal{F}$ in order to obtain suitable values for the SSCF parameters. With this strategy, the original problem (6) is transformed into the optimization problem:

$$\underset{\mathbf{A}_i,\mathbf{b}_i}{minimize} \; [\varphi(\mathcal{F}, w_i(\mathcal{F}))] \qquad (i = 1, \dots, n) \tag{7}$$

In this paper we focus on the SSCF problem and solve the optimization problem (7) through the cuckoo search algorithm described in next section.

3 The Cuckoo Search Algorithm

Cuckoo search (CS) is a powerful metaheuristic algorithm originally proposed by Yang and Deb in 2009 [17]. Since then, it has been successfully applied to difficult optimization problems [5,15,16,18]. The algorithm is inspired by the obligate interspecific brood-parasitism of some cuckoo species that lay their eggs in the nests of host birds of other species to escape from the parental investment in raising their offspring and minimize the risk of egg loss to other species, as the cuckoos can distributed their eggs amongst a number of different nests.

This interesting and surprising breeding behavioral pattern is the metaphor of the cuckoo search metaheuristic approach for solving optimization problems. In the cuckoo search algorithm, the eggs in the nest are interpreted as a pool of candidate solutions of an optimization problem while the cuckoo egg represents a new coming solution. The ultimate goal of the method is to use these new (and potentially better) solutions associated with the parasitic cuckoo eggs to replace the current solution associated with the eggs in the nest. This replacement, carried out iteratively, will eventually lead to a very good solution of the problem.

Table 1. Cuckoo search algorithm via Lévy flights as originally proposed in [17,18].

Algorithm: Cuckoo Search via Lévy Flights

```
begin
    Objective function f(x), x = (x_1, ..., x_D)^T
    Generate initial population of n host nests x_i (i = 1, 2, ..., n)
    while (t < MaxGeneration) or (stop criterion)
        Get a cuckoo (say, i) randomly by Lévy flights
        Evaluate its fitness F_i
        Choose a nest among n (say, j) randomly
        if (F_i > F_j)
            Replace j by the new solution
        end
        A fraction (p_a) of worse nests are abandoned and new ones
                are built via Lévy flights
        Keep the best solutions (or nests with quality solutions)
        Rank the solutions and find the current best
    end while
    Postprocess results and visualization
end
```

In addition to this representation scheme, the CS algorithm is also based on three idealized rules [17,18]:

1. Each cuckoo lays one egg at a time, and dumps it in a randomly chosen nest;
2. The best nests with high quality of eggs (solutions) will be carried over to the next generations;
3. The number of available host nests is fixed, and a host can discover an alien egg with a probability $p_a \in [0, 1]$. In this case, the host bird can either throw the egg away or abandon the nest so as to build a completely new nest in a new location. For simplicity, this assumption can be approximated by a fraction p_a of the n nests being replaced by new nests (with new random solutions at new locations).

The basic steps of the CS algorithm are summarized in the pseudocode shown in Table 1. Basically, the CS algorithm starts with an initial population of n host nests and it is performed iteratively. The initial values of the jth component of the ith nest are determined by the expression $x_i^j(0) = rand.(up_i^j - low_i^j) + low_i^j$, where up_i^j and low_i^j represent the upper and lower bounds of that jth component, respectively, and $rand$ represents a standard uniform random number on the interval $(0, 1)$. With this choice, the initial values are within the search space domain. These boundary conditions are also controlled in each iteration step.

For each iteration g, a cuckoo egg i is selected randomly and new solutions $\mathbf{x}_i(g+1)$ are generated by using the Lévy flight. According to the original creators of the method, the strategy of using Lévy flights is preferred over other simple random walks because it leads to better overall performance of the CS. The general equation for the Lévy flight is given by:

$$\mathbf{x}_i(g+1) = \mathbf{x}_i(g) + \alpha \oplus levy(\lambda) \tag{8}$$

where g indicates the number of the current generation, and $\alpha > 0$ indicates the step size, which should be related to the scale of the particular problem under study. The symbol $\oplus$ is used in Eq. (8) to indicate the entry-wise multiplication. Note that Eq. (8) is essentially a Markov chain, since next location at generation $g+1$ only depends on the current location at generation g and a transition probability, given by the first and second terms of Eq. (8), respectively. This transition probability is modulated by the Lévy distribution as:

$$levy(\lambda) \sim g^{-\lambda}, \qquad (1 < \lambda \leq 3) \tag{9}$$

which has an infinite variance with an infinite mean. From the computational standpoint, the generation of random numbers with Lévy flights is comprised of two steps: firstly, a random direction according to a uniform distribution is chosen; then, the generation of steps following the chosen Lévy distribution is carried out. The authors suggested to use the Mantegna's algorithm for symmetric distributions (see [18] for details), which computes the factor:

$$\hat{\phi} = \left(\frac{\Gamma(1+\hat{\beta}).sin\left(\frac{\pi.\hat{\beta}}{2}\right)}{\Gamma\left(\left(\frac{1+\hat{\beta}}{2}\right).\hat{\beta}.2^{\frac{\hat{\beta}-1}{2}}\right)} \right)^{\frac{1}{\hat{\beta}}} \tag{10}$$

where Γ denotes the Gamma function and $\hat{\beta} = \dfrac{3}{2}$ in the original implementation by Yang and Deb [18]. This factor is used in Mantegna's algorithm to compute the step length ς as: $\varsigma = \frac{u}{|v|^{\frac{1}{\hat{\beta}}}}$, where u and v follow the normal distribution of zero mean and deviation σ_u^2 and σ_v^2, respectively, where σ_u obeys the Lévy distribution given by Eq. (10) and $\sigma_v = 1$. Then, the stepsize ζ is computed as $\zeta = 0.01\,\varsigma\,(\mathbf{x} - \mathbf{x}_{best})$. Finally, $\mathbf{x}$ is modified as: $\mathbf{x} \leftarrow \mathbf{x} + \zeta.\mathbf{\Psi}$ where $\mathbf{\Psi}$ is a random vector of the dimension of the solution $\mathbf{x}$ and that follows the normal distribution $N(0,1)$. The CS method then evaluates the fitness of the new solution and compares it with the current one. In case the new solution brings better fitness, it replaces the current one. On the other hand, a fraction of the worse nests (according to the fitness) are abandoned and replaced by new solutions so as to increase the exploration of the search space looking for more promising solutions. The rate of replacement is given by the probability p_a, a parameter of the model that has to be tuned for better performance. Moreover, for each iteration step, all current solutions are ranked according to their fitness and the best solution reached so far is stored as the vector $\mathbf{x}_{best}$.

4 The Proposed Method

In this section, we describe the proposed method to solve the SSCF problem. The input of our problem is a self-similar fractal image, $\mathcal{F}$, and the number of contractive functions, n. Currently, our method does not compute the optimal value for n. This task will be part of our future work in the field.

4.1 Our Cuckoo Search-Based Method

Our approach is based on the application of the cuckoo search algorithm with Lévy flights described in Sect. 3 to our initial input. We consider an initial population of size n_p where each individual $\mathcal{P}_k^i$ is a real-valued vector of the free variables of the i-th contractive function w_i according to Eq. (1), that is:

$$\mathcal{P}_k^i = (a_{1,k}^i, b_{1,k}^i, \ldots, f_{1,k}^i) \tag{11}$$

These individuals are initialized with uniform random values in the interval $[-1, 1]$ for the variables in $\mathbf{A}_i$ and in the interval $[-5, 5]$ for the elements in $\mathbf{b}_i$. After this initialization step, we compute the contractive factor s_i and remove all functions w_i with $s_i \geq 1$ to ensure that only contractive functions are included in the initial population. Before applying the cuckoo search, we also need to define a suitable fitness function. Different metrics can be used to this purpose. As indicated in Sect. 2, the most natural choice is the Haussdorff distance. Unfortunately, it becomes computationally expensive and inefficient for this problem. For computational reasons, in this paper we use the Hamming distance instead. To this aim, we treat the fractal images as bitmap images on a grid of pixels for a given resolution driven by a parameter called the mesh size, m_s. Then, we generate their corresponding matrices with 0s and 1s, where 1 means that the corresponding pixel is activated and 0 otherwise. Then, we count the number of mismatches between both matrices to determine the rate of pixels with different values for the given resolution. This yields a reasonable measure of the similarity of both images. The procedure is repeated for a given number of iterations.

4.2 Parameter Tuning

It is well-known that the parameter tuning of metaheuristic methods is troublesome and problem-dependent. Fortunately, the cuckoo search is specially advantageous in this regard, as it depends on only two parameters: the population size, n_p, and the probability p_a. We carried out some numerical trials for different values of these parameters and found that $n_p = 40$ and $p_a = 0.25$ are very adequate for our problem. Moreover, the method is executed for n_{iter} iterations. In our simulations, we found that $n_{iter} = 1500$ is enough to reach convergence in all cases. In addition to the CS parameters, we also need two more parameters for our method: the number of contractive functions n and the mesh size, m_s. The analysis of the role of these parameters is not included here because of limitations of space. In this work, they are set to 3 and 40, respectively.

5 Experimental Results

The proposed method has been applied to several examples of fractals. Only two are included here because of limitations of space: the Sierpinsky gasket and the gather, depicted in left and right columns of Fig. 1, respectively. The original images are shown on the top. The application of our method to these examples yields the set of self-similar contractive functions minimizing the functional (7), which are then used to reconstruct the fractal image. The reconstructed images are shown in Fig. 1 (middle). Finally, a combination of both pictures is shown in Fig. 1 (bottom) for better visual comparison between both images.

As the reader can see, although the matching is not optimal, our method captures the underlying structure of the fractal image with good visual qual-

Fig. 1. Best results for the SSCF fractal problem for the Sierpinsky gasket (left) and the gather (right): (top) original self-similar fractal images; (middle) reconstructed images; (bottom) combination of the original image (in red) and the reconstructed image (in blue) for better visual comparison.

Table 2. Similarity rates for the contractive functions of the examples in Fig. 1.

Example	Similarity of w_1	Similarity of w_2	Similarity of w_3
Sierpinsky gasket	0.53	0.59	0.58
Gather	0.65	0.57	0.53

ity. This is a remarkable result because our initial population is totally random, meaning that their corresponding initial images are all very far from the target image. Our method is able to select the best contractive functions in each iteration and improve them over the generations until reaching a final image that replicates well the source image and is also a self-similar fractal. This good visual appearance is confirmed by our numerical results. The similarity rate for the two examples is reported in Table 2. These results indicate that our method is very promising, being able to provide a suitable solution for the SSCF problem as a good starting point to the more challenging IFS inverse problem.

6 Conclusions and Future Work

In this paper we address the SSCF problem: to compute a suitable set of self-similar contractive functions for the IFS inverse problem. Our method is based on the cuckoo search algorithm, a powerful swarm intelligence method for continuous optimization. The method has been applied to various examples of 2D self-similar fractal images with satisfactory results. Although the resulting IFS are not optimal yet, the method is able to recover the underlying shape of the source image with good visual quality and acceptable numerical accuracy.

The present method can be improved in several ways. On one hand, we want to modify our fitness function to obtain a better measure of the quality of the reconstructed fractal. On the other hand, we plan to apply another algorithm (proposed in [11,13] and known as *multifractal algorithm*) for a better choice of the probabilities p_i. We also wish to determine the optimal value for n along with the role of the different parameters of the method at full extent. Finally, this work must be extended to solve the general IFS inverse problem.

Acknowledgements. This research has been kindly supported by the Computer Science National Program of the Spanish Ministry of Economy and Competitiveness, Project Ref. #TIN2012-30768, Toho University, and the University of Cantabria.

References

1. Barnsley, M.F.: Fractals Everywhere, 2nd edn. Academic Press, San Diego (1993)
2. Elton, J.H.: An ergodic theorem for iterated maps. Ergodic Theor. Dynam. Syst. **7**, 481–488 (1987)
3. Falconer, K.: Fractal Geometry: Mathematical Foundations and Applications, 2nd edn. Wiley, Chichester (2003)
4. Gálvez, A.: IFS Matlab generator: a computer tool for displaying IFS fractals. In: Proceedings of ICCSA 2009, pp. 132–142. IEEE CS Press, Los Alamitos (2009)
5. Gálvez, A., Iglesias, A.: Cuckoo search with Lévy flights for weighted Bayesian energy functional optimization in global-support curve data fitting. Sci. World J. **2014**, 11 page (2014). Article ID 138760
6. Gálvez, A., Iglesias, A., Takato, S.: Matlab-based KETpic add-on for generating and rendering IFS fractals. CCIS **56**, 334–341 (2009)
7. Gálvez, A., Iglesias, A., Takato, S.: KETpic Matlab binding for efficient handling of fractal images. Int. J. Future Gener. Comm. Netw. **3**(2), 1–14 (2010)
8. Gálvez, A., Kitahara, K., Kaneko, M.: *IFSGen4* : interactive graphical user interface for generation and visualization of iterated function systems in. In: Hong, H., Yap, C. (eds.) ICMS 2014. LNCS, vol. 8592, pp. 554–561. Springer, Heidelberg (2014). doi:10.1007/978-3-662-44199-2_84
9. Graf, S.: Barnsley's scheme for the fractal encoding of images. J. Complex. **8**, 72–78 (1992)
10. Gutiérrez, J.M., Iglesias, A.: A mathematica package for the analysis and control of chaos in nonlinear systems. Comput. Phys. **12**(6), 608–619 (1998)
11. Gutiérrez, J.M., Iglesias, A., Rodríguez, M.A.: A multifractal analysis of IFSP invariant measures with application to fractal image generation. Fractals **4**(1), 17–27 (1996)
12. Gutiérrez, J.M., Iglesias, A., Rodríguez, M.A., Burgos, J.D., Moreno, P.A.: Analyzing the multifractal structure of DNA nucleotide sequences. In: Chaos and Noise in Biology and Medicine, vol. 7, pp. 315–319. World Scientific, Singapore (1998)
13. Gutiérrez, J.M., Iglesias, A., Rodríguez, M.A., Rodríguez, V.J.: Generating and rendering fractal images. Math. J. **7**(1), 6–13 (1997)
14. Hutchinson, J.E.: Fractals and self similarity. Indiana Univ. Math. J. **30**(5), 713–747 (1981)
15. Iglesias, A., Gálvez, A.: Cuckoo search with Lévy flights for reconstruction of outline curves of computer fonts with rational Bézier curves. In: Proceedings of Congress on Evolutionary Computation-CEC 2016. IEEE CS Press, Los Alamitos (2016)
16. Yang, X.-S.: Nature-Inspired Metaheuristic Algorithms, 2nd edn. Luniver Press, Frome (2010)
17. Yang, X.S., Deb, S.: Cuckoo search via Lévy flights. In: Proceedings World Congress on Nature and Biologically Inspired Computing (NaBIC), pp. 210–214. IEEE Press, New York (2009)
18. Yang, X.S., Deb, S.: Engineering optimization by cuckoo search. Int. J. Math. Model. Numer. Optim. **1**(4), 330–343 (2010)

Swarm Intelligence Scheme for Pathfinding and Action Planning of Non-player Characters on a Last-Generation Video Game

Guillermo Díaz[1] and Andrés Iglesias[2,3(✉)]

[1] Master Program in Creation of Video Games, University Pompeu Fabra, Balmes Building, Balmes 132-134, 08008 Barcelona, Spain
[2] Department of Information Science, Faculty of Sciences, Toho University, 2-2-1 Miyama, Funabashi 274-8510, Japan
[3] Department of Applied Mathematics and Computational Sciences, University of Cantabria, Avda. de los Castros, s/n, 39005 Santander, Spain
iglesias@unican.es

Abstract. Swarm intelligence is an emerging subfield of artificial intelligence (AI) where the sophisticated collective intelligence arising from a swarm of simple, unsophisticated individuals cooperating together is used to solve difficult problems. In our opinion, video games can be dramatically improved through swarm intelligence. As an illustration, we introduce a swam intelligence-based system for the representation and animation of some behavioral routines for the AI of the non-player characters (NPCs) of the last-generation first-person shooter video game "Isolated". In this work we focus on the problems of pathfinding and action planning of the NPCs. Some computer experiments have been conducted to analyze the feasibility and performance of this approach.

Keywords: Swarm computation · Video game · Non-player characters · Behavioral routines · Pathfinding · Action planning

1 Introduction

1.1 Artificial Intelligence and Computer Games

Game playing has always been a very important area of research in artificial intelligence (AI). Computer games pose a wide variety of difficult challenges that typically require some sort of intelligence for solving them. They are also simpler, cheaper, and faster than other traditional fields of application of artificial intelligence (e.g., robotics, natural language processing). In addition, they are very versatile and highly engaging. As a consequence, they have been highly regarded as excellent testbeds for AI techniques for many years.

Nowadays, the most classical application of AI in video games is the behavioral animation of their virtual characters, particularly the NPCs [9–12,15,16]. They are virtual characters not controlled by the player, so their AI must be fully

J. Del Ser (ed.), *Harmony Search Algorithm*, Advances in Intelligent Systems and Computing 514, DOI 10.1007/978-981-10-3728-3_34

specified by the programmer. A classical example of NPCs appears in shooter games, where the player takes the role of a virtual character fighting or competing against a number of computer-controlled enemies. The NPCs are not always enemies; they can also be allies (e.g., in squad games) or simply bystanders, such as in strategy or role-playing games (RPGs). Whatever the case, they need an AI providing a set of rules, methods, and procedures defining their behavioral routines: how the NPCs move, evolve, and react to player's actions and the game dynamics. For many years, the AI of NPCs was mostly based on scripts, leading to self-replicating patterns and simple and repetitive behavioral routines commonly seen in many video games (even blockbuster productions). This problem has recently become more noticeable with the increasing popularity of the massively multiplayer online (MMO) video games.

However, the extraordinary advances in artificial intelligence during the last decades have given rise to a new generation of powerful methods and techniques that go far beyond the classical scripts. Now, the game mechanics can be more varied and complex, with sophisticated behavioral routines for the NPCs. An illustrative example is given by the increasing use of CPU for AI tasks. While in old 8-bit days of gaming only about 1–2% of total CPU time was devoted to AI, now games are routinely allocating about 10–35% of CPU time for the AI system, if not more [18, 19]. Still, although more complex approaches have been recently reported, the field has a lot of potential for further improvement.

1.2 Aims and Structure of the Paper

In this paper we support the notion that the artificial intelligence of video games can be dramatically improved through the adoption of swarm intelligence. To back up our claims, we developed a swam intelligence-based system for the representation and animation of some behavioral routines for the NPCs. Because of limitations of space, the paper describes only a portion of this system, particularly its application to the problems of pathfinding and action planning of NPCs by using the last-generation FPS (first-person shooter) video game *"Isolated"*.

The structure of this paper is as follows: the main concepts and ideas about swarm intelligence and particle swarm optimization (the technique used in this paper) are outlined in Sect. 2. Then, Sect. 3 describes briefly our benchmark, the FPS video game *"Isolated"*. Some relevant issues of video games related to pathfinding and action planning are also discussed in this section. The implementation of our system is briefly described in Sect. 4. Finally, our experimental results are briefly reported in Sect. 5.

2 Swarm Intelligence

Swarm intelligence (SI) is an emerging subfield of artificial intelligence providing a number of powerful methods to solve very difficult problems [2, 7]. It has been defined as *"the property of a system whereby the collective behaviors of (unsophisticated) agents or boids interacting locally with one another and with their*

Algorithm 1. Particle Swarm Optimization

```
for each particle do
    Initialize particle                              //Initialization phase
end for
while (stop condition = false) do
    for each particle do
        Fitness ← fitness of the particle
        if (Fitness is better than BestIndividualFitness) then
            BestIndividualFitness ← Fitness      //update the memory individual fitness
            BestIndividualPosition ← Position    //update the memory individual solution
        end if
        if (Fitness is better than BestGlobalFitness) then
            BestGlobalFitness ← Fitness          //update the current best global fitness
            BestGlobalPosition ← Position        //update the current best global solution
        end if
    end for
    for each particle do
        Update velocity                              //eqn. (1)
        position ← position + velocity               //eqn. (2)
    end for
end while
return bestGlobalPosition
```

environment cause coherent functional global patterns to emerge" [2]. In SI there is not a centralized intelligence determining the evolution of the population; instead, the collective behavior of the swarm arises from decentralized systems comprised of simple mobile agents with the ability to communicate with each other. The key feature of SI are these local interactions among agents and with the environment. Because of these appealing features, SI is clearly a promising field regarding its potential applications to video game technology, particularly for the artificial intelligence of the NPCs.

2.1 Particle Swarm Optimization

One of the most popular SI techniques is *Particle Swarm Optimization* (PSO) [6,7]. Basically, PSO is a global stochastic optimization algorithm for dealing with problems where potential solutions can be represented as vectors in a n-dimensional space (the *search space*). In PSO, particles representing potential solutions are distributed over such space and provided with an initial velocity and the capacity to communicate with other neighbor particles, even the entire swarm. Particles "flow" through the solution space and are evaluated according to some fitness function after each instance. Particles evolution is regulated by two memory factors: their memory of their own best position and knowledge of the global or their neighborhood's best (these two cases are known as *global PSO* and *local PSO*, respectively). Particles of a swarm communicate good positions to each other and adjust their own position and velocity based on these good positions. As the swarm iterates, the fitness of the global best solution improves so the swarm eventually reaches the best solution. The original PSO algorithm was first reported in [6]. See also [1,7] for further information about PSO.

The global PSO procedure is briefly sketched in the pseudocode of Algorithm 1. It starts by choosing a population (swarm) of random candidate solutions in a n-dimensional space, called *particles*. Then they are displaced throughout their domain looking for an optimum taking into account global and local influences, the latest coming from the neighborhood of each particle. To this purpose, all particles have a position and a velocity and evolve all through the hyperspace according to two essential reasoning capabilities: a memory of their own best position and knowledge of the global or their neighborhood's best. The meaning of the "best" must be understood in the context of the problem to be solved. For instance, in a minimization problem that means the position with the smallest value for the target function [3,4].

The dynamics of the particle swarm is considered along iterations: each particle modifies its position P_i, keeping track of its best position in the variables domain implied in the problem. This is made by storing for each particle the coordinates P_i^b associated with the best solution (fitness) it has achieved so far along with the corresponding fitness value, f_i^b. These values account for the *memory* of the best particle position. In addition, members of a swarm can communicate good positions to each other, so they can adjust their own position and velocity according to this information. To this purpose, we also collect the best fitness value among all the particles in the population, f_g^b, and its position P_g^b from the initial iteration. This is a global information for modifying the position of each particle. Finally, the evolution for each particle i is given by:

$$V_i(k+1) = w\,V_i(k) + \gamma_1 R_1[P_g^b(k) - P_i(k)] + \gamma_2 R_2[P_i^b(k) - P_i(k)] \qquad (1)$$

$$P_i(k+1) = P_i(k) + V_i(k) \qquad (2)$$

where $P_i(k)$ and $V_i(k)$ are the position and the velocity of particle i at time k respectively, w is called *inertia weight* describing how much the old velocity will affect the new one and coefficients γ_1 and γ_2 are constant values called *learning factors*. In particular, γ_1 is a weight that accounts for the "social" component, while γ_2 represents the "cognitive" component, accounting for the memory of an individual particle along the time. Two random numbers, R_1 and R_2, with uniform distribution on $[0,1]$ are included to enrich the searching space. Finally, a fitness function must be given to evaluate the quality of a position. This procedure is iterated until a stopping condition is reached. Common terminating criteria are that a solution is found that satisfies a lower threshold value, or that a fixed number of generations has been reached, or that successive iterations no longer improve the results.

3 Our Benchmark: The Video Game Isolated

First/third-person shooters (FTPS) are one of the major video game genres more closely related to AI techniques. The main reason is that the action and dynamics of FTPS are much closer to real-world situations than other classical AI fields of application. Their graphical output and the simplicity of storyboard (compared to other games) facilitates the tracking and understanding of important AI features such as action planning, squad strategies, cooperation between agents, and others. For these reasons, we decided to use a FTPS in this paper.

Our benchmark is based on the video game *Isolated*, a last-generation first-person shooter developed by the first author and other developers on the powerful game engine *Unreal 4*. In the game, the player assumes the role of a *Special Operation Command* (SOC) member who, in close cooperation with other team members, is assigned to difficult missions involving hostage rescue and/or terrorist group deactivation. These missions can be performed on a variety of indoor/outdoor environments (see Fig. 1) and with no information available about the enemies and the environment. To fulfill the mission, many different AI tasks must be carried out. For instance, the player's character has to interact with the other squad members (all NPCs behaving autonomously according to their own AI) in a cooperative and synchronized way. This introduces realistic temporal and spatial constraints. Furthermore, our bots exhibit a more sophisticated behavior and a wider variety of skills than their counterparts in other types of video games. Bots can explore the environment autonomously, read and navigate maps, identify enemies and allies, cooperate between them for synchronized attacks, develop advanced action planning (e.g. espionage, stealth, simulation, counter-attack, or defensive strategies) and so on. For the sake of limitations of space, in this paper we restrict to two behavioral routines: pathfinding and action planning, briefly described in next paragraphs.

Fig. 1. Some screenshots of the video game: (left) indoor scene; (right) outdoor scene.

3.1 Pathfinding

In video games terminology, *pathfinding* refers to the collection of methods and tools to determine how to move a NPC from one location in a map to another, taking into account a number of features and constraints, such as the terrain environment, static and/or dynamic obstacles, lack of knowledge about the surrounding environment, and others. The classical approach for pathfinding in video games is based on standard graph-search methods (such as DFS, BFS, Dijkstra's algorithm, and the most popular one, A^* algorithm and all its variants: beam search, bandwidth search, bidirectional search, lifelong planning A^*, jump point search, and so on). The ultimate reason for this approach is that many video games are based on a tile system where the map is comprised of little squares that can easily be mapped onto a graph. Although these algorithms have some important limitations (they are based on graphs, and do not perform well with dynamic environments), several extensions and modifications have been described to take advantage of important GPU features such as multithreading.

FTPS pathfinding is usually more complex than that for other video games, since it typically requires to deal with many in-game elements such as elevators, choke points (e.g., doorways, bridges, tunnels, narrow streets), sniper locations, battle areas, and the like. This problem becomes more critical for squad games, where multiple units are moving at the same time within the same region. Pathfinding algorithms such as A^* are not designed for multiple paths being computed simultaneously, so they simply ignore other agents when computing the shortest path for a particular agent or just treat them as static obstacles. Consequently, they do not modify the path until the agent is colliding with other agent, a point when the agents have to re-search and select a new path. This leads to pathfinding behaviors that appear odd or stupid to human players, such as agents planning to pass through a narrow straight corridor and running in circles instead, just trying to avoid to collide with each other.

3.2 Action Planning

In video games, *action planning* refers to the set of techniques applied for anticipatory decision making: the development of preliminary decisions that may be adjusted and changed by circumstances before an action is actually executed [8,13,14,17]. Typical approaches for action planning are finite state machines (FSMs) or Behavior Trees (BTs), implementing STRIPS or, if decisions are based on goals, a GOAP (goal-oriented action planning) system. The first system, STRIPS, is an automated planning method that searches through possible situations or states by applying operators or actions, while GOAP is a simplified STRIPS-like planning architecture designed for real-time control of the behavior of autonomous characters in video games, first applied in *F.E.A.R.* game in 2005.

Another popular approach for action planning is given by the HTN (hierarchical task networks) planners, which are based on hierarchies of tasks that can be broken down recursively. HTNs are becoming increasingly popular in video games but have also some important drawbacks regarding the low number of actions and NPCs simultaneously handled by the planning system [5].

4 Implementation of the System

We applied the PSO algorithm described in Sect. 2.1 to the NPCs of the SOC so that they can cooperate with the human player in solving the missions. Of course, this is only a part of the complex artificial intelligence system of our bots. Therefore, it was created as a module of a more sophisticated NPC behavioral system. From a computational standpoint, the PSO algorithm is a node in the *Behavior Tree* provided by the *Blueprint* visual scripting system in *Unreal Engine*. First step is to initialize the variables for the PSO. Since the engine does not provide a specific method to create global variables, we use the class *SaveGame* to perform this task:

```
AAI_Character* npc = Cast<AAI_Character>(OwnerComp->GetOwner());
bool goalReached = npc->goalReached;
FVector currentPosition = npc->GetActorLocation();
float inertiaDifference = npc->inertiaDifference;
FVector velocity = npc->particleVelocity;
float r1, r2; // Random coefficients
float w = 0.7f; // Weight coefficient
float c1=1.4f, c2=1.4f; // Social and cognitive coefficients
```

Then, we check all the pre-conditions. If they are met, the algorithm will be executed. To this purpose, we calculate the current fitness for the particle and update the individual and global best fitness and position accordingly:

```
if (!goalReached){
        // Calculate current fitness
        float fitness = calculateFitness(currentPosition);
        // Update individual fitness if better
        if (fitness < npc->bestIndividualFitness){
                npc-> bestIndividualFitness = fitness;
                npc->bestIndividualPosition = currentPosition;
        }
        // Update global fitness if better
        if (fitness < bestGlobalFitness){
                bestGlobalFitness = fitness;
                bestGlobalPosition = currentPosition;
        }
```

Then, we check whether our agent has reached its goal. If so, the algorithm stops; otherwise, a new velocity and position are computed according to (1)–(2).

```
    // If particle is close enough
    if (fitness <= acceptanceRadiusThreshold){
            npc->goalReached = true; // Update the variable
            return EBTNodeResult::Succeeded;
    }
    else
        {
            // Re-calculate velocity
            r1 = FMath::FRandRange(0.0f, 1.0f);
            r2 = FMath::FRandRange(0.0f, 1.0f);
            npc->inertiaDifference = fmodf(inertiaDifference + 0.025f,w-0.025f);
            npc->particleVelocity = velocidad*(w - npc->inertiaDifference)+
                    (npc->bestLocalPosition - currentPosition)*r1*c1+
                    r2*c2*(bestGlobalPosition - currentPosition);
            // Calculates the new position and moves the character in game
            currentPosition = currentPosition + npc->particleVelocity;
            npc->SetActorLocation(currentPosition);
            return EBTNodeResult::Succeeded;
    }
}
else{return EBTNodeResult::Succeeded;
}
```

5 Experimental Results

Two computer experiments have been conducted to analyze the performance of our approach. The first one tests the performance of PSO for pathfinding. To this aim, 2000 executions have been carried out in a maze-like grey model scenario shown in Fig. 2 with the NPCs arising at different locations. The mission is to find a hidden object placed in an unknown location of the scenario. The second experiment is more complex and involves our behavioral routines for pathfinding and action planning. In this case, we focus on a particular mission: to find a bomb that a terrorist group has placed in a hidden location of the map. The mission is hampered by the fact that the members of the terrorist group are scattered (and often hidden) in the environment and can attack the command at any time. To make things even more challenging, the terrorists retain some hostages used as human shields in case of feeling potentially in danger. The mission goal is to minimize the time required to find the bomb and rescue the hostages while simultaneously minimizing the casualties of our platoon and the hostages. Two different environments have been used to test the algorithm under different conditions: the first one is located inside a multi-storey building with several rooms and corridors, while the second one shows an open-air scenery, with different roads through a forest and a cave (see Fig. 3).

The tests have been divided in series of 100 executions. For each series, two parameters are varied: the inertia value and number of particles. The former is varied from a very low value $w = 0.1$ to a large one $w = 1$ with step-size 0.3. To enrich the discussion, we also consider the case of a dynamic inertia decreasing linearly from 1 to 0.1 as the number of iterations. The population size is initially chosen as 16 and then doubled until 128 particles. A threshold of 300 seconds has been set to finish the search. Table 1 shows the average CPU time (in seconds) for

Fig. 2. Screenshots of the pathfinding test for different viewpoints.

Fig. 3. Screenshots of the two environments for the second experiment.

the fastest agent to reach the goal, with the best results for each population size highlighted in bold. As you can see, the CPU time improves as the population increases for all values of w. Regarding the inertia, the best results are obtained for $w = 0.7$ (even better than with the dynamic inertia), while $w = 0.1$ is the worst value in all cases. Our computer results have been very promising: the cooperation of the NPCs in the swarm leads to a surprising wealth of plausible motion patterns and realistic intelligent behaviors. Furthermore, the swarm of NPCs was able to complete the mission satisfactorily in all cases.

Table 1. Average CPU time (in seconds) for the fastest NPC to find the target.

Inertia	16 particles	32 particles	64 particles	128 particles
$w = 0.1$	95.581	77.673	60.100	55.476
$w = 0.4$	88.940	60.894	55.011	46.675
$w = 0.7$	**66.961**	**44.521**	**41.857**	**30.787**
$w = 1.0$	94.011	60.833	43.180	42.583
Decreasing w	70.891	48.732	44.115	34.560

Regarding our future work, the integration of this swarm intelligence system with other AI components of the NPCs not described here will be fully exploited to solve other challenging problems. An in-depth analysis of our results under different conditions and other parameter values is also part of our future work.

Acknowledgements. This research has been kindly supported by the Computer Science National Program of the Spanish Ministry of Economy and Competitiveness, Project Ref. #TIN2012-30768, Toho University, and the University of Cantabria.

References

1. Eberhart, R.C., Shi, Y.: Particle swarm optimization: developments, applications and resources. In: Proceedings of the IEEE Congress on Evolutionary Computation, CEC 2001, pp. 81–86. IEEE Computer Society Press, Los Alamitos (2001)
2. Engelbretch, A.P.: Fundamentals of Computational Swarm Intelligence. Wiley, Chichester (2005)
3. Gálvez, A., Iglesias, A.: Efficient particle swarm optimization approach for data fitting with free knot B-splines. Comput. Aided Des. **43**(12), 1683–1692 (2011)
4. Gálvez, A., Iglesias, A.: Particle swarm optimization for non-uniform rational B-spline surface reconstruction from clouds of 3D data points. Inf. Sci. **192**(1), 174–192 (2012)
5. Jacopin, E.: Game AI planning analytics: evaluation and comparison of the AI planning in three first-person shooters. In: Proceedings of the Tenth Annual AAAI Conference on Artificial Intelligence and Interactive Digital Entertainment, AIIDE 2014, pp. 119–124. AAAI Press, Palo Alto (2014)
6. Kennedy, J., Eberhart, R.C.: Particle swarm optimization. In: Proceedings of the IEEE International Conference on Neural Networks, Perth, Australia, pp. 1942–1948. IEEE Computer Society Press, Los Alamitos (1995)
7. Kennedy, J., Eberhart, R.C., Shi, Y.: Swarm Intelligence. Morgan Kaufmann Publishers, San Francisco (2001)
8. Iglesias, A.: A new framework for intelligent semantic web services based on GAIVAs. Int. J. Inf. Technol. Web Eng. **3**(4), 30–58 (2008)
9. Iglesias, A., Luengo, F.: Intelligent agents for virtual worlds. In: Proceedings of the CyberWorlds, CW 2004, Tokyo, Japan, pp. 62–69. IEEE Computer Society Press, Los Alamitos (2004)

10. Iglesias, A., Luengo, F.: A new based-on-artificial-intelligence framework for behavioral animation of virtual actors. In: Proceedings of the Computer Graphics, Imaging and Visualization, CGIV 2004, Penang, Malaysia, pp. 245–250. IEEE Computer Society Press, Los Alamitos (2004)
11. Luengo, F., Iglesias, A.: Framework for simulating the human behavior for intelligent virtual agents. Part I: framework architecture. In: Bubak, M., Albada, G.D., Sloot, P.M.A., Dongarra, J. (eds.) ICCS 2004. LNCS, vol. 3039, pp. 229–236. Springer, Heidelberg (2004). doi:10.1007/978-3-540-25944-2_29
12. Luengo, F., Iglesias, A.: Framework for simulating the human behavior for intelligent virtual agents. Part II: behavioral system. In: Bubak, M., Albada, G.D., Sloot, P.M.A., Dongarra, J. (eds.) ICCS 2004. LNCS, vol. 3039, pp. 237–244. Springer, Heidelberg (2004). doi:10.1007/978-3-540-25944-2_30
13. Iglesias, A., Luengo, F.: New goal selection scheme for behavioral animation of intelligent virtual agents. IEICE Trans. Inf. Syst. **E88–D**(5), 865–871 (2005)
14. Iglesias, A., Luengo, F.: AI framework for decision modeling in behavioral animation of virtual avatars. In: Shi, Y., Albada, G.D., Dongarra, J., Sloot, P.M.A. (eds.) ICCS 2007. LNCS, vol. 4488, pp. 89–96. Springer, Heidelberg (2007). doi:10.1007/978-3-540-72586-2_12
15. Luengo, F., Iglesias, A.: A new architecture for simulating the behavior of virtual agents. In: Sloot, P.M.A., Abramson, D., Bogdanov, A.V., Dongarra, J.J., Zomaya, A.Y., Gorbachev, Y.E. (eds.) ICCS 2003. LNCS, vol. 2657, pp. 935–944. Springer, Heidelberg (2003). doi:10.1007/3-540-44860-8_97
16. Luengo, F., Iglesias, A.: Animating Behavior of Virtual Agents: The Virtual Park. In: Kumar, V., Gavrilova, M.L., Tan, C.J.K., L'Ecuyer, P. (eds.) ICCSA 2003. LNCS, vol. 2669, pp. 660–669. Springer, Heidelberg (2003). doi:10.1007/3-540-44842-X_67
17. Luengo, F., Iglesias, A.: Designing an action selection engine for behavioral animation of intelligent virtual agents. In: Gervasi, O., Gavrilova, M.L., Kumar, V., Laganà, A., Lee, H.P., Mun, Y., Taniar, D., Tan, C.J.K. (eds.) ICCSA 2005. LNCS, vol. 3482, pp. 1157–1166. Springer, Heidelberg (2005). doi:10.1007/11424857_124
18. Schwab, B.: AI Game Engine Programming, 2nd edn. Course Technology, Boston (2009)
19. Woodcock, S.: Game AI: the state of the industry 2000–2001: it's not just art, it's engineering. Game Dev. **8**, 36–44 (2001)

Simulated Annealing and Natural Neighbor for Rational Bézier Surface Reconstruction from Scattered Data Points

Carlos Loucera[1], Andrés Iglesias[2,3(✉)], and Akemi Gálvez[2,3]

[1] Department of Communications Engineering (DICOM), University of Cantabria, Avda. de los Castros, s/n, 39005 Santander, Spain
[2] Department of Information Science, Faculty of Sciences, Toho University, 2-2-1 Miyama, Funabashi 274-8510, Japan
[3] Department of Applied Mathematics and Computer Science, University of Cantabria, Avda. de los Castros, s/n, 39005 Santander, Spain
iglesias@unican.es

Abstract. Surface reconstruction is a very important problem in fields such as geometric modeling and processing, and CAD/CAM. Most of the methods proposed to solve this problem rely on parametric polynomial schemes. However, there are shapes that cannot be described by using a strictly polynomial approach. In this paper we introduce a new method to address the surface reconstruction problem from scattered data points through rational Bézier surfaces. Our approach is based on the combination of simulated annealing, the natural neighbor interpolation method, and least-squares minimization to perform data parameterization, data fitting, and weight computation. Some computer experiments carried out for both organized and unorganized data sets show the good performance of our approach.

Keywords: Surface reconstruction · Rational Bézier surface · Simulated annealing · Natural neighbor · Scattered data points

1 Introduction

Surface reconstruction from data points is a very important issue in fields such as geometric modeling and processing, and computer-aided design and manufacturing (CAD/CAM). Most of the methods reported in the literature to solve this problem rely on parametric polynomial schemes such as Bézier or B-splines. Although adequate for several cases [2,4,9] these methods are also limited, because some shapes (e.g., the quadric surfaces) cannot be properly described through a strictly polynomial approach. A way to overcome this limitation is to consider rational surfaces, as they are a natural extension of the polynomial surfaces and include the quadratic shapes in a canonical way.

In this paper we introduce a new method to address the surface reconstruction problem from scattered data points through rational Bézier surfaces. In this

J. Del Ser (ed.), *Harmony Search Algorithm*, Advances in Intelligent Systems and Computing 514, DOI 10.1007/978-981-10-3728-3_35

approach, extra degrees of freedom (the weights) are introduced for more flexibility and better performance. The resulting optimization problem becomes more difficult, since we have additional unknown parameters to be computed. Furthermore, the case of scattered data points is very challenging, because the classical parameterization techniques cannot deal with unorganized data. This means that we have to solve the data parameterization, data fitting, and weight computation problems simultaneously. Our approach is based on the combination of simulated annealing for data parameterization and weight computation and the natural neighbor interpolation method to account for the unorganized data points. Once these steps are fully accomplished, data fitting is carried out by standard least-squares minimization.

The structure of this paper is as follows: previous work in the field is briefly reported in Sect. 2. Then, the mathematical background and the problem to be solved are presented in Sect. 3. The proposed method is described in detail in Sect. 4. Our method has been applied to some examples of organized and unorganized data sets in Sect. 5. The paper closes with the main conclusions and our plans for future work.

2 Previous Work

Surface reconstruction with splines is a very difficult, highly non-linear problem. Main approaches to solve it can be grouped into two categories: *incremental methods*, which typically follow a predefined sequence of incremental improvements by deciding which part is fixed and which is optimized, and *global methods*, which attempt to reconstruct the whole surface at once.

At their turn, the incremental methods can be either *variational* [20], where an hybrid functional is optimized to obtain a good trade-off between the approximation fidelity and its smoothness, or *fairing* [6,17], where an initial coarse approximation is obtained and subsequently refined through an iterative smoothing procedure. Both approaches are strongly limited because they require some subjective decisions from the user.

Global methods try to compute all required parameters without any assumption about data, the underlying function, or even the smoothness. This is an extremely difficult task where traditional mathematical methods tend to fail. The rising popularity of nature-inspired metaheuristics has led to new optimization techniques that have been applied to the shape reconstruction problem, e.g., particle swarm optimization [3], the firefly algorithm [4], cuckoo search [5], or the bat algorithm [8]. See [7] for a in-depth review on nature-inspired metaheuristic methods for data fitting.

Our approach combines the strengths of incremental and global methods. For organized data, the method applies simulated annealing to compute all required parameters in a single step. Otherwise, our method follows three major steps: firstly, construct an organized point cloud via natural neighbor interpolation; then, compute all rational Bézier surface parameters by minimizing a functional over the structured mesh and finally, find the original parameterization by optimizing a less complex function with fewer unknowns. A detailed description of the method will be given in Sect. 4.

3 Mathematical Background

A *rational Bézier surface of degree* (m, n) is mathematically represented as:

$$\mathbf{S}(u,v) = \sum_{i=0}^{m}\sum_{j=0}^{n} \mathbf{P}_{i,j} R_{i,j}(u,v) = \frac{\displaystyle\sum_{i=0}^{m}\sum_{j=0}^{n} \omega_{i,j}\mathbf{P}_{i,j} B_i^m(u) B_j^n(v)}{\displaystyle\sum_{i=0}^{m}\sum_{j=0}^{n} \omega_{i,j} B_i^m(u) B_j^n(v)} \tag{1}$$

where $R_{i,j}(u, v)$ are the *blending* functions, $u, v \in [0, 1]$ are the surface parameters, $\mathbf{P}_{i,j} \in \mathbb{R}^3$ and $\omega_{i,j} \in \mathbb{R}^+$ are the control points and their weights, respectively, and $B_k^l(t)$ are the *Bernstein polynomials of index* k *and degree* l, given by:

$$B_k^l(t) = \binom{l}{k} t^k (1-t)^{l-k} \quad \text{with} \quad \binom{l}{k} = \frac{l!}{k!(l-k)!} \quad \text{and } 0! = 1 \tag{2}$$

Let now $\{\mathbf{Q}_{k,l}\}_{k=1,\ldots,p;l=1,\ldots,q}$ be a given set of data points in $\mathbb{R}^3$. We seek to find the rational Bézier surface that approximates the given data better in the least-squares sense. This means that, for a given degree (m, n), the problem consists of finding the rational Bézier surface $\mathbf{S}$ given by (1) that minimizes the following least-squares functional $\mathcal{E}$:

$$\mathcal{E} = \sum_{k=1}^{p}\sum_{l=1}^{q} \left(\mathbf{Q}_{k,l} - \frac{\displaystyle\sum_{i=0}^{m}\sum_{j=0}^{n} \omega_{i,j}\mathbf{P}_{i,j} B_i^m(u_k) B_j^n(v_l)}{\displaystyle\sum_{i=0}^{m}\sum_{j=0}^{n} \omega_{i,j} B_i^m(u_k) B_j^n(v_l)} \right)^2 \tag{3}$$

To do so, our method must perform data parametrization, i.e., find the (u_k, v_l) associated to $\mathbf{Q}_{k,l}$, then compute the control points $\mathbf{P}_{i,j}$ and their weights $\omega_{i,j}$ and, finally, the method must deal with the model complexity to choose the values of (m, n). From (2) and (3), the problem can be formulated as an over-constrained non-linear system of equations with a large number of unknowns: the surface degree, data parameters, control points, and weights.

Least-squares problem (3) can be represented as the solution of the following system of equations $vec(\mathbf{Q}) = vec(\mathbf{P}).\mathcal{R}$, where the $vec(\cdot)$ symbol refers to the vectorization operator, column-wise stacking, and $\mathcal{R}$ is given by: $\mathcal{R} = \left\{vec(\mathbf{R}^T(u_k, v_l))\right\}_{k,l}$ where $vec(\mathbf{R}(u,v)) = \left\{vec(R_{i,j}^T(u,v))\right\}_{i,j}$ and $(\cdot)^T$ represents the transpose of a vector or matrix.

4 The Proposed Method

The core of our method consists of performing data parameterization and weight computation by means of the metaheuristic optimization algorithm simulated annealing. Then, the control points are obtained by least-squares minimization of the functional $\mathcal{E}$ in (3). Finally, a suitable surface degree is selected through the use of the Bayesian information criterion (BIC).

4.1 Simulated Annealing

The *simulated annealing* (SA) algorithm, originally introduced in [13] to solve large-scale combinatorial optimization problems, is inspired by the annealing physical phenomenon: the process of refining a metal inner structure by first exposing the material to extreme temperatures which are slowly decreased until a state of minimal energy is achieved.

Simulated annealing could be seen as a family of metaheuristic optimization methods that share a common architecture [14], namely: an iterative stochastic-driven process to search for the optimum, an annealing schedule which dictates when and how a transition occurs, and a function to accept or reject new candidates. Ideally, at the initial stages the system is allowed to freely explore the solution space, accepting bad transitions with ease; as the algorithm continues, this probability decreases until the search strategy behaves like a hill climbing algorithm towards the end. One of the most successful implementations to date is the *adaptive simulated annealing* (ASA) [11], which is actually the inspiration for our implementation. Next paragraph describes the algorithm in detail.

Let $f : \mathbb{R}^n \to \mathbb{R}^+$ be a real-valued function. The algorithm starts with a randomly chosen state $\boldsymbol{x}_0 \in \mathbb{R}^n$, $f_0 \equiv f(\boldsymbol{x}_0)$, and an initial temperature $T_0 \in \mathbb{R}^+$. Then, at each iteration k, a new candidate is generated via a neighborhood function $\mathfrak{N} : \mathbb{R}^n \times \mathbb{R}^+ \to \mathbb{R}^+$ by taking into account the system temperature, the previous candidate, and the solution space intrinsic characteristics. Each new proposed state is accepted or rejected in accordance with a modified Metropolis criterion [15] $\mathfrak{A}$: a better proposal is always accepted; a worse one is only accepted with a probability that depends on the system temperature, and the energy transition between the previous and current state. Then, the temperature is reduced by a monotonically decreasing function $\mathfrak{T} : \mathbb{R}^+ \to \mathbb{R}^+$ and a check for stagnation is made, i.e., to see if the average change in the objective function after N_s iterations is below a threshold ϵ_s. In case of stagnation, the algorithm triggers a local search with the current candidate as an initial guess. In either case, if the stopping criterion is not met, a new candidate is generated, effectively restarting the annealing cycle.

A flowchart of our SA implementation for real-valued optimization problems is shown in Fig. 1(right). The main components of the algorithm are:

1. *Temperature reduction:* $\mathfrak{T} : T_{k+1} \leftarrow \frac{T_0}{k}$
2. *Neighborhood function:* $\mathfrak{N} : \boldsymbol{x}_{k+1} \leftarrow \boldsymbol{x}_k + \Delta\boldsymbol{x}_k$, where $\Delta\boldsymbol{x}$ is a random variable sampled from the Cauchy distribution given by:

$$g_k\left(\Delta\mathbf{x}\right) = \frac{T_k}{\left(\|\Delta\mathbf{x}\|^2 + T_k^2\right)^{(D+1)/2}}$$

   where D is the dimension of the search space.
3. *Acceptance function:* $\mathfrak{A} = \min\left\{1, \left(1 + \exp\left(\frac{\Delta f}{T_k}\right)\right)^{-1}\right\}$

4.2 Natural Neighbor Interpolation

The natural neighbor interpolation [19], also known as the *Sibson* or *region stealing* method, is a powerful general-purpose spatial-data estimation technique based on combining a local weighted-average interpolation over a query point via its neighbors (in a Voronoi sense). In its most basic form the interpolant can be written as $\mathbf{I}(u_0, v_0) = \sum_{\mu=1}^{D} \xi_{\mu,0}\mathbf{Q}_u$. For a given (u_0, v_0) the method computes a Delauney triangulation $\blacktriangle(\mathbf{Q})$ of the data in order to find the closest nodes that form a convex hull around the query point, then the associated weights $\xi_{\mu,0}$ are calculated by finding how much area could be *stolen* when inserting the point into $\blacktriangle$.

4.3 Implementation

The implementation of our algorithm for selecting the optimal rational Bézier surface is illustrated in Fig. 1. The flowchart on the left describes the general methodology, whereas the flowchart on the right describes our simulated annealing implementation. From now on, η represents the number of free variables. In our method, two cases are considered:

1. *Organized data:* As above-mentioned, if the data is structured the simulated annealing performs all required computations, namely: the surface parameters and weights via the SA neighborhood function and the control points by solving Eq. (3). This SA process is referred to as $SA^{(1)}$. The solution encoding for the $SA^{(1)}$ algorithm is given by column-wise stacking the following vectors $\boldsymbol{u}, \boldsymbol{v}$ and $\boldsymbol{w}$, where $\boldsymbol{w} = vec\left(\left(\{w_{i,j}\}\right)^T\right)$.
2. *Scattered data:* Equation (3) assumes an organized set of points. When dealing with scattered data $\{\mathbf{Q}_\mu\}_{\mu=0}^{N}$, we first compute the natural neighbor interpolant $\mathbf{I}$ of the data. Then, evaluate it at an evenly spaced mesh which generates an structured set of points $\left\{\mathbf{I}(\hat{u}_{\hat{k}}, \hat{v}_{\hat{l}}) = \hat{\mathbf{Q}}_{\hat{p},\hat{q}} \mid \hat{k} = 1, \ldots, \hat{p},\ \hat{l} = 1, \ldots, \hat{q}\right\}$.

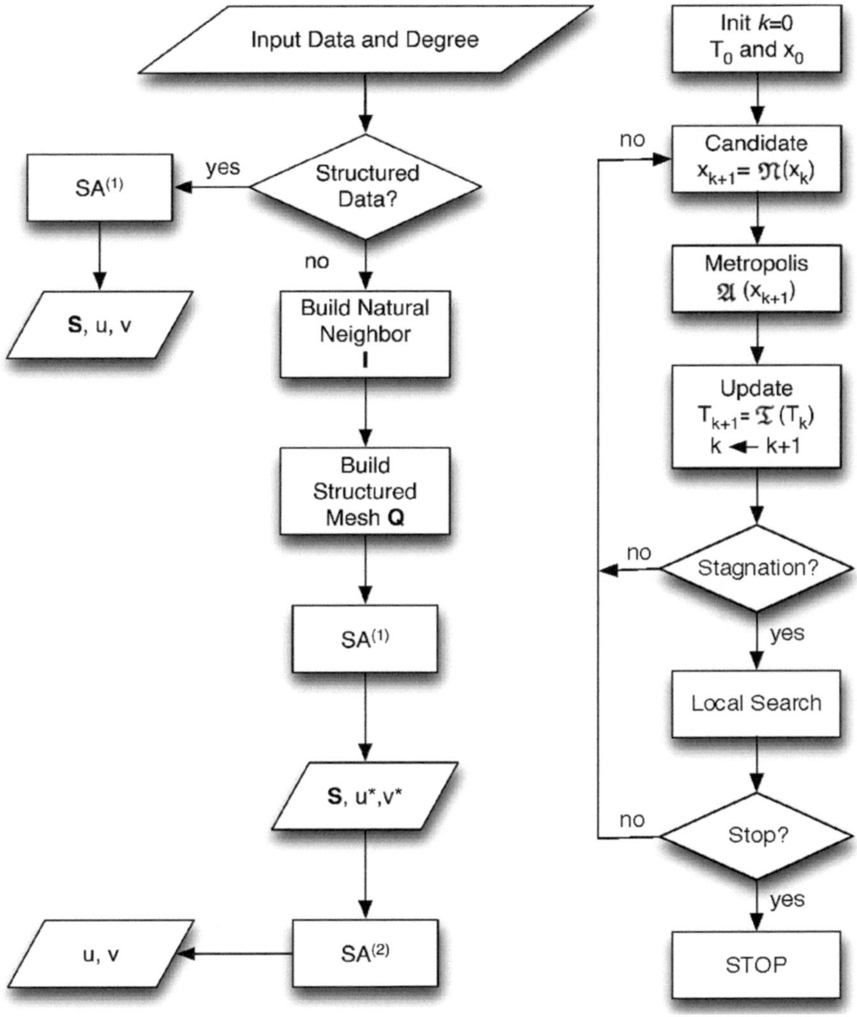

Fig. 1. Implementation of our algorithm: (left) the general methodology; (right) the simulated annealing algorithm.

Now we can approximate $\hat{\mathbf{Q}}_{\hat{p},\hat{q}}$ using the previously outlined method, i.e. solving Eq. (3) to obtain a rational Bézier surface $\mathbf{S}(\hat{u}_{\hat{k}}, \hat{v}_{\hat{l}})$. Finally, we find the associated parametrization (u_μ, v_μ) to the original data $\mathbf{Q}_\mu$ by minimizing (4) by means of the simulated annealing process referred to as $SA^{(2)}$:

$$\mathcal{E} = \sum_{\mu=1}^{N} \left(\mathbf{Q}_\mu - \mathbf{S}\left(u_\mu, v_\mu\right)\right)^2 \tag{4}$$

Note, however, that the only unknowns in (4) are the parameters $\{(u_\mu, v_\mu)\}$. The solution encoding for the $SA^{(2)}$ process is given by the columnwise stacking of the vectors $\boldsymbol{u} = \{u_\mu\}$ and $\boldsymbol{v} = \{v_\mu\}$.

Finally, the *Bayesian information criterion* [18] has been computed from the modified version given by: $BIC = N \log(\mathcal{E}) + \eta \log(N)$.

4.4 Parameter Tuning

Regarding the choice of values for the parameters of our method, they have been chosen as follows:

1. *Initial temperature:* it has been selected as: $T_0 = \max\{1, 0.8f^*\}$, where f^* is the maximum distance between the evaluation of 100 random points.
2. *Stagnation:* given by $N_s = 10\eta$ and $\epsilon_s = 10^{-4}$.
3. *Stopping criterion:* an iteration budget of $N_{end} = 10^4$ is considered.
4. *Local search:* we refine each candidate by means of the well known Nelder-Mead simplex optimization algorithm [16] as implemented in the `NLopt` library (see [12] for details).

5 Numerical Experiments

Our methodology for surface fitting has been tested against three different datasets providing a broad range of challenging features for surface reconstruction. Our experiments were run on a AMD-FX^{TM}-4100 Quad-Core Processor at 3600 *Mhz* with 8GB DDR3 RAM running *Linux 3.14.x LTS kernel* and *MATLAB 2012a*. Each experiment has been executed 16 times; then, we removed the three best and three worst executions in order to provide statistical evidence for the results presented and assert the experiment reproducibility. Each example is reconstructed for surfaces of degrees (m, n) with $m, n \in \{3, \ldots, 20\}$. In this paper we shown only the best results according to their BIC value.

5.1 Franke's Test Function

First example is constructed by evaluating the Franke's bivariate test function:

$$\begin{aligned} g(x,y) = &\frac{3}{4} Exp\left(-\frac{9y}{10} - \frac{(9x+1)^2}{49} - \frac{1}{10}\right) - \frac{1}{5} Exp\left(-(9x-4)^2 - (9y-7)^2\right) \\ &+ \frac{3}{4} Exp\left(-\frac{(9x-2)^2}{4} - \frac{(9y-2)^2}{4}\right) + \frac{1}{2} Exp\left(-\frac{(9x-7)^2}{4} - \frac{(9y-3)^2}{4}\right) \end{aligned} \tag{5}$$

at an evenly spaced grid of size 51×51, i.e. 2601 points. Then, a Gaussian noise of intensity 0.03 is applied to every point and 100 points are randomly removed. Finally we permute 100 randomly chosen points. As a consequence, the resulting point cloud is noisy, dense, and unorganized. But even under these very adverse conditions, our method is able to reconstruct the underlying shape with high fidelity as shown in Fig. 2(right). The degree of the resulting surface is $(6, 5)$ with an *RMSE* (root-mean-square error) of 0.0162.

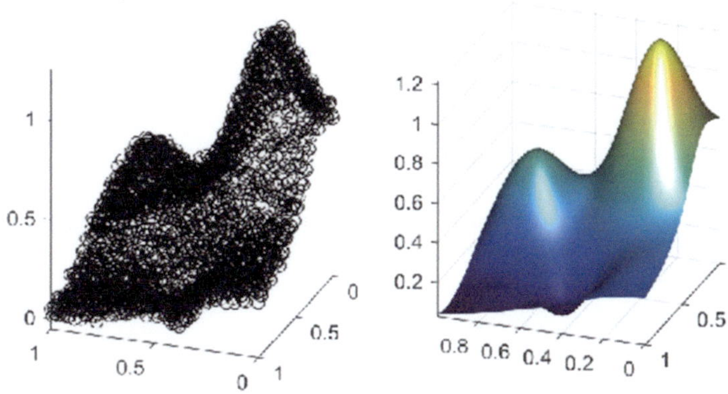

Fig. 2. Franke's function: data set (left) and best reconstructed surface (right).

5.2 Pipe Elbow

This example uses a data point set coming from a NURBS surface, so it is also a good test for our method. It consists of 10,000 points generated from a NURBS surface of degree $(4, 4)$ with three and four free knots, evaluated at an evenly spaced mesh of 100×100 points. Since the data is structured, we apply the $SA^{(1)}$ process to compute the data parametrization, the weights, and the control points. The best reconstructed rational Bézier surface is shown in Fig. 3(left). The surface degree is $(7, 9)$: as expected from the original NURBS surface, the degree is higher on the v direction. The fitting $RMSE$ value of the reconstructed surface is: $RMSE = 0.0040$.

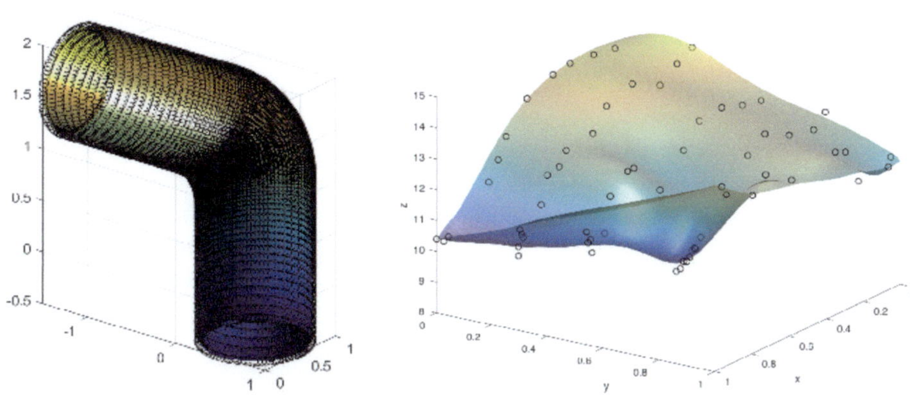

Fig. 3. Best reconstructed rational Bézier surfaces of two datasets: (left) pipe elbow; (right) Big Sur.

Big Sur Data. The final example corresponds to the Big Sur dataset [1], a point cloud generated by taking water temperature measurements from a boat. This low density (64 points), rapid varying and unorganized dataset poses a challenge for most interpolation and approximation methods. Our method automatically reconstructs the underlying shape without the need of any user input or subjective parameter.

Because we are dealing with a very low density dataset where we typically require the interpolation/approximation to be extremely accurate at the vertical component, we have computed the normalized mean square error (a goodness of fit metric) for each spatial component as:

$$NMSE_{\bullet} = 1 - \left\| \frac{\boldsymbol{Q}_{\bullet} - \boldsymbol{S}_{\bullet}}{\boldsymbol{Q}_{\bullet} - mean(\boldsymbol{Q}_{\bullet})} \right\|^2 \tag{6}$$

and we got $NMSE_x = 0.85, NMSE_y = 0.83, NMSE_z = 0.96$. Note that this metric vary between $-\infty$ (bad fit) and 1 (perfect fit). The best reconstructed rational Bézier surface is displayed in Fig. 3(right).

6 Conclusions and Future Work

In this paper we introduced a new method to solve the surface reconstruction problem from scattered data points through rational Bézier surfaces. Our method consists of the hybridization of SA, natural neighbor interpolation method, and least-squares minimization to perform data parameterization, data fitting, and weight computation. This approach is based on our previous work with simulated annealing for rational Bézier curves in [10]. However, data points for curves were always organized, so that methodology cannot be extended to the case of surfaces without further modification. In this work, we apply a modified version to solve the problem of surface approximation from noisy scattered data. Our new method is able to automatically reconstruct the underlying shape of unorganized point clouds without any subjective parameter. The addition of the natural neighbor interpolation technique provides our method with the ability to compute all relevant parameters even for point clouds not dense enough, noisy, or extremely dense. Some computer experiments shown that the method performs very well for both organized and unorganized data sets.

Our future work includes the extension of this approach to B-splines, NURBS, and other free-form parametric surfaces. A way to enhance the method is to fully hybridize natural neighbor and SA by embedding the region stealing approach into the neighbor generating function, hence removing the sequential strategy. We are also planning to extend the capacities of the SA by coupling the SA cooling schedules with the Differential Evolution (DE) optimization algorithm. The idea behind our hybridization is to improve the DE diversity mechanism by coupling a set of acceptance probabilities interlinked by one annealing temperature per DE individual.

Acknowledgments. This work has been supported by the Spanish Ministry of Economy and Competitiveness (MINECO) under grants TEC2013-47141-C4-R (RACHEL) and #TIN2012-30768 (Computer Science National Program) and Toho University.

References

1. Foley, T.A.: Interpolation and approximation of 3-D and 4-D scattered data. Comput. Math. Appl. **13**(8), 711–740 (1987)
2. Gálvez, A., Cobo, A., Puig-Pey, J., Iglesias, A.: Particle swarm optimization for Bézier surface reconstruction. In: Bubak, M., Albada, G.D., Dongarra, J., Sloot, P.M.A. (eds.) ICCS 2008. LNCS, vol. 5102, pp. 116–125. Springer, Heidelberg (2008). doi:10.1007/978-3-540-69387-1_13
3. Gálvez, A., Iglesias, A.: Particle swarm optimization for non-uniform rational B-spline surface reconstruction from clouds of 3D data points. Inf. Sci. **192**, 174–192 (2012)
4. Gálvez, A., Iglesias, A.: Firefly algorithm for polynomial Bézier surface parameterization. J. Appl. Math. **2013**, 1–9 (2013)
5. Gálvez, A., Iglesias, A., Cabellos, L.: Cuckoo search with Lévy flights for weighted Bayesian energy functional optimization in global-support curve data fitting. Sci. World J. **2014**, 1–11 (2014). Article ID 138760
6. Greiner, G., Hormann, K.: Interpolating and approximating scattered 3D-data with hierarchical tensor product B-splines. In: Surface Fitting and Multiresolution Methods, pp. 163–172. Vanderbilt University Press (1997)
7. Iglesias, A., Gálvez, A.: Nature-inspired swarm intelligence for data fitting in reverse engineering: recent advances and future trends. Stud. Comput. Intell. **637**, 151–175 (2016)
8. Iglesias, A., Gálvez, A., Collantes, M.: Bat algorithm for curve parameterization in data tting with polynomial Bézier curves. In: Proceedings of Cyberworlds 2015, Visby, Sweden, pp. 107–114. IEEE Computer Society Press, Los Alamitos (2015)
9. Iglesias, A., Gálvez, A., Loucera, C.: A simulated annealing approach for data fitting with Bézier surfaces. In: Proceedings of International Conference on Intelligent Information Processing, Security and Advanced Communication, IPAC 2015, pp. 4:1–4:6. ACM Press (2015)
10. Iglesias, A., Gálvez, A., Loucera, C.: Two simulated annealing optimization schemas for rational Bézier curve fitting in the presence of noise. Math. Problems Eng. **2016**, 1–17 (2016). Article ID 8241275
11. Ingber, L.: Adaptive simulated annealing (ASA): lessons learned. Control Cybern. **25**, 33–54 (1996)
12. Johnson, S.G.: The NLopt nonlinear optimization package. http://ab-initio.mit.edu/nlopt
13. Kirkpatrick, S., Gelatt, C.D., Vecchi, M.P.: Optimization by simulated annealing. Science **220**(4598), 671–680 (1983)
14. van Laarhoven, P.J.M., Aarts, E.H.L.: Simulated annealing. In: van Laarhoven, P.J.M., Aarts, E.H.L. (eds.) Simulated Annealing: Theory and Applications, pp. 7–15. Springer, Dordrecht (1987)
15. Metropolis, N., Rosenbluth, A.W., Rosenbluth, M.N., Teller, A.H., Teller, E.: Equation of state calculations by fast computing machines. J. Chem. Phys. **21**(6), 1087 (1953)

16. Nelder, J.A., Mead, R.: A simplex method for function minimization. Comput. J. **7**(4), 308–313 (1965)
17. Sarkar, B., Menq, C.H.: Smooth-surface approximation and reverse engineering. Comput. Aided Des. **23**(9), 623–628 (1991)
18. Schwarz, G.: Estimating the dimension of a model. Ann. Stat. **6**(2), 461–464 (1978)
19. Sibson, R.: A brief description of natural neighbor interpolation. In: Barnet, V. (ed.) Interpreting Multivariate Data, pp. 21–36. Wiley, Chichester (1981)
20. Zhao, H.K., Osher, S., Merriman, B., Kang, M.: Implicit, nonparametric shape reconstruction from unorganized points using a variational level set method. Comput. Vis. Image Underst. **80**, 295–319 (1998)

Author Index

J. Del Ser (ed.), *Harmony Search Algorithm*, Advances in Intelligent Systems and Computing 514, DOI 10.1007/978-981-10-3728-3

MIX
Papier aus verantwortungsvollen Quellen
Paper from responsible sources
FSC® C105338

If you have any concerns about our products,
you can contact us on
ProductSafety@springernature.com

In case Publisher is established outside the EU,
the EU authorized representative is:
Springer Nature Customer Service Center GmbH
Europaplatz 3, 69115 Heidelberg, Germany

Printed by Libri Plureos GmbH
in Hamburg, Germany